Springer
Proceedings in Physics 74

Springer Proceedings in Physics

Managing Editor: H. K. V. Lotsch

Volumes 1–45 are listed at the end of the book

A. Lau F. Siebert W. Werncke (Eds.)

Time-Resolved Vibrational Spectroscopy VI

Proceedings of the Sixth International Conference
on Time-Resolved Vibrational Spectroscopy
Berlin, Germany, May 23-28, 1993

With 118 Figures

Springer-Verlag
Berlin Heidelberg New York
London Paris Tokyo
Hong Kong Barcelona
Budapest

Dr. Albrecht Lau
Dr. Wolfgang Werncke

Max-Born-Institut für Nichtlineare Optik und
Kurzzeitspektroskopie im Forschungsverbund Berlin e. V.
Rudower Chaussee 6, D-12474 Berlin, Germany

Prof. Dr. Friedrich Siebert

Institut für Biophysik und Strahlenbiologie
Albert-Ludwigs-Universität, Albertstrasse 23
D-79104 Freiburg, Germany

ISBN-13:978-3-642-85062-2 e-ISBN-13:978-3-642-85060-8
DOI: 10.1007/978-3-642-85060-8

Library of Congress Cataloging-in-Publication Data. International Conference on TRVS (6th: 1993: Berlin, Germany) Time-resolved vibrational spectroscopy VI: proceedings of the Sixth International Conference on Time-Resolved Vibrational Spectroscopy, Berlin, Germany, May 23-28, 1993/A. Lau, F. Siebert, W. Werncke, eds. p. cm. - (Springer proceedings in physics; v.74) Includes bibliographical references and index. ISBN-13:978-3-642-85062-2

1. Vibrational spectra-Congresses. 2. Raman spectroscopy-Congresses. 3. Time-resolved spectroscopy-Congresses. I. Lau, A. (Albrecht), 1933- . II. Siebert, F. (Friedrich) III. Werncke, W. (Wolfgang), 1944- . IV. Series. QD96.V53I57 1993 543'.0858-dc20 93-40610

Softcover reprint of the hardcover 1st edition 1994

Typesetting: Camera ready copy from the authors/editors
54/3140 - 5 4 3 2 1 0 - Printed on acid-free paper

Preface

The Sixth International Conference on Time-Resolved Vibrational Spectroscopy was held from May 23 to 28, 1993 in Berlin, Germany. It continued the series of biennial conferences initiated in 1982 by Prof. George Atkinson (University of Arizona) at Lake Placid, USA, followed by conferences which were chaired by Prof. Alfred Laubereau (University of Bayreuth) and Dr. Manfred Stockburger (Max-Planck-Institut, Göttingen) at Bayreuth-Bischofsgrün, Germany, in 1985, by Prof. Joop D.W. Van Voorst (University of Amsterdam) at Amersfoort, The Netherlands, in 1987, Prof. Thomas G. Spiro (Princeton University) at Princeton, USA, in 1989, and by Prof. Hiroaki Takahashi (Waseda University) at Tokyo, Japan, in 1991.

The Berlin conference attracted 120 participants from 19 different countries, representing the most active scientific groups of the world in this field. Since 1982 the field has benefited from the development of lasers with shorter pulses and of reliable tunable light sources in the infrared. Now, the main activities are focused on the primary photo-induced processes and their excited-state dynamics and on detailed investigations in photochemistry and photobiology. The high quality of the contributions given at this conference is reflected in this proceedings volume and will provide all scientists interested in this field with current state-of-the-art results.

We gratefully acknowledge the financial support of Deutsche Forschungsgemeinschaft, Senatsverwaltung von Berlin, Deutsche Physikalische Gesellschaft, Commission of the European Communities, Max-Born-Institut, Gesellschaft zur Förderung angewandter Optik, Optoelektronik und Spektroskopie e.V. (GOS), and the grants from Coherent GmbH, Quantronix GmbH, Soliton GmbH, Optilas GmbH, Sopra GmbH and Siemens AG Berlin.

Furthermore, we wish to thank the international program committee for the selection of the scientific presentations, the authors of this proceedings volume, and the members of the local organizing group.

Berlin — *A. Lau*
Freiburg — *F. Siebert*
August 1993 — *W. Werncke*

Contents

Part I Theory and Vibrational Dynamics

Part III Excited States

Part IV Chemical, Photochemical, and Photophysical Reactions

Part V Photobiology

Part I

Theory and Vibrational Dynamics

Quantum Vibrational Motion in Polyatomic Molecules from Femtoseconds to Days

M. Quack

Laboratorium für Physikalische Chemie, ETH Zürich (Zentrum), CH-8092 Zürich, Switzerland

Abstract. We provide a brief review of the work from the Zürich group on the primary processes of molecular motion and kinetics during and after excitation of vibrational motion by intense coherent infrared radiation. Very short (fs and below) and very long (Ms) time scales are considered. We emphasize the fundamental concepts and provide a comparative discussion of the "direct" kinetic approach, say, by time resolved pump-probe spectroscopy and the "indirect" spectroscopic approach, emphasized in our work. Some recent results on wavepacket motion in functional groups of polyatomic molecules are mentioned and a list of key references is provided.

1. Introduction

The physical primary processes of quantum vibrational motion in polyatomic molecules are central to our understanding of chemical reaction dynamics in general and laser chemistry after IR-multiphoton excitation in particular [1–4]. We have previously provided general reviews of various aspects of this field [3–5], as well as more specific reviews addressed to work of our own research group [1,2,6–8]. We shall not duplicate these here and refer the reader to extensive lists of references in these articles.

The goal of the present brief review is to address the conceptual background of our approach and mention some of the more pertinent recent results. Both experimental and theoretical aspects are discussed and we refer here also to an earlier report of some theoretical work from our group in this series of conferences [9].

2. Time scales for primary quantum intramolecular dynamics

We distinguish time scales of primary events from complicated, combined processes. The latter usually can range from the primary process time scale to many order of magnitude longer times (by almost arbitrary amounts), whereas the former can usually be characterized by typical, natural time ranges [7]. The fastest primary process in ordinary chemistry is related to electronic motion and electron transfer. We can characterize it by a typical electronic time scale

$$t_e = h/E_h \simeq 1.52 \times 10^{-16}\,\mathrm{s} \qquad (1)$$

that is about 150 attoseconds. Electronic (photochemical) rearrangements clearly can occur on this time scale, which is somewhat shorter than reached by the fastest current pump-probe spectroscopic experiments (a few fs). The next typical time scale in molecules concerns the rearrangement of "atoms" or nuclei in molecules, which is usually necessary to render a chemical reaction irreversible, with a definite change of molecular structure. The time scale is then of the order of vibrational periods, say 10 fs for a high frequency mode with a fundamental wavernumber $\tilde{\nu}_f$ near 3300 cm^{-1}, such as an acetylenic CH stretch or about 100 fs for a low frequency molecular mode, even longer for torsional motion.

$$t_v \simeq 1/(c\,\tilde{\nu}_f) \simeq 10 \text{ to } 100\text{ fs} \qquad (2)$$

This vibrational time scale smoothly continues from the 100 fs to the ns time range, if we consider typical processes of intramolecular vibrational redistribution (IVR), Whereas in former times IVR was considered to occur only on the subpicosecond time scale, it is now clear from both direct, time resolved experiments, and indirect line shape analyses, that the time scale may extend into the nanoseconds [8]. Even longer times may apply to intramolecular vibrational-rotational redistribution (IVRR). The fundamental equation in both cases refers to coupling terms $\Delta\tilde{\nu}_c$ between states characterized by a simple (separable) vibrational rotational motion

$$t_{IVR(R)} = 1/(c\,\Delta\tilde{\nu}_c) \simeq 100 \text{ fs to } 1 \text{ ns (and longer?)} \tag{3}$$

Even longer primary times scales may apply for rearrangements of molecular structure, if we include tunneling processes, which may extend into (almost) arbitrarily long times, if we consider primary processes such as interconversion between chiral enantiomers by tunneling [7,10].

The time scales for the various IVR(R)-processes are ideally suited for time resolved vibrational spectroscopy of polyatomic molecules from the femtosecond to the nanosecond time scale. They are also the typical time scales for kinetic spectroscopy of chemical reactions, for which nanosecond time resolution is often convenient [11] and subpicosecond time resolution is rarely necessary [12,13].

One may wonder, whether this exhausts the possibilities for physical-chemical primary processes. This is clearly not the case. If we consider coupling between rovibrational motion and nuclear spin, time scales extend into the millisecond to second domain, and if we include parity violation [7,10], time scales of ks (days and even longer) may apply. With the parity violating energy difference ΔE_{pv} in chiral molecules one has, for instance [10]:

$$t_{pv} = h/\Delta E_{pv} \simeq 10^4 \text{ s} \tag{4}$$

We are reaching here the borderline between the primary processes of high energy elementary particle physics and molecular physics. The slow time scale for molecular parity violation results from the indirect influence of the weak nuclear force on molecular processes, characterized by very weak coupling. On the other hand, configuration interaction will mix molecular and highly excited nuclear states, resulting in a coupling between intranuclear nucleon motion with a characteristic time scale $\simeq 10^{-20}$ s, i.e. about 10 zeptoseconds (zs), and molecular motion. Estimates for the fastest processes that relate to such couplings can be obtained also by the time a high energy photon or a neutrino needs to travel through a molecule ($\simeq$ 0.3 nm) or a nucleus ($\simeq$ 3 fm), for the former one has times of about 1 as, for the latter times of 10 yoctoseconds (1 ys = 10^{-24} s [14]. For instance, we may consider chemical coupling to capture of high energy neutrinos

$$^{37}\text{Cl} + \nu_e \longrightarrow {}^{37}\text{Ar}^+ + e^- \tag{5}$$

which happens on characteristic time scales of zs to ys.

Particularly for the very short time scales, the direct time resolved experiments may become difficult in practice. We thus have systematically developed an approach to fast primary dynamics of molecules, which is independent of any method of time resolved experiment as it entirely relies upon the derivation of full hamiltonian molecular dynamics and statistical mechanics from non-time resolved experiments.

3. The "indirect" spectroscopic and the "direct" kinetic approaches to time dependent molecular motion

The progress in time resolved experiments has been enormous and at this conference the direct kinetic experiments clearly dominate. However, there are inherent limitations to this approach and it may be useful to search for alternatives. Figure 1 shows the scheme for our "spectroscopic approach".

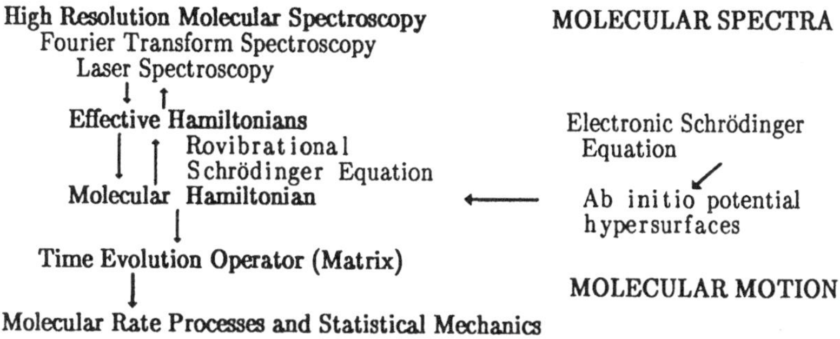

Figure 1 Scheme of the combined experimental-theoretical approach "Molecular Spectra and Molecular Motion"

From high resolution molecular spectra one derives in several, highly nontrivial steps the molecular hamiltonian, or relevant parts thereof, and the time evolution operator, which allows us to describe all the time dependent molecular processes at any desired level of detail from wavepacket motion to statistical mechanics and rate constants. We shall not discuss details of the evaluation and analysis of spectra (see [5–7]), but it is clear that the end result of the time dependent dynamics thus derived depends upon the proper understanding and analysis of the spectra in terms of the hamiltonian, the following steps being sometimes difficult, computationally, but straight-forward conceptually. Because of the necessary step of intermediate "interpretation" of the spectra, the spectroscopic approach to time dependent molecular dynamics may be termed "indirect", as opposed to the allegedly "direct" approach of time resolved observation of molecular processes. However, in kinetic experiments *one observes some time-dependent signal but not a time dependent molecular process.* There is usually a considerable amount of interpretation necessary to analyze the time-dependent signal in terms of a certain, well defined molecular process. The interpretations in kinectics, more often even than in time independent spectroscopy, require highly speculative, uncertain assumptions to establish the desired information. It is therefore by no means clear that the currently so much more popular approach of time resolved picosecond and femtosecond experiments in molecular dynamics is in any sense more "direct" than the spectroscopic approach outlined in Fig. 1 and actively pursued by the Zürich group among very few. The somewhat bad image of non-time-resolved spectral analysis in terms of kinetic data in the past has in part been due to inadequate methods such as oversimplified line width interpretations or trivial Fourier transform approaches, which are either approximate, of a very limited range of validity, or sometimes plainly wrong, indeed.

In contrast, the hamiltonian molecular dynamics approach outlined in Fig. 1 is theoretically absolutely safe, subject only to the validity of quantum mechanics in addition to the requirement that the data interpretation step has been taken correctly and unambiguously, which may sometimes be difficult or impossible, resulting in corresponding sources for error. Figure 2 illustrates the symmetry of the various approaches to fundamental dynamical properties.

The figure should demonstrate very clearly that the spectroscopic as the kinetic approaches are both experimental. They both make use of theory in various degrees at the level of interpretation. Indeed, in both approaches the best degree of certainty that can be achieved in the interpretation usually results from a complete quantum dynamical simulation (calculation) of the experimental data on the basis of the model that has been derived from the data by their interpretation. If there is a complete match between experiment and calculation or simulation, we tend to believe the interpretation to be correct, even more so if the model is able to predict future observations. This is true for both observational approaches (type I and type II), and it is really a matter of taste, whether we wish to call one more direct or indirect than the other. Both, the spectroscopic and the kinetic approaches are either "direct" or "indirect" to perfectly similar degrees and one should judge their merits upon the quality and usefulness of the results, not on the basis of personal preference.

Figure 2 compares observations of type I and type II not only for the current situation in kinetics, but also for the now largely historical situation in molecular

Observations type I
(frequency resolved spectra, line intensities, etc.)

analysis | understanding | interpretation ↓ ↑ calculation (simulation)

Theory
Logical structures, Laws summarized from previous investigations ⟶ ⟵ **Fundamental dynamical molecular properties**
(laws, nature and parameters of hamiltonian, symmetries, molecular structure, thermochemistry, open questions)

analysis | understanding | interpretation ↑ ↓ calculation (simulation)

Observations type II
(e.g., time resolved spectra and intensities in kinetics)
(specific heats, entropies, reaction enthalpies in thermodynamics)
(electron diffraction patterns in electron diffraction of gases)
(Röntgen diffraction pattern in crystal structure analysis)

Figure 2 Illustration of various experimental approaches to molecular properties in kinetics, thermodynamics and structure analysis

thermodynamics and molecular structure research. Molecular structure can be derived either from observations of type I using typically microwave and infrared spectra and so forth [15,16] and calculating structures using a hamiltonian model for the observed frequencies. Alternatively, one may use "direct" observations of type II, such as electron diffraction of gases or Röntgen diffraction of crystals.

In the end what matters is, of course, which technique provides accurate molecular structures. In practice, for different types of structural problems, different methods have merits. For accurate structural data on small molecules in the gas phase, microwave and infrared spectroscopy will be the methods of choice, whereas for the structure of some crystallized large biomolecule one will choose Röntgen diffraction. There is no unique answer which method is "best". Very similar considerations apply to obtaining thermodynamic data such as absolute entropies either from specific heat measurements and the third law (also from reaction equilibria and electrochemical cells), or else from molecular spectra in conjunction with statistical thermodynamics.

Without deprecating the time resolved techniques represented so beautifully at this conference, we wish to stress here the potential usefulness of the non-time resolved spectroscopic approach. For some complementary discussion of the evolution of methods for studying primary photoinduced dynamics, we refer to the summary remarks by R.N. Zare in the Faraday discussion on the dynamics of molecular photofragmentation [17].

We have applied the spectroscopic approach to a variety of problems in intramolecular kinetics and quantum dynamics as recently reviewed [6,7]. Among the more general recent results we may mention the quantum "chaotic" delocalization at large anharmonic couplings observed for CHX_3 molecules [18]. Of potential future importance for general chemical aspects is also the discovery of the characteristic femtosecond to nanosecond quantum dynamics of functional groups in organic compounds as exemplified by quantum wavepacket motion in R–CHO, $R_1R_2R_3CH$ and R–C≡C–H [19]. We refer to the reviews and papers cited for more details.

Acknowledgment: The many contributions of the members of my group are acknowledged as well as help from I. Kipfer. Our work is supported financially by the Schweizerischer Schulrat and the Schweizerischer Nationalfonds.

References

1. M. Quack: Advances in Chemical Physics **50**, 395 (1982)
2. M. Quack: Phil. Trans. Roy. Soc. London, **A 332**, 203 (1990)
3. D.W. Lupo, M. Quack: Chem. Rev. **87**, 181 (1987)
4. M. Quack: Infrared Physics **29**, 441 (1989)
5. M. Quack: Annu. Rev. Phys. Chem. **41**, 839 (1990)
6. M. Quack: Jerusalem Symp. **24**, 47 (1991)
7. M. Quack: J. Mol. Struct. **292**, 171 (1993)
8. M. Quack, NATO ASI Series **299**, 293 (1992)
9. M. Quack, J. Stohner, E. Sutcliffe: **Time-resolved vibrational spectroscopy**, Springer Proceedings in Physics **4**, 284 (1985)
10. M. Quack: Angewandte Chemie (Intl. Ed.) **28**, 571 (1989)
11. H. Groß, Y. He, M. Quack, G. Seyfang: at this conference (this volume)
12. L.R. Khundkhar, A.H. Zewail: Annu. Rev. Phys. Chem. **41**, 15 (1990)
13. T. Elsässer: at this conference (this volume)
    T.P. Dougherty, E.J. Heilweil: at this conference (this volume)
    R. Mathies: at this conference (this volume)
    R.M. Hochstrasser: at this conference (this volume)
14. **The international system of units (SI)**, (Imprimerie Durand, Paris 1991), CIPM 1990 resolution to 19th CGPM, p. 105 and p. 77
15. G. Herzberg: **Molecular Spectra and Molecular Structure**, Vols. I,II,III (van Nostrand, Toronto 1950, 1945, 1966)
16. H.W. Kroto: **Molecular Rotation Spectra** (Wiley, New York 1975)
17. R.N. Zare: Faraday Disc. Chem. Soc. **82**, 391 (1986)
18. R. Marquardt, M. Quack: J. Chem. Phys. **95**, 4854 (1991)
19. M. Quack, J. Stohner: J. Phys. Chem. (1993) in press

Dynamics and Spectroscopies in Four-Wave Mixing with Coherent and Incoherent Light

S. Schaertel, S. Smith, and A.C. Albrecht

Department of Chemistry, Cornell University, Ithaca, New York

Abstract. "Interferometric Coherent Raman Scattering" (ICRS) involving incoherent light is characterized by fringes (Rabi detuning oscillations (RDO's)) or strongly down-converted Raman Bohr frequencies. New examples of RDO's are shown involving both resonant and nonresonant components. Along the frequency axis the ICRS signal reveals new "zero linewidth" (ZLW) features that promise ultrahigh frequency resolved Raman spectroscopy. New examples of these features are shown. Neither RDO's nor ZLW features are seen when using femtosecond light pulses. The latter, however, permit the resolution of stochastic line-shaping processes in real time. Thus a time resolved (25fs) CARS study of the ring breathing mode in liquid benzene has been completed. The line-shaping event is analyzed as a Kubo process. It is suggested that the spinning benzene ring may be principally responsible for the dephasing of this mode.

1. Introduction

The ultrashort correlation times of phase noise in nanosecond incoherent light fields and the ultrashort amplitude modulation of femtosecond fields can probe ultrafast material dynamics each in their own way. It was in the early 1980's when two centers in Japan and one in America independently appreciated that phase noisy light could offer experimentally useable ultrashort time gating comparable to that available from femtosecond light sources. Applications, however, have not been widespread. Typically, nanosecond cavity-spoiled dye lasers are used (in which gratings are either in zero-order or are entirely absent and replaced by a mirror). Such nanosecond noisy fields may offer time gates as short as ~35 fs, though more typically ~100 fs is the rule. However, amplitude modulated light pulses on the femtosecond scale are the conventional tool for studying fast dynamics. The nanosecond sources involve nanosecond accumulation times while the femtosecond studies are measurements in real time - an important distinction for many applications.

We continue to explore the application of phase noisy nanosecond light to four-wave-mixing (4WM) experiments - the coherent Raman spectroscopies (CRS - CARS and CSRS) in particular. Called "Interferometric Coherent Raman Scattering" (ICRS), it is briefly summarized and new applications are presented in Section 2.1. By way of contrast, dephasing in liquid benzene is followed by femtosecond (25 fs) CARS experiments in real time and very briefly summarized in Section. 2.2. Appropriate literature citations are to be found within the referenced material.

Springer Proceedings in Physics, Vol. 74 Time-Resolved Vibrational Spectroscopy VI
Editors: A. Lau, F. Siebert, W. Werncke

2.1 Interferometric Coherent Raman Scattering with Incoherent Light (ICRS)

ICRS has been described in detail both from the theoretical [1] and experimental [2] points of view. In ICRS three beams are incident upon the sample in a 'box' configuration, two are the twin noisy fields 2 and 2' derived from a single incoherent source, one is a "monochromatic" field 1. The experimental time parameter derives from the interferometer setting, τ, that displaces 2 from 2'. The frequencies are chosen such that $|\omega_1 - \omega_2| \approx \omega_A$, the Bohr frequency of a resonant vibration. One has either Coherent Stokes Raman Spectroscopy (CSRS or ICSRS) ($\omega_1 > \omega_2$), or Coherent Anti Stokes Raman Spectroscopy (CARS or ICARS) ($\omega_1 < \omega_2$). Provided the signal (the fourth wave) is sharply spectrally filtered, damped oscillations ("fringes") are seen as the interferometer setting (τ) is varied. These fringes represent a new class of higher order Rabi detuning oscillations (the RDO's) which is unrelated to the more common Bohr frequency oscillations seen in other 4WM experiments. The latter are oscillations due to material coherences that are set up when the incident light has sufficient bandwidth. Detuning oscillations are the difference beat between a radiation oscillator and a material oscillator that are mutually coupled. The RDO's at the 4WM level are simply higher order versions of these. In ICRS the period of the RDO's contains no reference whatsoever to the incoherent light. Its noise has been precisely nulled. In principle, one can elicit a RDO from any pair of eigenstates (electronic, vibrational, or phonon/librational) [1]. Thus the RDO's constitute a true down conversion of any Bohr frequency.

In CSRS and CARS beams 1 and 2 (or 2') act out of phase and represent a radiation difference oscillator at $\omega_1 - \omega_2$. Another radiation difference oscillator at $\omega_{2'} - \omega_S$ (for CSRS) is well defined by beams 2' (or 2) and the signal beam, S, provided the latter is sharply defined by monochromatic detection. Each of these radiation difference oscillators may be detuned from the Raman Bohr frequency, ω_A, but may still drive the vibrational coherence since beams 2 (2') are spectrally broad (~150 cm^{-1}). One then encounters radiation/matter oscillators at the two detuning frequencies: $\Delta_1 = \omega_A - (\omega_1 - \omega_2)$ and $\Delta_2 = \omega_A - (\omega_{2'} - \omega_S)$. These two radiation/matter difference oscillators can beat at their sum, $\Delta_+ = \Delta_1 + \Delta_2 = (2\omega_A - \omega_1 + \omega_S)$, and their difference $\Delta_- = \Delta_1 - \Delta_2 = (2\omega_2 - \omega_1 - \omega_S)$. (We have taken $\omega_{2'} = \omega_2$, the central frequency of the twin incoherent light beams.) Δ_+ exactly represents the RDO. The RDO represents a downtuning of the Raman frequency by $(\omega_1 - \omega_S)/2$. It also clearly has no reference to the incoherent light and it accurately determines the Raman frequency once ω_1 and ω_S are specified. These RDO's decay at one-half the dephasing time of the two-level system. RDO's are sensitive to polarization conditions because the nonresonant part of $\chi^{(3)}$, $\chi^{(3)}_{NR}$, alone is incapable of producing RDO's, there being no material resonance in this component. A polarization configuration that suppresses $\chi^{(3)}_R$ also suppresses the RDO's. RDO's from cross-terms between $\chi^{(3)}_{NR}$ and $\chi^{(3)}_R$ do exist and are discussed for the first time here.

With $\chi^{(3)} = \chi^{(3)}_R + \chi^{(3)}_{NR}$ the theoretical task is to properly perform the stochastic averaging (at the signal level) over the noise of the incoherent fields 2 and 2', $\langle |\chi^{(3)}|^2 \rangle$. This has just recently been accomplished (using a Lorentzian spectral density for the noise) [3]. It is found that the RDO from the cross-term is essentially a sine function, while that from $\langle |\chi^{(3)}|_R^2\rangle$ is primarily a cosine function. This distinction provides a new approach for separating the $\chi_{NR}^{(3)}$ and $\chi_R^{(3)}$ components of $\chi^{(3)}$. In the following section we give an abbreviated form for $\langle |\chi^{(3)}|^2\rangle$ and show RDO's from two very different systems to which the equation applies.

2.1.1 RDO's for $\chi^{(3)} = (\chi_R^{(3)} + \chi_{NR}^{(3)}) \neq \chi_R^{(3)}$

We discuss two classes of systems in which both resonant and nonresonant susceptibilities can play a role: neat systems where $\chi_{NR}^{(3)}$ and $|\chi_R^{(3)}|$ happen to be comparable, and mixtures where the solute at mole fraction x_R contributes $\chi_R^{(3)}$ and the solvent at mole fraction x_N is the major provider of $\chi_{NR}^{(3)}$. For both classes we have to perform

$$\langle |\chi^{(3)}|^2\rangle = \langle |\chi^{(3)}|_R^2\rangle + 2\langle (Re\ \chi_R^{(3)})\ \chi_{NR}^{(3)}\rangle + \langle |\chi_{NR}^{(3)}|^2\rangle .$$

With $\chi^{(3)}$ in units of the usual macroscopic susceptibility per total number density, per circular frequency, and for a Lorentzian spectral density of the phase noise, we have found [3] that the purely resonant part is given by:

$$\langle |\chi_R^{(3)}|^2\rangle = x_R^2\ |\gamma^{(R)}|_0^2 F_R \left(\tfrac{1}{2}\cos(\Delta\,\tau) + \tfrac{\gamma}{\Delta}\sin(\Delta\,|\tau|)\right) \exp[-2\gamma|\tau|] + G_{R/R},$$

the cross-term by:

$$2\langle (Re\ \chi_R^{(3)})\ \chi_{NR}^{(3)}\rangle = 2\, x_R x_{NR}\, |\gamma^{(R)}|_0\, (\gamma^{(NR)})\, F_R \sin(\Delta\,|\tau|) \exp[-2\gamma|\tau|] + G_{R/NR}$$

and the purely nonresonant part by:

$$\langle |\chi_{NR}^{(3)}|^2\rangle = x_N^2\ (\gamma^{(NR)})^2 F_N\ \exp[-2\Gamma|\tau|] .$$

In the above, $|\gamma^{(R)}|_0$ is the peak magnitude of resonant third order molecular susceptibility of the molecule carrying the resonance, $\gamma^{(NR)}$ is the nonresonant third order molecular susceptibility, $F_R = (\Gamma^2\gamma)/((\Delta_1^2+\Gamma^2)(\Delta_2^2+\Gamma^2))$, $F_N = (\Gamma)/(\Delta_-^2 + 4\Gamma^2)$, and $\Delta \equiv \Delta_{+}$,. $G_{R/R}$ and $G_{R/NR}$ are normally weaker terms that contribute to the coherent spike. There are two damping parameters - Γ, the HWHM spectral width of the noisy light, and γ, the HWHM of the Raman line.

Fig. 1a) displays RDO's seen by ICSRS in neat mesitylene. Unlike the case of neat benzene which only shows RDO's in ICSRS [2], one now detects both a coherent spike and RDO's, which in fact are found to contain cross-term contributions. In neat mesitylene the $\chi_R^{(3)}$ does not dominate the $\chi_{NR}^{(3)}$ and both play a role in its ICSRS. Fig. 1b) displays RDO's seen in a solution.by ICSRS when benzene is sufficiently diluted in cyclohexane (like x_R (benzene) = 0.12). Now $|\chi_R^{(3)}|$ (benzene) ~ $\chi_{NR}^{(3)}$ (cyclohexane)) and a coherent spike appears with the RDO's and the cross term is important. In neat benzene and even at $x_R = x_{NR} = 0.5$, only RDO's are seen and cross-term effects are insignificant [2]. The above equation applies to both classes of systems (however for a neat sample both x_R and x_{NR} are equal to unity).. We are currently quantitating $\chi_R^{(3)}$ relative to $\chi_{NR}^{(3)}$ for both classes of

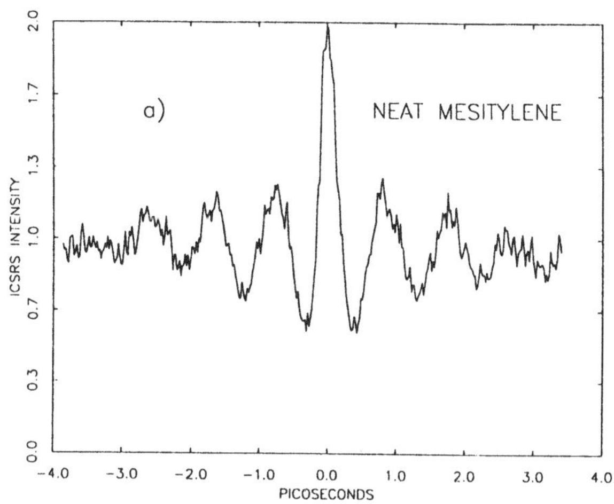

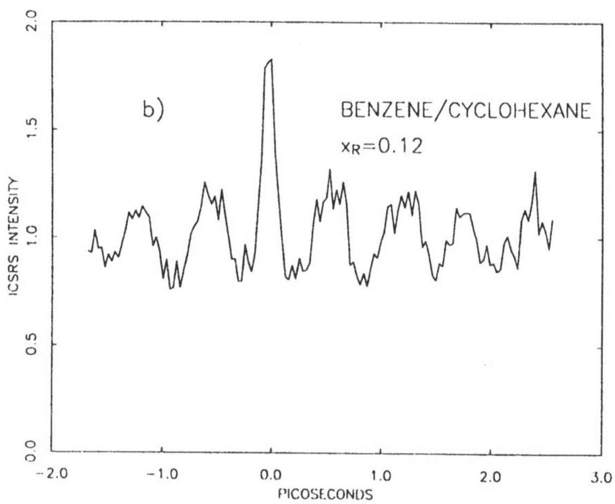

Fig. 1. RDO's with both resonant and nonresonant contributions: a) neat mesitylene. b) benzene-h6 at $x_R = 0.12$ in cyclohexane. See [3].

samples. Raman cross-sections for benzene can provide its $|\gamma^{(R)}|_0$ so that $\gamma^{(NR)}$ of the solvent is then determined [3]. The reverse of this procedure is also possible.

2.1.2 The Zero-Line-Width (ZLW) Features

A second new feature is seen in ICRS when the spectrum of the signal intensity at fixed τ (usually zero) is analyzed under high resolution. Surprising "zero-linewidth" features (ZLW's) are predicted [1] provided the "monochromatic" field 1 and the monochromator resolution of the signal field, S, are not infinitely narrow. The ZLW features appear to be a new kind of dephasing induced extra resonance in which the dephasing originates from the (narrow band) radiation. Its width is fixed by the

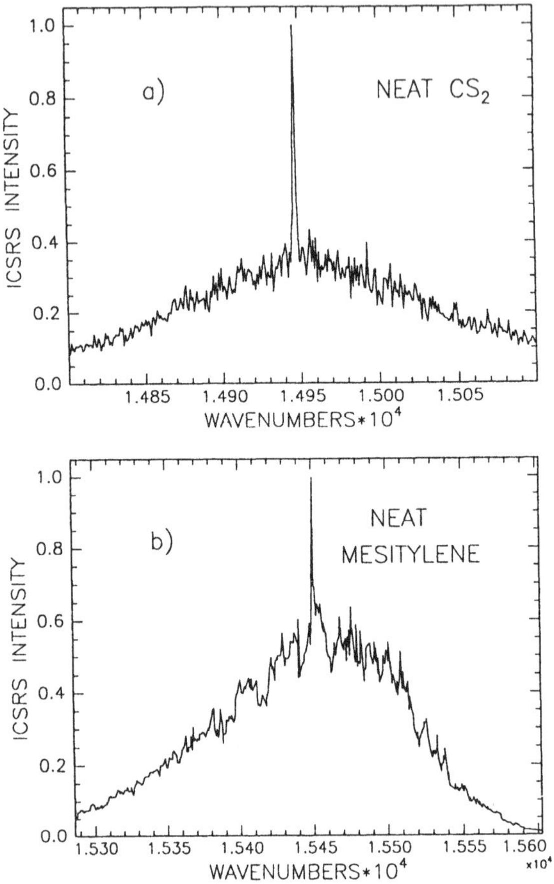

Fig. 2. ZLW features seen exactly at $\Delta_+ = 0$ in: a) neat carbon disulfide and b) neat mesitylene. See [4].

naturally non-zero spectral widths of fields 1 and S and not by the natural linewidth. The ZLW feature appears exactly at $\Delta_+ = 0$. Fig. 2a) shows its sighting in neat mesitylene and Fig. 2b) in neat carbon disulfide [4]. Theory says [1] that such ultra-sharp ZLW features should be available from any two-level coherence regardless of the natural width provided it is not inhomogeneously broadened. It will be interesesting to see whether such ZLW features can be found not only from ground electronic state vibrational modes, but also from those in an excited electronic state, as theory predicts [1].

2.2 Dephasing in Real Time by Femtosecond CARS

An ultrafast (25 fs) real-time CARS study of vibrational dephasing of the ring breathing mode of liquid benzene [5] has managed to resolve the dephasing event in its non-exponential regime. A Kubo function anaylsis says that the responsible stochastic process has a correlation time of about 550 fs and causes an ~ 3 cm^{-1} (rms) modulation of the breathing mode. It is argued that spinning motion around the C_6 axis is the likely dephasing process. Insight of this type into dephasing dynamics is unavailable through ICRS. Its intrinsic long time scale of averaging provides information comparable to that contained in cw line shape analysis - but in an entirely new way.

Acknowledgements

We are grateful for support from the National Science Foundation (CHE-9002234) and the Materials Science Center of Cornell University.

References

1. M. A. Dugan and A. C. Albrecht, Phys. Rev. **A** **43**, 3877(1991)
2. M. A. Dugan and A. C. Albrecht, Phys. Rev. **A** **43**, 3922(1991)
3. S. Schaertel and A. C. Albrecht, (to be published)
4. S. Smith and A. C. Albrecht, (to be published)
5. T. Joo and A. C. Albrecht, J. Chem. Phys., **99**, 0000(1993)

Influences of Internal Rotation on Vibrational Predissociation Dynamics

H.J. Elston, B.D. Gilbert, C.S. Parmenter, H.K. Oh, T.A. Stone, M.C. Su, and Z.Q. Zhao

Indiana University, Department of Chemistry, Bloomington, IN 47405, USA

Abstract. Characteristics that influence the nonstatistical dissociation of van der Waals complexes between aromatic molecules and Ar have been explored by comparing five complexes of which p-difluorobenzene-Ar (pDFB-Ar) is the prototype.

1. Introduction

D.H. Levy and co-workers [1] showed in their early explorations of supersonic jet S_1-S_0 spectroscopy that the van der Waals molecule s-tetrazine-Ar formed between the aromatic and the carrier gas was an instructive prototype for unimolecular dissociation. One could initiate dissociation by exciting any of several known S_1 ring vibrational levels in the complex with a UV laser. It was then possible to monitor the subsequent flow of vibrational energy within the S_1 complex and its deposition in the S_1 dissociation product by S_1-S_0 fluorescence. The state-to-state view of vibrational energy flow from reactant to product gave a detailed view of vibrational dynamics.

This work inspired explorations of dissociation in many large molecule complexes from which a common characteristic has emerged. All exhibit nonstatistical dissociation. This behavior is seen most clearly in the product vibrational state distribution. Only a few levels are populated among the many accessible in the free polyatomic product.

This report summarizes studies of a series of aromatic complexes chosen to explore molecular characteristics that influence this state selectively. The complexes are displayed in Fig. 1. The benchmark is pDFB-Ar [2,3] whose dissociation almost certainly occurs by the sequential process described in Fig. 1. Vibrational energy is initially deposited in a ring level without excitation of van der Waals modes. Prior to dissociation, the van der Waals modes are first excited by IVR. The product vibrational state distributions have now been determined for the four additional van der Waals clusters displayed in Fig. 1. The level selectivity of dissociation is reminiscent of similar selectivity in collisional vibrational energy transfer (VET) in aromatic systems, including the pDFB + Ar system [4].

Springer Proceedings in Physics, Vol. 74 Time-Resolved Vibrational Spectroscopy VI
Editors: A. Lau, F. Siebert, W. Werncke

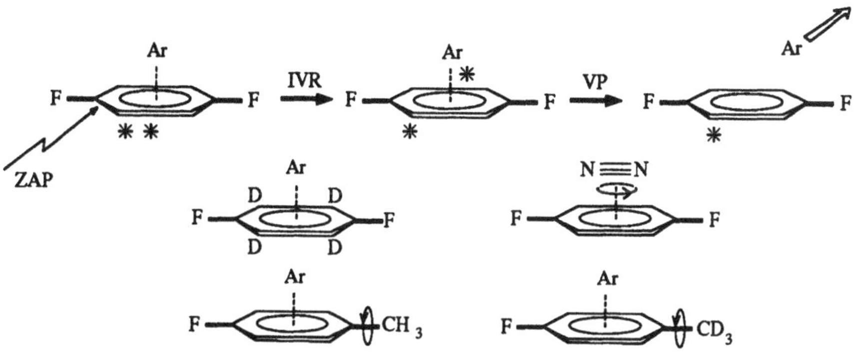

Figure 1. Five van der Waals complexes whose dissociation characteristics have been studied. The dissociation is thought to occur by the illustrated serial process in which IVR within the complex precedes the dissociation, here labelled as vibrational predissociation (VP).

2. Results and Discussion

The dissociation selectivity in pDFB-Ar is illustrated in Fig. 2 for the complex level 5^1 with $\epsilon_{vib} \approx 800$ cm^{-1}. Only two of more than a dozen accessible levels are populated by dissociation [2a]. Equally selective dissociation has been seen from six other initial levels [2b].

One cannot account for the observed product levels by assuming direct vibrational predissociation (VP) from the initially pumped complex levels [2a,7]. As an alternative, it has been proposed that an accidental two-level Fermi resonance between the pumped complex level and a lower ring level combined with high van der Waals mode excitation may drive the selectivity [2a]. Deuterating pDFB to change the accidental level coincidences provides a test. As shown in Fig. 2, the same final levels are still selected by pDFB-d_4-Ar dissociation. Thus Fermi resonances within the complex are not controlling the dissociation channels.

There remains inference, therefore, that dissociation occurs by a serial process involving first IVR within the complex as is directly observed in s-tetrazine-Ar [8] and other systems. Spectroscopic evidence for IVR exists also for pDFB-Ar when pumping low complex levels.

Comparisons of pDFB-Ar and pDFB-N_2 dissociation concern this IVR. Aside from the nearly free rotation of N_2 above the ring, an additional van der Waals mode is introduced by the change from Ar to N_2. This mode increases the density of van der Waals states so that the N_2 IVR rate should be greater than that of the Ar complex. The fluorescence spectra after pumping 5^1 in the Ar and the N_2 complex bear out this prediction.

Comparison of the dissociation channels following excitation of the 5^1 level in pDFB-Ar, pDFB-d_4-Ar and pDFB-N_2 is shown in Fig. 2. Other than the appearance of an extra channel for pDFB-N_2 (dashed line), the selectivity remain the same.

Our modelling is similar to that used for s-tetrazine-Ar [5]. Channel selectivity begins with IVR to van der Waals levels built on a single lower ring level. The hot van der Waals levels so produced dissociate by a competition between direct VP [6]

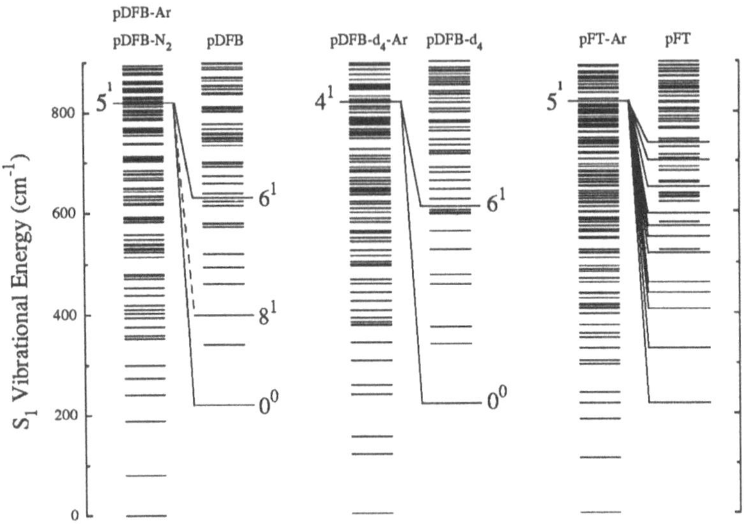

Figure 2. The S_1 ring vibrational level structure of complexes and their free molecule dissociation products. None of the internal rotation states is shown for pFT. Dissociation channels following excitation of the 5^1 level in the complexes (4^1 for pDFB-d_4-Ar) are shown.

and an RRKM dissociation [5]. The modelling is fairly successful in predicting the correct channels in all three complexes.

A qualitative propensity rule appears in the modelling. Selectivity in IVR is driven in part by a propensity for small changes in ring mode vibrational quantum numbers. Similarly, selectivity in VP is also driven, again in part, by this propensity. Such a propensity rule is familiar to those modelling the full-collision aromatic molecule VET [4].

We finally turn to pFT-Ar and pFT-d_3-Ar complexes. Two changes in the level structure occur in comparison with pDFB-Ar. First, the ring level structure changes. The change is, however, modest, similar to that produced by deuterating the ring, and should have little effect on the dissociation.

Second, a nearly free internal rotation is introduced by the CH_3 group. Studies of pFT demonstrate that this additional degree of freedom effectively mixes ring levels of different vibrational character [9]. If this characteristic persists in the pFT-Ar complex, one would expect a large perturbation of the dissociation. Two effects may occur. First, the increased state mixing should accelerate IVR within the complex. More importantly, it will destroy the high ring mode selectivity of the IVR. If so, this change would reduce the selectivity in final product states.

The predictions are borne out. The most impressive effect of the internal rotation is in the dissociation channels. As shown in Fig. 2, pFT-Ar dissociation is now completely different. The final level selectivity is largely gone. Similar behavior still occurs if we deuterate the internal rotor. These examples are the first cases (of which we are aware) in which the level selectivity of aromatic complex dissociation is largely destroyed.

There also is direct evidence of the internal rotation-vibration interaction. The S_1-S_0 fluorescence contains emission from the free molecule dissociation product with quanta of its internal rotation excited. Since the rotor is not excited in the initial complex state, the observation is sure evidence for the participation of internal rotation-vibration coupling in the dissociation dynamics.

We are grateful for the financial support of the National Science Foundation.

References

[1] a) D.V. Brumbaugh, J.E. Kenny and D.H. Levy, J. Chem. Phys. **78**, 3415 (1983); b) R.E. Smalley, L. Wharton, D.H. Levy, D.W. Chandler, J. Chem. Phys **68**, 2487 (1978).

[2] a) K.W. Butz, D.L. Catlett, Jr., G.E. Ewing, D. Krajnovich, C.S. Parmenter, J. Phys. Chem. **90**, 3533 (1986); b)H.-K. O, C.S. Parmenter, M.C. Su, Ber. Bunsenges. Phys. Chem. **92**, 253 (1988); c) M.-C. Su, H.-K. O, C.S. Parmenter, Chem. Phys. **156**, 261 (1991);

[3] B.A. Jacobson, S. Humphrey, S.A. Rice, J. Chem. Phys. **89**, 5624 (1988).

[4] a) D.L. Catlett, Jr., C.S. Parmenter, J. Phys. Chem. **95**, 2864 (1991); b) D.J. Krajnovich, C.S. Parmenter, D.L. Catlett, Jr., Chem. Rev. **87**, 237 (1987).

[5] D.F. Kelly, E.R. Bernstein, J. Phys. Chem. **90**, 5164 (1986).

[6] a) G.E. Ewing, J. Phys. Chem. **90**, 1790 (1986); b) G.E. Ewing, J. Phys. Chem. **91**, 4662 (1987).

[7] A.R. Tiller, A.C. Peet, D.C. Clary, Chem. Phys. **129**, 125 (1989).

[8] J.J.F. Ramaekers, H.K. van Dijk, J. Langelaar, R.P.H. Rettschnick, Faraday Discuss. Chem. Soc. **75**, 183 (1983)

[9] a) D.B. Moss, C.S. Parmenter, G.E. Ewing, J. Chem. Phys. **86**, 51 (1987); b) D.B. Moss, C.S. Parmenter, J. Chem. Phys **98**, 6897 (1993); c) Z.Q. Zhao, Ph.D. Thesis, Indiana University (1992).

Measurement of Rotational-State Relaxation by Time-Resolved Infrared Double-Resonance Spectroscopy

B. Abel, S. Coy, C. Flannery, J.J. Klaassen, and J.I. Steinfeld*

Department of Chemistry and G.R. Harrison Spectroscopy Laboratory, Massachusetts Institute of Technology, Cambridge, MA 02139, USA
*Present address: Institut für physikalische Chemie, University of Göttingen, Germany

Abstract. Many trace species of critical importance in atmospheric processes, notably ozone and methane, can be monitored by remote sensing of their infrared absorption and emission bands. Recovery of accurate column densities and mixing ratios from field measurements requires accurate knowledge of band parameters, viz., assigned line frequencies, line intensities, and pressure-broadened linewidths. The last-named quantities are generally derived from line-broadening models which have been tested against only a limited experimental data base. We have been using time-resolved infrared double-resonance spectroscopy to measure inelastic collision rates, which are a key component of broadening models. A pulsed infrared laser is used to prepare selected rovibrational levels of the species of interest; these are probed by a tunable semiconductor diode laser which provides the resolution necessary for resolution of energy-transfer processes at the state-to-state level. Results obtained for several systems indicate that in many cases the first-order collision models used to account for rovibrational relaxation and collision broadening are inadequate to describe the inelastic processes in detail, particularly for large J and K changes in collisions.

1. Introduction

The severe [1] and apparently accelerating [2] depletion of the ozone content of the earth's stratosphere has driven a renewed and urgent interest in chemical and physical processes affecting our atmosphere. A central component of such studies is determination of spatial and temporal variations in column densities and mixing ratios of atmospheric trace species, such as ozone, methane, nitrogen oxides, and chemically reactive free radicals. Such determinations are generally carried out using spectroscopic remote sensing techniques [3]. In order to interpret such measurements in terms of actual concentration values, basic spectroscopic data must be known for the species of interest, viz., assigned line frequencies, line intensities, and pressure-broadened linewidths. Atmospheric transmission models [4] require broadening coefficients for tens of thousands of individual transitions over a wide range of pressures, temperatures, and composition. Since it is impractical to determine all of these quantities by individual measurement, theoretical models are generally employed to generate the needed values.

The pressure broadened linewidth $\gamma(u,l)$ includes contributions from inelastic collisions in the lower state (l) and the upper state (u) of the spectroscopic transition, as well as possible additional contributions from "non-inelastic" processes such as

Springer Proceedings in Physics, Vol. 74 Time-Resolved Vibrational Spectroscopy VI
Editors: A. Lau, F. Siebert, W. Werncke

dephasing or elastic reorientation. These "non-inelastic" processes are generally neglected in standard pressure-broadening models, so that the collision-broadened linewidth is given as a simple average of rotational-state-changing collisions in the upper and lower states, i.e.,

$$\gamma(j_u\tau_u, j_\ell\tau_\ell) = \frac{n_2}{2\pi c} \cdot \frac{1}{2}\left[\sum_{j_u'\tau_u'} k\left(j_u\tau_u \rightarrow j_u'\tau_u'\right) + \sum_{j_\ell'\tau_\ell'} k\left(j_\ell\tau_\ell \rightarrow j_\ell'\tau_\ell'\right)\right]$$

In the preceding expression, j and τ are rotational quantum numbers [5,6], n_2 is the perturbing gas density, and the speed of light c converts γ to cm^{-1} units.

Evaluation of pressure-broadened linewidths (or broadening coefficients γ/p) thus comes down to finding a suitable inelastic collision model. Among the most widely used models are the Anderson-Tsao-Curnutte [7] and Quantum Fourier Transform with Improved Dynamics, or QFT-ID [8]. These models are generally tested against a limited range of experimentally measured broadening coefficients for selected values of ($j\tau$), temperature, pressure, and perturber gas identity. In order to validate these models and determine their range of applicability, however, it is necessary to examine the predicted values of individual state-to-state inelastic rate coefficients $k(j\tau \rightarrow j'\tau')$ in some detail. We have measured these rate coefficients using the technique of time-resolved infrared double-resonance (TRIRDR) spectroscopy, as described below. Since a full specification of the $k(j\tau \rightarrow j'\tau')$ matrix may involve thousands of individual rate coefficients, an alternative approach has been suggested which makes use of energy- or angular-momentum-based scaling laws [9,10]. These scaling laws contain a small number of parameters which are empirically adjusted to fit available pressure-broadening data and, if available, state-to-state relaxation data.

2. Time-Resolved Infrared Double-Resonance Spectroscopy

In the TRIRDR technique [11,12], a pulsed infrared laser is used to prepare selected rovibrational levels of the species of interest. Since many important species possess infrared absorption in the neighborhood of the CO_2 laser output (9 - 11 μm), this reliable and widely available laser has been used for many of our experiments. Of course, many species do not absorb in this region, and so we have been developing a Raman-shifted, Nd:YAG-pumped Ti:sapphire laser system as a TRIRDR pump source [13]. This system provides tunable, narrow-bandwidth infra-red pulses between 1 and 3.3 μm (10,000 - 3,000 cm^{-1}), which covers the hydrogen stretching fundamental region as well as numerous overtone and combination bands. The Ti:sapphire laser also affords an improvement in pulsewidth-limited time resolution, viz., 5 - 10 nsec as compared with the 80 - 100 nsec from a TEA-CO_2 laser.

The time-dependent absorption from specific rovibrational levels of the target molecule is monitored using a tunable semiconductor diode laser, which provides the ca. 10^{-4} cm^{-1} resolution necessary to distinguish individual levels. Photoconductive mercury-cadmium-telluride detectors provide the time resolution and sensitivity required for isolating the state-to-state molecular relaxation processes.

3. Results for Individual Systems

Ozone. Total and state-to-state rovibrational energy transfer rates and propensity rules have been determined for ozone in collisions with itself and with atmospheric broadeners such as nitrogen [14,15]. The QFT-ID model [8] was adapted to compute the state-to-state inelastic contributions to the pressure-broadened linewidths. Comparisons of kinetic model simulations using the computed rates with experimental TRIRDR signals confirmed a propensity for first-order dipolar transitions, but also clearly demonstrated the importance of higher-order interactions, not comprehended by the model, leading to large changes in j in single collisions. Simple energy and angular-momentum scaling laws were also found to be inadequate for representing the experimental data.

Ammonia. NH_3 has been the subject of numerous investigations utilizing time-resolved spectroscopy to elucidate energy-transfer and relaxation processes occurring among the rotational and vibrational energy levels of this molecule. Each rotational energy level is split into symmetric (s) and antisymmetric (a) components by inversion doubling, and the splitting increases with excitation in the ν_2 mode. In the vibrational ground state, relaxation is dominated by dipole-like transitions between s and a inversion levels, but these become less dominant in the higher vibrational levels. Processes with $\Delta j > 1$ and $\Delta K > 0$ make significant contributions to rotational relaxation, and parity-conserving ($s \leftrightarrow s$, $a \leftrightarrow a$) transitions occur with a probability of around 70% of that of parity-changing collisions [16]. Neither of these observations is in accord with a first-order dipole interaction model.

Methane. Relaxation in spherical-top molecules such as methane is in principle more complicated than in the systems mentioned previously, because long-range dipole interactions are absent in spherical tops and inelastic collisions are dominated by short-range atom-atom interactions. This is manifested by smaller inelastic cross sections and a somewhat broader distribution of final angular momentum states, as compared with polar molecules. In addition, both time-resolved and linewidth data show a systematic variation of relaxation rate with the symmetry species (A, E, or F) of the molecule. Our TRIRDR measurements on methane-d_4 [17] and silane [18-20] show a further marked selectivity with respect to the C^n fine-structure states [6]. This selectivity has been interpreted by Parson [21] in terms of a correlation between the projection of the total angular momentum j along one of the molecule's symmetry axes and the parities of the initial and final states in the inelastic collison. The predictions of Parson's model are currently being tested by further measurements on methane [13].

4. Concluding Comments

The first-order collision models used to calculate pressure-broadened linewidths are in many cases inadequate to describe inelastic processes at the state-to-state level. Large changes in rotational angular momentum can occur in individual collisions, and the energy-level structure of the molecule is of comparable importance to the interaction potential in determining final-state distributions. Simple energy- or angular-momentum-based scaling laws provide a rough guide to these distributions, but in general are unable to reproduce the observed final-state selectivity. Further development of line-broadening models will need to take these higher-order interactions into account.

The study of these inelastic processes is a small but significant component of the ongoing effort to become aware of the impact of our actions on our environment, in this case by improving the accuracy of atmospheric monitoring techniques. Scientific questions of this type address only one part of the problem, however. What is required, in addition to improving our understanding of the complex interactions between environmental systems and human activities, is the determination to act on the basis of that understanding in order to remediate any damage which may have already occurred to these systems, and to protect them for the future.

Acknowledgments. This research was supported by grants NAGW-1667 from the NASA Upper Atmosphere Research Program and NAGW-2387 from the NASA Planetary Atmospheres Program. We also gratefully acknowledge National Science Foundation grants CHE89-14953 and CHE93-04251 to the G.R. Harrison Spectroscopy Laboratory, and a Traineeship Program in "Chemistry of the Environment" awarded to the M.I.T. Chemistry Department.

References and Notes

1. "Scientific Assessment of Ozone Depletion: 1991", World Meteorological Organization Global Ozone Research and Monitoring Project Report No. 25 (Geneva, Switzerland).
2. R.A. Kerr, Science **260**, 490 (1993); J.F. Gleason, P.K. Bhartia, J.R. Herman, R. McPeters, P. Newman, R.S. Stolarski, L. Flynn, G. Labow, D. Larko, C. Seftor, C. Wellemeyer, W.D. Komhyr, A.J. Miller, and W. Planet, Science **260**, 523 (1993).
3. E.D. Hinkley(ed.), "Laser Monitoring of the Atmosphere", Topics in Applied Physics No. 14, Springer-Verlag, Berlin (1976).
4. L.S. Rothman, R.R. Gamache, R.H. Tipping, C.P. Rinsland, M.A.H. Smith, D.C. Benner, V.M. Devi, J.-M. Flaud, C. Camy-Peyret, A. Perrin, A. Goldman, S.T. Massie, L.R. Brown, and R.A. Toth, J. Quant. Spectrosc. Rad. Transfer **48**, 469 (1992).
5. The quantum number j represents total angular momentum in all cases, but τ takes on different meanings depending on the symmetry of the molecule. τ is identical with j for a linear molecule such as N_2 or acetylene, $\tau = K$ for a symmetric rotor such as ammonia, and $\tau = (K_a,K_c)$ for an asymmetric rotor such as ozone. For a spherical top such as methane, τ includes an additional fine-structure index C^n [see ref. 6].
6. J.-P. Champion, M. Loete, and G. Pierre, in "Spectroscopy of the Earth's Atmosphere and Interstellar Molecules" (K.N. Rao and A. Weber, eds.), pp. 339 ff., Academic Press, New York (1992).

7. C.J. Tsao and B. Curnutte, J. Quant. Spectrosc. Rad. Transfer **2**, 41 (1962).
8. R.R. Gamache and L.S. Rothman, Appl. Opt. **24**, 1651 (1985); R.R. Gamache and L.S. Rothman, J. Mol. Spectroscopy **109**, 283 (1985).
9. G. Millot, J. Chem. Phys. **93**, 8001 (1990).
10. J.I. Steinfeld, P. Ruttenberg, G. Millot, G. Fanjoux, and B. Lavorel, J. Phys. Chem. **95**, 9638 (1991).
11. P.L. Houston and J.I. Steinfeld, in "Laser and Coherence Spectroscopy" (J.I. Steinfeld, ed.), pp. 1-123, Plenum Press, New York (1978).
12. L. Laux, D. Harradine, B.R. Foy, J.I. Steinfeld, and M. Dubs, Proc. 3rd Intl. Conf. on Infrared Physics (W. Ruegsegger and F. Kneubuhl, eds.), pp. 572-575 (Zurich, Switzerland, July 1984).
13. J.J. Klaassen and J.I. Steinfeld (to be published).
14. C.C. Flannery and J.I. Steinfeld, J. Chem. Phys. **96**, 8157 (1992).
15. C.C. Flannery, J.I. Steinfeld, and R.R. Gamache, J. Chem. Phys. (in press).
16. B. Abel, S.L. Coy, J.J. Klaassen, and J.I. Steinfeld, J. Chem. Phys. **96**, 8236 (1992).
17. B. Foy, J. Hetzler, G. Millot, and J.I. Steinfeld, J. Chem. Phys. **88**, 6838 (1988).
18. G. Millot, J. Hetzler, B. Foy, and J.I. Steinfeld, J. Chem. Phys. **88**, 6742 (1988).
19. J. Hetzler, G. Millot, and J.I. Steinfeld, J. Chem. Phys. **90**, 5434 (1989).
20. J.R. Hetzler and J.I. Steinfeld, J. Chem. Phys. **92**, 7135 (1990).
21. R. Parson, J. Chem. Phys. **93**, 8731 (1990); R. Parson, J. Chem. Phys. **95**, 8941 (1991).

Water Molecules in Liquid Solution: Vibrational and Reorientational Relaxation

H. Graener and G. Seifert

University of Bayreuth, Postfach 101251, D-95440 Bayreuth, Germany

Abstract: Using time-, frequency- and polarization resolved Infrared Double Resonance Spectroscopy with picosecond pulses we have investigated the vibrational and reorientational relaxation of water molecules in various solvents. Here we will present results on the relaxation of HDO molecules.

1. Introduction

The investigation of vibrational and reorientational relaxation processes can provide detailed information about interactions and dynamics in a liquid. To improve the understanding of these phenomena we have investigated the relaxation of H_2O and D_2O in various organic solvents [1,2]. The extension of our studies to a further water isotope (HDO) is of great interest. It allows to relate these investigations to a pure water system (HDO:D_2O) [3]; furthermore the separation of inter- and intramolecular contributions to the vibrational relaxation processes can be extended in comparison to the vibrational relaxation processes of H_2O and D_2O.

2. Experimental Results

The details of the Infrared Double Resonance Spectroscopy and the corresponding theoretical description were given recently [2]. In short: an intense, linear polarized laser pulse, tuned to a vibrational resonance, pumps some percent of the molecules to a vibrational excited state. The resulting transmission changes are monitored by a weak, delayable and independently tunable probing pulse. Using a proper polarization geometry for the probe the populational and orientational contributions to the transmission changes can be separated [2]. Measurements as a function of probe frequency (for fixed delay time) allow to identify the populated vibrational states, whereas experiments with variable delay time (at different frequency settings of the probe pulse) provide the desired information about the relaxation times.

The samples were prepared by mixing 1 mol H_2O with 1 mol D_2O, resulting in a 1:2:1 mixture of H_2O:HDO:D_2O. This mixture was diluted in the chosen organic solvent. Fig. 1a shows an example of a corresponding IR Spectrum (solvent 1,2-dichloroethane) in the region of the OD stretch absorption (solvent spectrum subtracted): clearly visible is the strong OD absorption line of HDO at 2673 cm^{-1}; the symmetric and antisymmetric OD stretching absorptions of D_2O at 2626 cm^{-1} and 2729 cm^{-1} are distinctly separated. The absorption frequencies

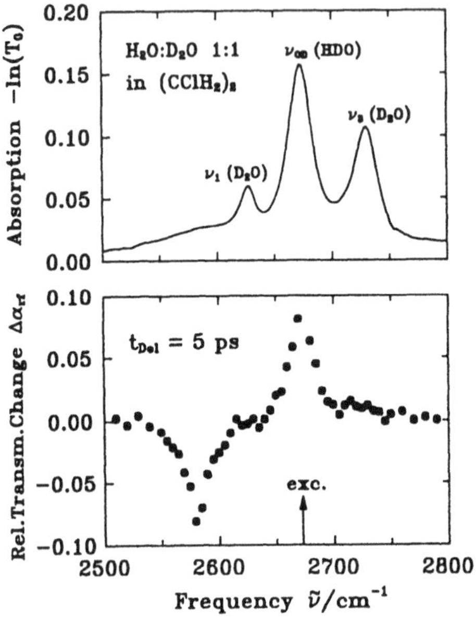

Fig. 1: IR absorption spectra of an equimolar $H_2O:D_2O$ mixture in 1,2-dichloroethane; overall water concentration 0.1 mol/l;
a) conventional spectrum in the region of the OD stretching absorptions;
b) transient spectrum taken 5 ps after an excitation pulse at 2673 cm^{-1}; the rotation free transmission change is plotted versus probe frequency

are found to be independent of the concentration indicating that the water molecules are isolated in the sample (no water-water hydrogen bonds). On the other hand the vibrational frequencies exhibit a significant variation for the different solvents. This can be explained by a strong solvent - solute interaction (hydrogen bonding).

Fig. 1b presents a transient spectrum taken 5 ps after an excitation pulse at the OD absorption frequency of HDO. The main features in this spectrum are the bleaching around the pump frequency and the red shifted induced absorption due to the excited state absorption. The OD stretching absorptions of D_2O (see Fig. 1a) do not contribute to the transient spectrum. Accordingly there is no HDO - D_2O interaction in this solution; furthermore the high selectivity of the transient spectroscopy should be noted.

Fig. 2 gives two examples of time resolved measurements for HDO in dichloroethane. The pump pulse was tuned to ν_{pu}= 3635 cm^{-1}, the OH stretch absorption of HDO. The upper curve (full points) shows the bleaching for a probe frequency $\nu = \nu_{pu}$. The lower curve (open circles) represents a measurement for a probe pulse tuned to the excited state absorption of the OH stretch vibration at 3465 cm^{-1}.
Both curves rise and fall synchronously, indicating that no long-lived intermediate state is involved in the relaxation process of the OH vibration. The falling wings can be fitted with a single exponential decay function, yielding a relaxation time of 73 ps. From similar measurements at the corresponding OD frequencies a population relaxation time of 24 ps was obtained. Additionally HDO in CD_3CN was investigated resulting in a relaxation time of 11 ps for the OH stretching vibration.

Fig. 3 shows the induced dichroism, which is the pure orientational signal, for measurements at the OH vibration (full points) and OD vibration (open circles). The solid lines are the results of model calculations [2,4] using a single exponential decay function with reorientational time constants

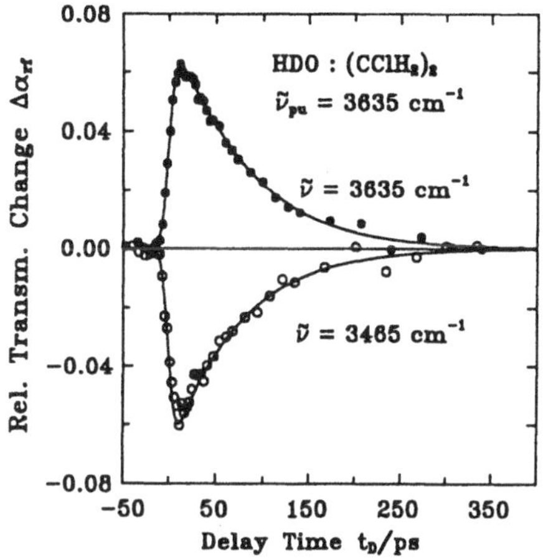

Fig. 2: Relative transmission change (rotation free) vs. delay time for HDO: $(CClH_2)_2$; excitation frequency ν_{pu}= 3635 cm^{-1}; full points: probe frequency $\nu = \nu_{pu}$; open circles ν = 3465 cm^{-1}; experimental points, calculated curves

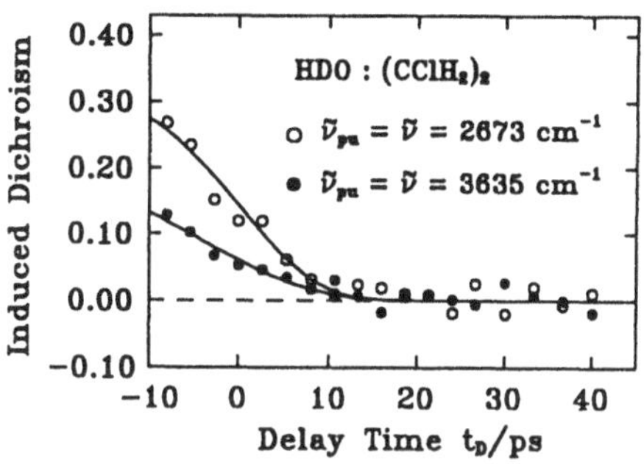

Fig. 3: Induced dichroism as a function of delay time for two frequency settings of pump and probe pulses; full points: $\nu_{pu} = \nu$ = 3635 cm^{-1} (OH vibration); open circles: $\nu_{pu} = \nu$ = 2675 cm^{-1} (OD vibration); sample: 0.1 mol/l HDO in $(CClH_2)_2$; experimental points, calculated curves.

τ of 2 ps (OH) and 4 ps (OD), respectively. From the analysis of the induced dichroism of HDO in CD_3CN a time constant of τ = 6 ps was derived.

3. Discussion

It is interesting to compare the numbers obtained for the vibrational relaxation of HDO to the corresponding values for H_2O in the same solvents [1,2]. The solvent dependence for both systems is very similar: the vibrational relaxation in the system with stronger hydrogen bonding (here acetonitrile) is much faster, whereas the reorientational relaxation shows the reciprocal behavior. Obviously the solvent - solute interaction plays an important role for both relaxation processes. A more detailed comparison yields that the relaxation pathways for these two water isotopes differ notably. In the case of H_2O (in $CClH_2)_2$ a rapid equilibration among the two OH stretching vibrations (time constant 4 ps) followed by a relaxation of this ensemble to the overtone of the bending vibration (time constant 26 ps; energy mismatch -436 cm^{-1}) was observed. For the relaxation of the OH vibration of HDO (time constant 73 ps) two pathways are possible: either via the higher lying combination vibration (OD stretch + bending; energy mismatch +443 cm^{-1}) or via the bending overtone (mismatch -831 cm^{-1}). For both situations the observed slower vibrational relaxation (compared to H_2O) would be expected. The most probable accepting mode for the OD vibration is the overtone of the bending vibration. The small energy mismatch of +131 cm^{-1} corresponds well to the rather short relaxation time.

A final remark to the reorientational relaxation: the experimental results (Fig. 3) indicate that the assumption of a single reorientational time constant for HDO is not justified. For the anisotropic rotational diffusion of HDO the induced dichroism is expected to decay with three exponentials with different amplitudes. Assuming the diagonal elements of the diffusion tensor to be inversely correlated to the corresponding moments of inertia, it can be calculated [5] that the decay of the induced dichroism at early delay times is dominated by single exponentials for both situation shown in Fig. 3; but the relevant time constants differ by approximately a factor of two for the investigated situations (OH- and OD vibration). This is in good agreement with the experimental findings.

References:

[1] H. Graener and G. Seifert, J. Chem. Phys., 98, 36 (1993)

[2] H. Graener, G. Seifert and A. Laubereau, Chem. Phys., in press

[3] H. Graener, G. Seifert and A. Laubereau, Phys. Rev. Lett., 66, 2092 (1991)

[4] G. Seifert and H. Graener, Proceedings of UPS '91, Eds. A. Laubereau and A. Seilmeier, I.O.P. Publishing LTD, Bristol and Philadelphia 1992, p. 539

[5] T.J. Chuang and K.B. Eisenthal, J. Chem. Phys., 57, 5094 (1972)

Femtosecond Molecular Dynamics in the Liquid Phase Studied by Time-Resolved CARS

H. Okamoto[1], *R. Inaba*[1,2], *K. Yoshihara*[2], and *M. Tasumi*[1]

[1]Department of Chemistry and Research Centre for Spectrochemistry, School of Science, The University of Tokyo, 7-3-1 Hongo, Bunkyo-ku, Tokyo 113, Japan

[2]Institute for Molecular Science, Myodaiji, Okazaki 444, Japan

Abstract. Femtosecond time-resolved CARS has been applied to studies of the dephasing of C≡N and C≡C stretching vibrations of *n*-alkane nitriles and alkynes and reorientational relaxation. The dephasing rates depend on the length of the alkyl chain, and the chain-length dependencies are different between nitriles and alkynes. Acceleration of the dephasing by adding hydrogen-bonding solvents has been observed for both nitriles and alkynes. Polarization measurements have been performed for the C≡N stretching of benzonitrile mixed with various solvents in order to obtain information on reorientational relaxation.

1. Introduction

Molecular vibrational motion generally occurs with a period in the femto- to picosecond regime. In order to observe various perturbations on molecular vibrations directly in the time-domain, it is necessary to use time-resolved spectroscopy with femtosecond time resolution. Since Leonhardt et al. first applied femtosecond laser pulses to the observation of time-resolved CARS [1], several groups have reported femtosecond time-resolved CARS measurements. We have constructed a femtosecond time-resolved CARS measuring system, which has the capability of performing polarization experiments [2]. Using this apparatus, we have systematically studied vibrational dephasing and reorientational motion of molecules in the liquid phase. In this paper, we describe the vibrational dephasing of nitriles [3-5] and alkynes [6] with lengths of alkyl chains mixed with various solvents, and reorientational relaxation of benzonitrile [7].

2. Experimental

The experimental setup for time-resolved CARS was already described in Ref. 2. The polarizations of incident and scattered radiation were so chosen that only the isotropic (for the dephasing and reorientation measurements) or anisotropic (for the reorientation measurements) decay component contributes to the signal [8].

3. Results and discussion

3.1 Dependence of the dephasing rates of nitriles and alkynes on the length of the alkyl chain

Temporal profiles of CARS signals for the C≡N stretching vibration (≈2250 cm^{-1}) of nitriles ($C_nH_{2n+1}CN$, n=2~17) show single-exponential decays, and the decay time constant (corresponding to half the dephasing time) depends on the number of carbon atoms in the chain (n). Figure 1A shows a plot of the dephasing rates (reciprocal of the dephasing time) of the C≡N stretching of nitriles against the square root of n. It is noted that the dephasing rates for the neat liquids (filled circles) are nearly proportional to $\sqrt{n}$ in the region of $\sqrt{n}$ >2. When methanol (strong proton-donating solvent) is mixed with the nitriles in 1:1 mole ratio, the dephasing rates increase by about 40 %, regardless of the chain length (filled triangles).

The decay profiles of CARS signals for the C≡C stretching (≈2120 cm^{-1}) of monoalkylacetylenes ($C_nH_{2n+1}C{\equiv}CH$, n=4~10) show slight deviations from the single-exponential functions, which may be caused by beat among several Raman transitions or inhomogeneous effects. However, the deviation is small enough for the purpose of the following qualitative discussion. In Fig. 1B, a plot of the dephasing rates against $\sqrt{n}$ is shown (filled circles). In contrast to the case of nitriles, the dephasing rates are almost independent of the chain length in the region of $\sqrt{n}$ >2. The dephasing rates increase by 9~20 % when dimethylformamide (strong proton-accepting solvent) is added to the sample in 1:1 mole ratio, and the degree of acceleration of dephasing seems to be independent of n.

It is believed that the vibrational dephasing of small molecules in the liquid state is predominantly controlled by intermolecular processes (e.g., collisional dephasing). However, the present results on nitriles suggest that intramolecular processes contribute to vibrational dephasing for relatively large nitriles. On the other hand, the results on alkynes may be consistent with the view that intermolecular processes determine the most part of vibrational dephasing. Such a difference may be attributed to the presence and absence of the dipole associated with the C≡N and C≡C bonds,

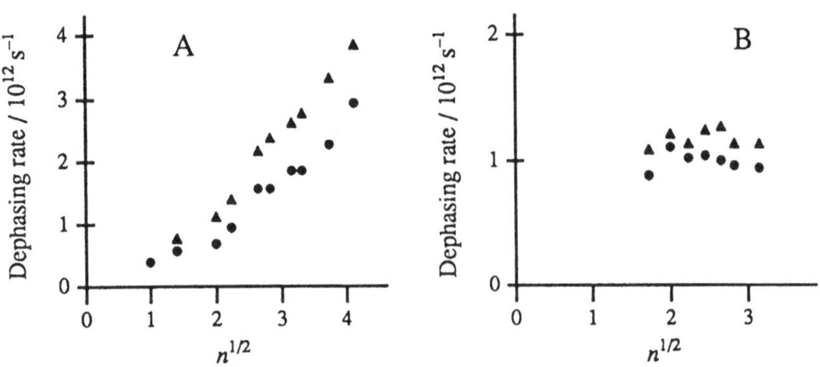

Figure 1. Plot of the dephasing rates of the C≡N stretching of nitriles (A) and the C≡C stretching of alkynes (B) against the square root of n. ●, neat liquids; ▲, 50 % mixtures with methanol (A) and 50 % mixtures with dimethyl formamide (B).

respectively. It is conceivable that the dephasing of the C≡N stretching is affected by interactions between its own dipole and dipoles associated with the methylene groups in the same molecule but located far from the C≡N group.

The effect of adding proton-accepting solvent in alkynes is reasonably understood in terms of intermolecular dephasing processes. Acceleration of dephasing by adding proton-donating solvents in nitriles also suggest an important contribution of intermolecular processes even in large nitriles, which seemingly contradicts the above view that the intramolecular processes have significant contributions to the dephasing of the C≡N stretching vibration. Further studies are needed in order to clarify the dephasing mechanism for nitriles.

3.2 Effects of hydrogen-bonding solvents on the dephasing of the C≡N stretching of acetonitrile and the C≡C stretching of 1-butyne

We have examined the correlation between the dephasing rate and the hydrogen-bonding ability of the solvent for the C≡N stretching of acetonitrile and the C≡C stretching of 1-butyne. It is found that the dephasing time constant is negatively correlated to the proton-donating (for both acetonitrile and 1-butyne) and accepting (for 1-butyne) abilities of the solvent. This result may be interpreted as follows.

The CN group of acetonitrile is strongly polar and acts as a proton-accepting site. Therefore, it is expected that an acetonitrile molecule interacts with a proton-donating molecule, and that the vibrational dephasing of the C≡N stretching is accelerated in the presence of a proton-donating solvent. The C≡CH group of 1-butyne is also polarized at the C–H bond: the H atom and the C≡C group are positively and negatively charged, respectively. Therefore 1-butyne may interact with both proton-donating and accepting molecules.

3.3 Reorientational relaxation of benzonitrile

From the measurements of isotropic and anisotropic components of time-resolved CARS, we can obtain the rotational correlation function for a totally-symmetric Raman transition, directly in the time domain. We have obtained the rotational correlation functions for the C≡N stretching of benzonitrile (≈2230 cm^{-1}) mixed with various solvents. All the mixtures examined gave single-exponential correlation functions. Neat benzonitrile showed slightly non-exponential behavior in short time regime (Δt<0.7 ps). The reorientational relaxation time has been obtained from the decay time constant of the rotational correlation function, and is plotted against the viscosity of the mixture in Fig. 2. The reorientational relaxation time should be linearly correlated to the viscosity within the framework of macroscopic Stokes-Einstein model [9]. In Fig. 2, a roughly linear relationship is found in the mixtures with viscosities <≈1.25 mPa s. From the slope of this linear relationship, ≈1.8 Å is obtained as the radius of molecule, when we apply the Stokes-Einstein relation. This may be a reasonable value for the benzonitrile molecule. The Stokes-Einstein model seems to hold for the low-viscosity mixtures of benzonitrile.

On the other hand, deviations from the prediction of the Stokes-Einstein model are observed for the mixtures with viscosities >≈1.25 mPa s. For these mixtures, no clear relationship seems to exist between the reorientational relaxation time and viscosity. In the present stage, we have no clear interpretation of this result.

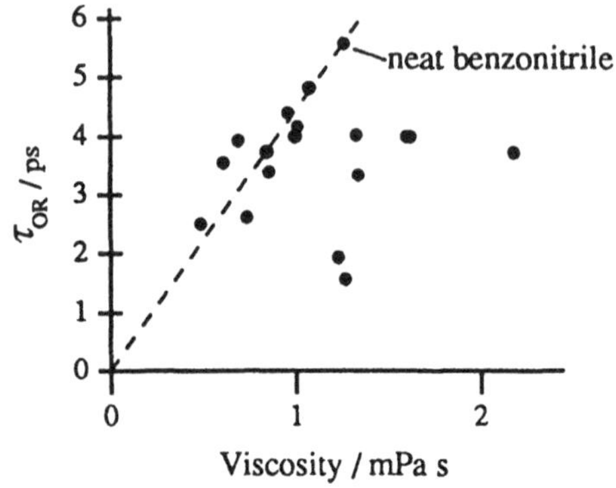

Figure 2. Plot of the reorientational relaxation times (τ_{OR}) of benzonitrile against the viscosities of various mixtures.

References

[1] R. Leonhardt, W. Holzapfel, W. Zinth, W. Kaiser: *Chem. Phys. Lett.*, **133**, 373 (1987).

[2] H. Okamoto, K. Yoshihara: *J. Opt. Soc. Am. B*, **7**, 1702 (1990).

[3] R. Inaba, H. Okamoto, K. Yoshihara, M. Tasumi: *Chem. Phys. Lett.*, **185**, 56 (1991).

[4] R. Inaba, H. Okamoto, K. Yoshihara, M. Tasumi: *J. Phys. Chem.*, **96**, 8385 (1992).

[5] H. Okamoto, R. Inaba, M. Tasumi, K. Yoshihara: *Chem. Phys. Lett.*, **206**, 388 (1993).

[6] R. Inaba, H. Okamoto, K. Yoshihara, M. Tasumi: *J. Phys. Chem.* (submitted)

[7] H. Okamoto, R. Inaba, K. Yoshihara, M. Tasumi: *Chem. Phys. Lett.*, **202**, 161 (1993).

[8] B. Dick: *Chem. Phys.*, **113**, 131 (1987).

[9] D. Kivelson, M. G. Kivelson, I. Oppenheim: *J. Chem. Phys.*, **52**, 1810 (1970).

Time-Resolved Double Raman-Raman Resonance

V.I. Fabelinskii[1], *L. Holz*[2], *V.V. Smirnov*[1], and *K.A. Vereschagin*[1]

[1]Institute of General Physics, Vavilov Street 38, 117942 Moscow, Russia
[2]Max-Born-Institute, Rudower Chaussee 6, D-12474 Berlin, Germany

Abstract. Biharmonic (Raman) excitation has been employed to create nitrogen molecules in the first vibrational state. Coherent Anti-Stokes Raman Scattering (CARS) of Q-branch of the vibrational band 1-2 of N_2 molecules has been used to probe the temporal behaviour of the first vibrational level population with temporal resolution of about 20 ns, which has allowed to measure the rate constant of the v-v exchange in nitrogen.

1. Biharmonic excitation

For a number of molecules without a dipole moment and for the nondipole-active transitions of many polyatomic molecules, biharmonic excitation allows to prepare a molecular ensemble with up to 30% molecules selectively excited into the first vibrational state. This method was used for the first time in [1] for vibrational excitation of H_2 molecules for investigating v-T-relaxation processes and was developed later in [2,3,4] for v-v exchange.

Energy redistribution over the vibrational mode in nitrogen molecules is mainly conditioned in by the processes of collisional single-quantum vibrational-vibrational (v-v) exchange:

$$N_n + N_{m+1} \xrightarrow[K_{(n+1,m)(n,m+1)}]{} N_{n+1} + N_m + \Delta E, \qquad (1)$$

where N_n is the population of the Nitrogen n = v - vibrational level; $K_{(n+1,m)(n,m+1)}$ are the rate constants of the v-v process; ΔE is the anharmonic energy defect. Density N_n of the vibrationally excited molecules in the zone of excitation can be obtained from equations that express the conservation law for the number of molecules in the volume unit:

$$dN_n(r,t)/dt+div(j_n(r,t)) = q_{1n}(r,t) + q_{2n}(r,t) + (dN_n(r,t)/dt)_{vv}. \qquad (2)$$

In the absence of mass-transfer and heating of the gas, $j_n = -D_p\, grad(N_n)$ in (2) is the density of the diffusion flow of the vibrationally excited molecules. The sum on the right side is the density of distributed sources characterizing the alteration of the n-th vibrational level population : $q_{1n}(r,t)$ describes the pump and $q_{2n}(r,t)$ - all other (including the v-T relaxation) process in the excitation zone; $(dN_n(r,t)/dt)_{vv}$ describes the vibrational levels population kinetics, where $(dN_n/dt)_{vv} = f_n$ $(K_{(n,m+1)(n+1,m)},$

$N_n, N_m, N_{n+1}, N_{m+1}$). With selective excitation of the first vibrational state by a short pump pulse ($\tau_p \sim$ 10-15 ns in duration) levels with n = 2 are not populated to values close to the stationary value determined by v-v exchange and in the initial formulation stage of the quasistationary vibrational distribution function (VDF), advantageous depopulation of the first level occurs. Under these conditions from the system (2) we obtain for $N_1(t)$ simple expression:

$$N_1(0)/N_1(t) = 4\,K_{(10)(01)}N_1(0)\,t + 1 \qquad (3)$$

Thus, relation $N_1(0)/N_1(t)$ must be near linear and the slope of these lines to the time axis is proportional to rate constant $K_{(10)(01)}$.

2. Experimental technique

In the experiment a pulsed CARS-spectrometer, as described in [5], was employed. Biharmonic excitation was provided by the second harmonic from a single-mode YAG:Nd^{3+} laser with an energy up to 40 mJ per pulse with a pulse duration of 15 ns and a spectral width of 0.5 cm^{-1} as well as emission from a dye laser with an energy up to 10 mJ and a spectral width of 2 cm^{-1}. The excitation beams at ω_1 and ω_2 and the probe beams at ω_1 and ω_2 were focused onto a cell; these beams were counterpropagating and had perpendicular polarizations. The beam dimensions in the constriction region were w_0 = 30 μm for the excitation beams and w_0 = 10 μm for the probe beams.

To determine the rate constant of vibrational-vibrational exchange in nitrogen a part of the molecules is raised to the first excited vibrational state over the duration of the biharmonic (at the transition v=0 - v=1) pumping pulses τ_p=15 ns. The change in population over time was recorded by means of pulsed CARS-probing the v=1 - v=2 transition over a delay of τ_d. In order to investigate the behaviour of the level population with v=1 as a function of time and density a family of spectra for various gas pressures and various delays (τ_d = 50 ns to 2.5 μs, accuracy 10 ns) was measured. At constant pressure the degree of excitation was determined by the ratio of the intensities of the rotational lines (J = 8-12) of the vibrational-rotation al transition v=1 - v=2 to the intensity of the nonresonance CARS signal.

3. Results and discussion

Experimentally we succeed to excite up to 30% of the molecules in the first excited vibrational state. The corresponding populations over various delays are shown in Fig.1 (points).

The temporal evolution of the CARS peak intensities of the Q-branch lines (v=1- v=2) (which was used to probe the vibrational populations) takes place not only because of changes of the VDF due to v-v exchange but also owing to diffusion of the vibrationally excited out of and unexcited molecules into the zone of interaction of the light beams. Really, the total alteration of the N_1 over the

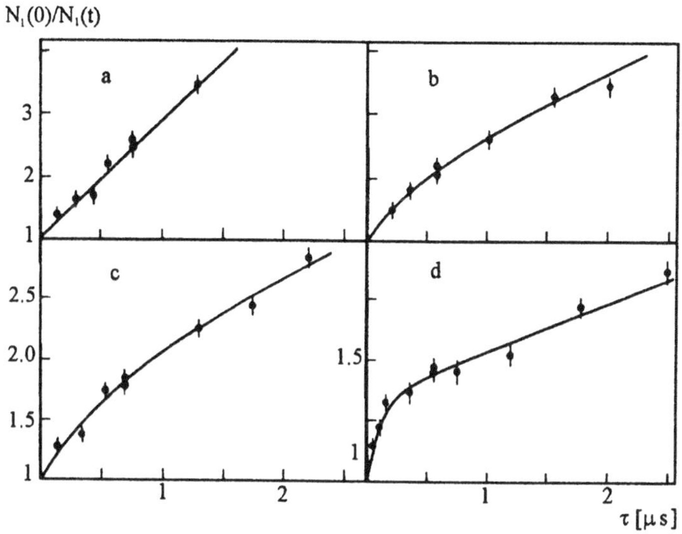

Fig. 1 Experimental (circles) and calculated (solid line) values of $N_1(0)/N_1(t)$; a) p = 0.25 atm, β = 0.12; b) p=0.45 atm, β = 0.3; c) p = 0.7 atm, β = 0.15; d) p=2.25 atm, β = 0.11; $\beta = N_1(0)/N_0(0)$

observation time is greater then it should be due to v-v process only at the experimentally achieved degrees of excitation; from 500 ns the experimental relations $N_1(0)/N_1(t)$ are nearly linear and the slope of these lines to the time axis is inversely proportional to pressure. Estimates show, that at gas pressure P_0 = 1 atm over the time t ~ 1 μsec the characteristic diffusion length is with about $l_0 = 2 (D_0 t)^{1/2}$ ~10 μm comparable to the focal radii of the laser beams.

The v-v exchange is accelerated by increasing the pressure, or, in contrast, the diffusion rate is slowed down by decreasing the spatial distribution gradients of the excited molecules. Both approaches have some shortages: the appearance of laser induced breakdown in the beam waist while increasing the pressure, and a decreased pump rate when using slightly focused laser beams only. Therefore we have resorted to numerical modeling of the temporal behaviour of the peak CARS intensities under conditions when both the v-v exchange and the diffusion of the excited molecules take place. The eq.s (2) were solved. The diffusion coefficient $D_p=D_0(P_0/P)$ of the vibrationally excited molecules is assumed to be independent of n. The numerical results are given in the form of solid curves in Fig. 1. In order to determine the v-v exchange rate constant of N_2 in the calculations, the quantity $K_{(10)(01)}$ as well as the radii of the excitation and probe beams were varied. The rate constants $K_{(10)(01)}$ measured at different pressures agree well with each other. The best agreement with experimental values was reached for $K_{(10)(01)}$= (10 ± 4) 10^{-14} cm^3/sec.

References

1. F. De Martini, Ducuing: Phys.Rev.Lett. **17**, 117-119 (1966)
2. R. Frey, J. Lukasik, J. Ducuing: J.Chem.Phys.Lett **14**, 514-517 (1972)
3. V.A.Arkhipov,N.G.Basov,E.M.Belenov et al.: Pis'ma v ZhETF **16**, 469-471 (1982)
4. S.M.Gladkov, M.G.Karimov, N.I. Koroteev: Pis'ma v ZhETF **35**, 381-383 (1982)
5. Laser-Induced Raman Spectroscopy in crystals and Gases, Proceedings of GPI, Vol 2 ed P.P. Pashinin, Nova Science Publishers, 1988

Time-Domain Coherent Polarization Spectroscopy of Inhomogeneously Broadened Transitions: Theory and Experimental Possibilities

N.I. Koroteev, D.V. Kolomoitsev, and S.Yu. Nikitin

International Laser Center, Moscow State University, 119899 Moscow, Russia

Introduction

Time-domain Coherent Antistokes Raman Spectroscopy (CARS) provides valuable information about interaction of particles in gaseous media [1]. Here we present brief outline of our theoretical models and give some quantitative parameters, relating to Hydrogen, Nitrogen Ammonia and Methane molecules, as well as to Thallium and Thulium atoms.

Theory

In experiments [2–10] on time-domain optical spectroscopy of molecular and atomic gases the most important role play the following mechanisms of molecular and electronic vibrations:

• Doppler dephasing caused by the thermal motion of particles in the gaseous medium,

• quantum dephasing due to fine structure of the transition under investigation,

• collisional dephasing caused by elastic and inelastic collisions.

Doppler dephasing is described by the following pulse-response function [1, 2, 11]:

$$h_{\rm D}(\tau) = \exp\left[-\beta(y - 1 + e^{-y})\right], \qquad (1)$$

where $\beta = (k_0 \sigma_v \tau_v)^2$, $y = \tau/\tau_v$, k_0 is the wave number of the transition under investigation, σ_v^2 is the dispersion of thermal velocities, τ_v is the thermal velocities correlation time. Formula (1) assumes the exponential correlation of thermal velocities.

Quantum dephasing arises as a result of splitting of energy levels, forming the optical transition. Examples of such splitting are rotational splitting of vibrational levels in molecules and hyperfine splitting of electronic levels in atoms. In the simplest case quantum dephasing can be described by formula

$$h(t) = \sum_J p_J \, e^{i\omega_J t}, \qquad (2)$$

where J is the quantum number, ω_J is frequency of optical transition, p_J is statistical weight of J-state. Pulse-response function (2) can be presented also

Springer Proceedings in Physics, Vol. 74 Time-Resolved Vibrational Spectroscopy VI
Editors: A. Lau, F. Siebert, W. Werncke

in the equivalent form as a set of independent differential equations:

$$h(t) = \sum_J h_J, \quad h_J(t=0) = p_J, \tag{3}$$

$$dh_J/dt \equiv \dot{h}_J = i\omega_J h_J. \tag{4}$$

Inelastic collisions of atoms or molecules in the gas cause the quantum number J to be random function of time: $J = J(t)$. In its turn, it leads to random fluctuations of the frequency of transition: $\omega = \omega(t)$. This effect can be called as frequency exchange or spectral exchange. To take it into account, we modify equations (4) to the set of coupled equations

$$\dot{h}_J = i\omega_J h_J + \sum_{J'} \Gamma_{J'J} h_{J'}, \tag{5}$$

where $\Gamma_{J'J}$ is the spectral exchange operator. Nondiagonal elements of this matrix have the sense of probability of transition $J' \to J$ per unit time due to inelastic collisions.

Diagonal elements

$$|\Gamma_{JJ}| = -\Gamma_{JJ} = 1/\tau_J, \tag{6}$$

where τ_J is the collisional life-time for J-state.

In general case spectral exchange operator must satisfy two conditions: detailed equilibrium principle

$$p_J \Gamma_{JJ'} = p_{J'} \Gamma_{J'J} \tag{7}$$

and conservation of a number of particles

$$\Gamma_{JJ} = -\sum_{J' \neq J} \Gamma_{JJ'} \tag{8}$$

In the strong collisions approximation the probability of transition is supposed to be proportional to the statistical weight of the final state:

$$\Gamma_{JJ'} = (p_{J'} - \delta_{JJ'})/\tau_0. \tag{9}$$

Here τ_0 is the mean time between inelastic collisions. In this case the set of kinetic equations takes the form [1, 12]:

$$\dot{h}_J + h_J/\tau_0 = i\omega_J h_J + (p_J/\tau_0) \sum_{J'} h_{J'}. \tag{10}$$

The same result was obtained by D'yakov with the help of the model of generalized random telegraph signal or "Kubo–Anderson process" [13].

Burshtein and Storozhev have proposed more complicated model of the spectral exchange for molecular nitrogen [14]. In this model the cross-section of rotational inelastic collision is described by formula ($J > J'$):

$$\sigma_{JJ'} = a(2J'+1)\ \exp\left[-c\sqrt{\varepsilon-\varepsilon'}\left(\sqrt{\varepsilon}+\sqrt{\varepsilon'}\right) - b(\varepsilon+\varepsilon')\right]\cdot I_0\left(2b\sqrt{\varepsilon\varepsilon'}\right), \quad (11)$$

where $\varepsilon = J(J+1)$, $\varepsilon' = J'(J'+1)$, J and J' are rotational quantum numbers, I_0 is the modified Bessel function. The shape of the surface proper $\sigma_{JJ'}$ is determined by two independent parameters. One of them, b, determines the strength of collisions, and the other, c their adiabaticity. Varying the parameter b, one can obtain a continuous transition from "strong" collisions ($b \ll 1$) to "weak" ones ($b \gg 1$).

Dephasing, caused by adiabatic (elastic) collisions, can be described by equation

$$\dot{h}_J + \alpha_J h_J = i(\omega_J + \beta_J)h_J, \quad (12)$$

where parameters α_J and β_J describe, correspondingly, spectral line broadening and shift due to adiabatic collisions. Unifying equations (10) and (12), we get the following kinetic equation [15]:

$$\dot{h}_J + (\alpha_J + 1/\tau_0)h_J = i(\omega_J + \beta_J)h_J + (p_J/\tau_0)\sum_{J'} h_{j'}. \quad (13)$$

Note, that in this model quantum and collisional dephasings are not, generally speaking, statistically independent.

Results

Hydrogen. Q-branch of molecular vibrations of hydrogen ($4155\,\mathrm{cm}^{-1}$) was investigated by time-domain CARS at room temperature and different pressures of gas [1, 2]. Doppler and adiabatic collisional dephasings are the most essential in this case. Experimental data agree well with the theory, based on the model (1). Using experimental data, we have estimated the following fitting parameters of our model: collisional dephasing time $T = A/\rho$ and correlation time of the thermal velocities $\tau_v = B/\rho$.

Here $A = (6.9 \pm 0.5)$ ns $B = (0.13 \pm 0,02)$ ns ρ is the density of gas in Amagat units. These estimations agree well with the data, obtained from the spectral measurements.

Nitrogen. Q-branch of molecular vibrations of nitrogen ($2330\,\mathrm{cm}^{-1}$) was investigated by time-domain CARS at low temperatures (supersonic free jet) [3] as well as for normal conditions [1, 4]. Rotational dephasing and spectral exchange are the most important for this system. Using the strong collisions approximation, we have estimated the cross-section of rotational inelastic collisions σ_R for different temperatures. We have found [16]: $\sigma_R(T = 295\ \mathrm{K})= 61\text{Å}^2$, $\sigma_R(T = 35\ \mathrm{K})= 208\ \text{Å}^2$, $\sigma_R(T = 25\ \mathrm{K})= 855\text{Å}^2$, Using the model (11), we have estimated all elements of the spectral exchange matrix for nitrogen

[14]. The cross-section of rotational energy relaxation for nitrogen at the room temperature is found to be $\sigma_E = 12$ Å^2.

Ammonia. Q-branch of molecular vibrations ν_1 mode 3334 cm^{-1} was studied by time-domain CARS at room temperature and different pressures of the gas [1, 5, 6]. Using experimental data and taking into account Doppler dephasing, rotational dephasing and spectral exchange, we have estimated the cross-section of rotational inelastic collisions [17]: $\sigma_R = 243$ Å^2.

Methane. Q-branch of ν_3 vibrational mode of methane was studied by Free Induction Decay Spectroscopy at room temperature and different pressures [7]. Using the same model as in the case of ammonia, we have estimated the cross-section of rotational inelastic collisions [17]: $\sigma_R = 38$ Å^2.

Thallium and Thulium. Atomic Thallium and Thulium were investigated by time-domain CARS at temperatures 1000 K and 1500 K and different pressures of buffer gas [8-10]. Using the model, taking into account hyperfine splitting of atomic energy levels, Doppler dephasing, collisional dephasing and spectral exchange [10, 18], we have explained some qualitative features in pulse-response functions, observed for Tl and Tm atoms. We also have estimated cross-sections of different dephasing processes for Thulium in the atmosphere of buffer gases He, Ne, Xe [10].

Conclusion

Time-domain optical spectroscopy and developed theoretical model, taking into account Doppler dephasing, quantum dephasing, collisional dephasing, spectral exchange and parameters of laser pulses enable one to measure kinetic, thermodynamic and nonlinear optical parameters of gaseous media. Predictions of the theory agree well with experimental data on time-domain optical spectroscopy for the following molecular and atomic gases: Hydrogen, Nitrogen, Ammonia, Methane, Thulium, Thallium.

Authors would like to thank Yu. E. D'yakov, V. G. Tunkin, V. B. Morozov, I. G. Konovalov, A. I. Burshtein, A. V. Storozhev and E. I. Lobodenko for fruitful cooperation.

References

1. D. V. Kolomoitsev, S. Yu. Nikitin, Proc. SPIE, 1402, 11 (1991).
2. Yu. E. D'yakov, S. A. Krikunov, S. A. Magnitskii, S. Yu. Nikitin, V. G. Tunkin, JETP, 84, 2013 (1983), in Russian.
3. S. A. Akhmanov, N. I. Koroteev, S. A. Magnitskii, V. B. Morozov, A. P. Tarasevich, V. G. Tunkin, J. Opt. Soc. Am. B, 2, 640 (1985).
4. A. P. Tarasevich, Ph.D. Dissert., Moscow State University, Moscow (1985), in Russian.

5. S. A. Magnitskii, Ph.D. Dissert., Moscow State University, Moscow (1983), in Russian.
6. D. V. Kolomoitsev, E. I. Lobodenko, S. A. Magnitskii, S. Yu. Nikitin, V. G. Tunkin, Optics and Spectroscopy, 70, 321 (1991), in Russian.
7. K. Bratengeier, H. -G. Purucker, A. Laubereau, Opt. Commun., 70, 393 (1989).
8. F. Sh. Ganikhanov, V. N. Kuliasov, I. G. Konovalov, V. B. Morozov, V. G. Tunkin, Optics and Spectroscopy, 70, 483 (1991), in Russian.
9. F. Sh. Ganikhanov, I. G. Konovalov, V. N. Kuliasov, V. B. Morozov, V. G. Tunkin, JETP Lett., 54, 433 (1991), in Russian.
10. F. Sh. Ganikhanov, D. V. Kolomoitsev, I. G. Konovalov, V. N. Kuliasov, V. B. Morozov, S. Yu. Nikitin, V. G. Tunkin. Russ. Acad. Sci. Bull., Ser. Phys., 57, 154 (1993), in Russian.
11. Yu. E. D'yakov, JETP Lett., 37, 14 (1983), in Russian.
12. D. V. Kolomoitsev, S. Yu. Nikitin, Optics and Spectroscopy, 61, 1201 (1986), in Russian.
13. Yu. E. D'yakov, Preprint N 43, Moscow State University (1987), in Russian.
14. A. I. Burshtein, D. V. Kolomoitsev, S. Yu. Nikitin, A. V. Storozhev, Chem. Phys., 150, 231 (1991).
15. D. V. Kolomoitsev, S. Yu. Nikitin, Optics and Spectroscopy, 71, 809 (1991), in Russian.
16. D. V. Kolomoitsev, S. Yu. Nikitin, Optics and Spectroscopy, 66, 286 (1989), in Russian.
17. D. V. Kolomoitsev, S. Yu. Nikitin, Inst. Phys. Conf. Ser. No 126: Section 3, p. 213, paper presented at Int. Symp. on Ultrafast Processes in Spectroscopy, Bayreuth, Germany (1991), IOP Publishing Ltd, 1992.
18. D. V. Kolomoitsev, S. Yu. Nikitin, Optics and Spectroscopy, 73, 862 (1992), in Russian.

Intermolecular Coherence and Vibrational Dephasing of BIC J-Aggregates

V.F. Kamalov[1,2], *R. Inaba*[2,3], *I.A. Struganova*[2,4], *M. Tasumi*[3], *and K. Yoshihara*[2]

[1]Photochemistry Department, Institute of Chemical Physics, Russian Academy of Sciences, Kosygin Str. 4, 117977 Moscow, Russia
[2]Institute for Molecular Science, Myodaiji, Okazaki 444, Japan
[3]Department of Chemistry, University of Tokyo, Tokyo 113, Japan
[4]Lasers for Photochemistry Ltd., 117977 Moscow, Russia

Abstract. BIC J-aggregates are studied by femtosecond CARS and fluorescence spectroscopy. Subpicosecond dephasing of Raman transitions of J-aggregates is observed at room temperature, which is accompanied by femtosecond beats of Raman transitions. Temperature dependence of BIC J-aggregate superradiant decay in glass is measured.

1. Introduction

Molecular J-aggregates are characterized by a red shifted sharp (relatively to monomer) absorption band originated by exciton resonance. This is described by delocalization of an electronic excitation in a well-aligned molecular aggregate [1]. The optical characteristics of J-aggregates have recently been studied with pseudoisocyanine (PIC) by several time-resolved spectroscopic techniques [2-4]. J-aggregates are the subject of high interest due to high values of nonlinear optical susceptibilities [5,6].

In this paper the properties of ground and excited electronic states of carbocyanine BIC J-aggregate are studied. This dye forms aggregates at much lower concentrations compare to PIC which allows to prepare samples of dilute solutions with low absorbance and scattering. We use time-resolved CARS to measure vibrational dephasing. Intermolecular coherence of molecules in J-aggregate is studied by picosecond time-resolved emission spectroscopy at very low temperature.

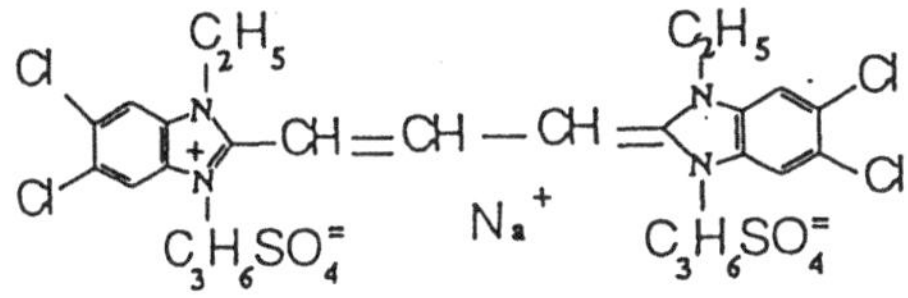

2. Time-resolved CARS

BIC J-aggregates formed in water/ethylene glycol(EG) mixture (v/v=9/1) had a sharp (300 cm^{-1})absorption band with the maximum at 595 nm. The intensity of CARS signal at the anti-Stokes frequency $\omega_a=2\omega_1-\omega_2$ was measured as a function of delay time Δt

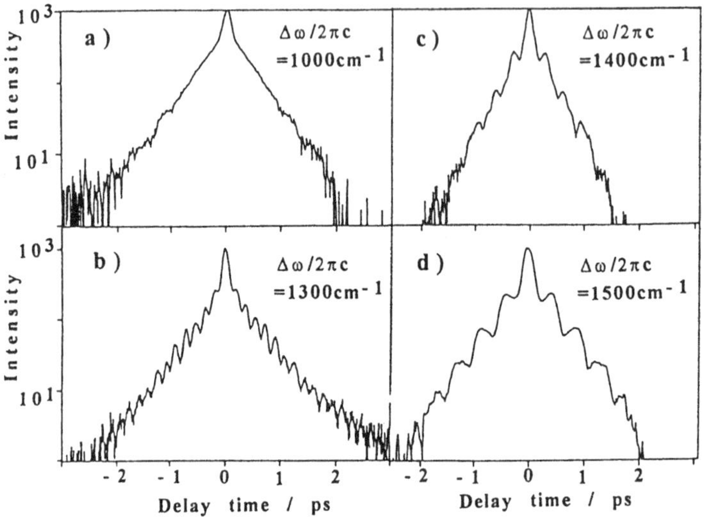

Fig.1.Time-resolved CARS of BIC J-aggregate at different $\Delta\omega$.

between two femtosecond pulses (85 fs FWHM) at 592 nm, in the resonance with the excitonic transition. Picosecond pulses (7 ps FWHM) of second dye laser were tunable in the 610-690 nm region [7]. Two beams ω_1 and ω_2, with peak powers 2 kW and 0.1 kW, respectively, and a probe beam ω_1 with 2 kW, were focused by a lens (f=4.0 cm) in a jet stream of J-aggregate solution with a thickness of 300 μm. CARS beams were linearly polarized. All polarization vectors were parallel to one another, which corresponded to measurement of CARS signal determined by χ_{1111} component of $\chi(\omega_a,\omega_1,\omega_1-\omega_2)$ tensor. The CARS signal intensity plotted against Δt at different $\Delta\omega$ is shown in fig.1 for 295K. CARS intensity at $\Delta\omega/2\pi c$= 1000 cm^{-1} presented in curve a) fig.1 shows a fast response at short Δt followed by a longer exponential decay. Two components in decay profile can be assigned to excitonic and Raman susceptibilities, correspondingly. Dephasing time of the excitonic transition is shorter than 100 fs [8]. The absolute value of signal increased significantly when dye laser tuned closer to excitonic transition due to resonant enhancement.

Assignment of longer component in curve a) to Raman dephasing is also based on the CARS decay measurements, which shows different behavior at different ω_2. CARS decay measured at $\Delta\omega/2\pi c$=1300 cm^{-1} is shown in curve b) fig.1. Here CARS decay is accompanied by many peaks with period of about 180 fs. This 180 fs period corresponds to high frequency (≃5.5 THz) beats. This beat is due to coherent excitation of two different vibrational modes of J-aggregate by a broad spectrum of femtosecond pulse (85 fs duration corresponds to 400 cm^{-1}). The decay time at this $\Delta\omega$ (1300 cm^{-1}) is 0.44 ps. CARS decay measured at $\Delta\omega/2\pi c$=1400 cm^{-1} (curve c)) and 1500 cm^{-1} (curve d)) also shows decay accompanied by ultrahigh frequency beats. Vibrational dephasing time and parameters estimated from these decay curves are presented in Table 1.

Table 1. Vibrational Dephasing and Ultrahigh Frequency Beats of BIC J-aggregates

Detuning $\Delta\omega/2\pi c$, cm^{-1}	Dephasing $T_2/2$, ps	Beats Period fs	Raman Modes Beating, $\nu_1-\nu_2$, cm^{-1}
1000	0.43	-	-
1300	0.44	180	185
1400	0.32	300	110
1500	0.39	275	120

The dephasing time of all Raman modes measured was on the subpicosecond time scale and changes slightly from mode to mode as can be seen from Table 1. The dephasing time of BIC monomer was not measured due to nonresonant conditions for the case of monomer. Assuming that the dephasing time of such a Raman modes of BIC monomer is typical for dephasing time of organic dyes in the subpicosecond range, changes of Raman dephasing due to aggregation seems not significant.

3.Temperature Dependence of Superradiant Decay

Time-correlated photon counting (FWHM=70ps) used for emission decay measurements. J-aggregates were excited by a pulsed dye laser (FWHM 5 ps) at 582 nm. Low pump intensity was used to prevent annihilation of singlet excitons in J-aggregate [4].

Emission decay of BIC J-aggregate in water/EG glass (v/v=1/1) measured at emission maximum at 595 nm at T = 8 K and 65 K is presented in fig.2. It is clearly seen that emission decay at T = 65 K is faster than at T = 8 K. Assuming exponential decay we obtained lifetime of 100 ps at 8K.

To estimate the number (N) of molecules forming exciton at 4K, the nonradiative decay processes was assumed to be negligible in BIC J-aggregate at 4K. Radiative lifetime of BIC J-aggregate is τ_{rad}=100 ps under such an assumption. J-aggregate radiative rate constant is related with the radiative rate constant of monomer (k^m_{rad}) and number N of molecules locked to form an exciton [9]:

$$k^J_{rad} = N\, k^m_{rad} \cos^2\alpha \qquad (1)$$

N is estimated to be N=12-13 (here we used the radiative rate constant of monomer $k^m_{rad}=7.8\cdot10^8 s^{-1}$, which we measured for BIC monomer in methanol at 4 K and assumed $\cos\alpha^2=1$).

The temperature dependence of BIC decay (curve a), fig.3) shows an increase of lifetime at lower temperature. It suggests directly that nonradiative relaxation channel exists in BIC J-aggregate. Emission lifetime changes mainly in the temperature range of 20-60K, therefore the activation energy ΔE of this nonradiative channel can be roughly estimated (ΔE = kT) as 6.0 x 10^{-22} J, which corresponds to 30 - 40 cm^{-1}.

Several possible mechanisms can be proposed for the nonradiative decay channel. i) Internal conversion directly to the ground electronic state with the transformation of about 2

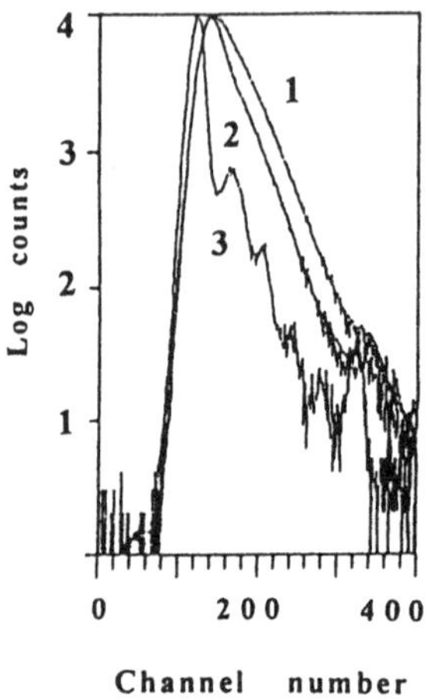

Fig.2. Emission decay of BIC J-aggregates in glass at T = 8 K (1), 65 K (2). Curve 3 - response function. 1 channel- 5.15 ps. Excitation 582nm.

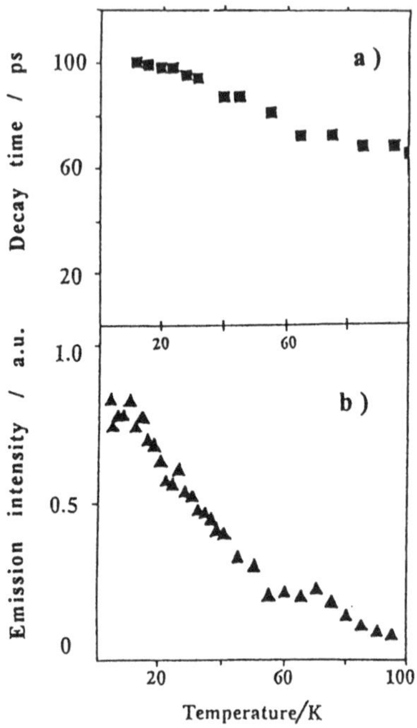

Fig.3. BIC J-aggregate emission decay time (a), emission intensity (b) temperature dependence. Excitation 582nm

eV electronic energy into vibrational energy. ii) Nonradiative decay to another excited electronic state of J-aggregate located close to initially excited state. The symmetry of this state should be the same as that of ground electronic state so that the single photon transition is forbidden. iii) Intra-aggregate electron transfer to a trap at the special site of aggregate (for example at the end of J-aggregate chain). A more detailed study to choose one of these mechanisms is necessary.

Our data shows (curve b) in fig.3) that the increase of emission intensity at lower temperature is much larger in extent compare to decay times. Thus, it can be supposed that the radiative rate which is given by (quantum yield of emission)/(emission lifetime) decreased with increasing of temperature. The above-stated tendency is the same as that observed by de Boer and Wiersma [3]. In [3], however, we can see a region (at temperature below 50K) where the radiative rate was temperature independent. This data were used as a test of Mukamel theory [10]. It was found in [10] that this data can be best fitted in the suggestion that the optical phonon with 240 cm^{-1} frequency is primary responsible for destruction of intermolecular coherence in PIC J-aggregate. In the case of BIC J-aggregate both lifetime and intensity were temperature dependent at 10 - 100 K, so we can suppose that the radiative rate constant is also temperature dependent in this region. It can be explained in favor of exciton interaction with acoustic phonon rather than with an optical phonon [10,11].

4. Conclusion

Dephasing of Raman modes of BIC J-aggregate was measured by time-resolved CARS resonantly enhanced by excitonic transition with dephasing time of subpicosecond. Ultrahigh frequency beats of J-aggregate Raman modes are observed. The characteristics of superradiant emission of BIC J-aggregate showed the intermolecular coherence of about 12-13 molecules at 4K. Fluorescence lifetime increased at lower temperature in the range 20-60K.

5. References

1. E.W.Knapp, Chem.Phys. 85 (1984) 73.
2. S.de Boer, K.J.Vink, D.A.Wiersma, Chem.Phys.Lett. 137 (1987) 99.
3. S.de Boer, D.A.Wiersma, Chem.Phys.Lett. 165 (1990) 45.
4. V.Sundstrom, T.Gillbro, R.A.Gadonas, A.Piskarskas, J.Chem.Phys. 89 (1988) 2754.
5. V.L.Bogdanov, E.N.Victorova, S.V.Kulya, A.S.Spiro, JETP Lett. 53 (1991) 105.
6. Y.Wing, J.Opt.Soc.Am. 8B (1991) 981.
7. H.Okamoto, K.Yoshihara, J.Opt.Soc.Am. 7B (1990) 1703.
8. V.F.Kamalov, R.Inaba, K.Yoshihara, Ultrafast Phenomena 8, Eds.J.-L.Martin, A.Migus, G.A.Mourou, A.H.Zewail, Springer - Verlag, 1993, p.87.
9. F.C.Spano, S.Mukamel, J.Chem.Phys. 91 (1989) 683.
10. F.C.Spano, J.R.Kuklinski, S.Mukamel, Phys.Rev.Lett. 65 (1990) 211.
11. K.Kemnitz, K.Yoshihara, T.Suzumoto, T.Tani, M.Lindrum, J.Moll, S.Daene, Lecture at the 2 Int.Conference on Solar Energy Storage and Appl. Photochem., Cairo, Egypt, 1993.

Vibrational Relaxation of Hydroxyls in Zeolite Y: H-Bonding, Dipole-Dipole Coupling and Accepting Modes

M.J.P. Brugmans[1], *A.W. Kleyn*[1], *A. Lagendijk*[1], *W.P.J.H. Jacobs*[2], *and R.A. van Santen*[2]

[1]FOM-Institute for Atomic and Molecular Physics, Kruislaan 407, 1098 SJ Amsterdam, The Netherlands

[2]Schuit Institute of Catalysis, Eindhoven University of Technology, P.O. Box 513, 5600 MB Eindhoven, The Netherlands

Abstract. Excited vibrational population lifetimes of hydroxyls in zeolite Y are shown to provide insight in the (dynamic) coupling of the hydroxyls to the zeolitic lattice as well as to other O–H oscillators.

Time-resolved saturation spectroscopy on O–H in zeolites was performed to gain new information about the relaxation dynamics of the O–H stretch vibration in this crystalline solid and about the structural environment of the acid O–H groups. The presence of these acid sites in proton loaded zeolites is exploited widely in the petrochemical industry, where they are used as catalysts in the hydrocarbon cracking process.[1] The dynamical chemical and physical interactions of the protons with the zeolitic lattice are of interest to understand the catalytic properties.

Zeolites Y with silicon to aluminum ratios Si/Al = 2.4 and 2.8 and different proton concentrations were studied. Conventional FTIR spectra show two resolved absorption bands that are attributed to O–H stretch vibrations of bridging oxigen atoms: a low frequency band (LF) near 3550 cm^{-1} and a HF band near 3640 cm^{-1}.[2]

We have measured O–H vibrational population relaxation times T_1 for both LF and HF absorption bands, using intense ($\sim$ 220 μJ) picosecond (20 ps) tunable infrared pulses (the experimental setup is described in ref.[3]). After excitation by a strong pump pulse that bleaches the O–H absorption, the decay of the excited population difference between $v = 1$ and $v = 0$ is monitored by the transmission of a weak probe pulse, whose time-delay with respect to the pump pulse is varied. From this decay the energy relaxation time T_1 can be deduced.

In Fig.1 results of typical pump-probe experiments are shown for both the LF and the HF absorption bands of zeolite Y with a proton to T (T = Si or Al) ratio of 0.22 and with Si/Al=2.4. Note that the population lifetime for the LF band is significantly smaller than T_1^{HF} (see also ref.[4]). When the laser frequency is scanned through the LF absorption band, the relaxation lifetime increases from $T_1^{\mathrm{LF}} = 29 \pm 10$ ps at 3518 cm^{-1} to 83 ± 14 ps at 3567 cm^{-1} for this particular zeolite. No frequency dependence of T_1^{HF} was observed. The same observations were made for zeolites with different proton concentrations, but both T_1^{LF} and T_1^{HF} decrease with increasing proton concentration.

Springer Proceedings in Physics, Vol. 74 Time-Resolved Vibrational Spectroscopy VI
Editors: A. Lau, F. Siebert, W. Werncke

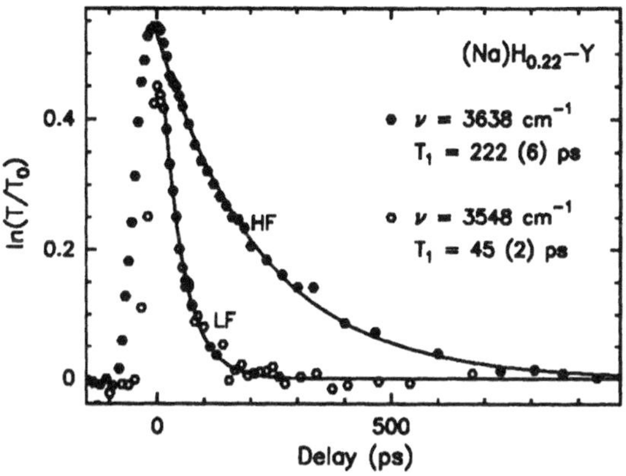

Figure 1: *Relative transmission of a probe pulse as function of the delay between pump and probe, for both the LF and the HF absorption bands of zeolite Y.*

The smaller T_1^{LF} compared to T_1^{HF} and the dependence of T_1^{LF} on the laser frequency are explained by hydrogen bonding of the protons in the LF band. Both LF and HF stretch vibrations relax to very similar accepting modes, but H-bonding of the protons in the LF band increases the coupling to the accepting modes.

To study the effect of proton concentration we decreased the proton concentration in a fully exchanged zeolite by partial deuteration. After correction for laser heating effects in the focus (temperature increases up to 250 K were estimated) we found that replacing 80% of O–H by O–D increases T_1^{LF} and T_1^{HF} for the O–H stretch vibration with a factor 1.5. This indicates that a (nearly) resonant deactivation mechanism like dipole-dipole coupling increases the relaxation rate at higher proton concentrations.

The relatively long O–D lifetimes (only a factor ≤ 3 smaller than O–H lifetimes) indicate that the O–H bending vibrations are important accepting modes for O–H stretch relaxation. Both from this and from the the temperature dependence of T_1^{HF} for O–H the relaxation scheme for O–H stretch relaxation is conluded to be a five-mode process with at least three O–H bending vibrations.

References

[1] Sir J.M. Thomas, Scientific American **April 1992**, 82 (1992).

[2] W.P.J.H. Jacobs, H. Jobic, J.H.M.C. van Wolput, and R.A. van Santen, Zeolites **12**, 315 (1992).

[3] H.J. Bakker, P.C.M. Planken, L. Kuipers, and A. Lagendijk, J. Chem. Phys. **94**, 1730 (1991).

[4] J. Kubota, M. Furuki, Y. Goto, J. Kondo, A. Wada, K. Domen, and C. Hirose, Chem. Phys. Lett. **204**, 273 (1993).

Restrictions to Homodyne Time-Resolved Techniques: Asymmetric Lineshapes of the Vibrons in α-$Ar_x(N_2)_{1-x}$

J. De Kinder, A. Bouwen, and D. Schoemaker

Physics Department, University of Antwerp (U.I.A.),
B-2610 Wilrijk (Antwerp), Belgium

The dephasing signals of the $^{14}N_2$ vibron in pure α-N_2 [1] and in mixed crystals with $^{15}N_2$ [1] or Ar, [2] were measured using time-resolved coherent anti-Stokes Raman scattering (TR-CARS). Our experimental results could be well fitted [1,2] starting from the ad-hoc assignement of the homogeneous and the inhomogeneous component in the frequency domain respectively as a Lorentzian and a Gaussian lineshape. However, recently we performed high-resolution spontaneous Raman scattering measurements using Fabry-Perot interferometry on a mixed crystal of α-$(^{15}N_2)_{0.013}(^{14}N_2)_{0.937}Ar_{0.05}$. The small concentration of $^{15}N_2$ molecules admixed in the crystal allows us to compare the excitonic lineshapes with these of localized vibrations, while it has only a minor effect [1] on the $^{14}N_2$ vibron lineshape. As can be seen in Fig. 1, both the A_g mode and the T_g mode of the $\mathbf{k} \simeq 0$ vibron have asymmetrical lineshapes. This is in clear contrast to what was assumed for fitting our time domain measurements.

Both techniques i.e., TR-CARS and spontaneous Raman scattering, can within the Born-Oppenheimer approximation [3] be expressed in terms of the nuclear response function $d(t) = \langle[\chi(t), \chi(0)]\rangle$, with χ the linear optical susceptibility. The typical form for the nuclear response function is given by $d(t) = A(t)\exp[i\Omega t + \Phi(t)] + c.c.$, with Ω the vibrational frequency and $A(t)$ and $\Phi(t)$ slowly varying amplitude and phase shift functions. For a symmetrical lineshape, $\Phi(t)$ will be constant during the decay, while for an asymmetrical lineshape, it exhibits a more complex behaviour.

The spontaneous Raman scattering is given by

$$I(\Omega) \propto B(\Omega, T) \int_0^{+\infty} \cos(\Omega t) d(t) dt,$$

where $B(\Omega, T)$ is the Boltzmann factor governing the population of the vibrational states (with an energy $\hbar\Omega$) for a given temperature T. It contains information on the phase and amplitude modulation of the nuclear vibration.

Contrary to this, the TR-CARS signal $S(t) \propto |\chi^{(3)}|^2 \propto |A(t)|^2$ and the TR-CARS measurements lack the information as to whether the vibrational bands under study are symmetric or asymmetric. In order to verify the consistency of our explanation, we have Fourier transformed our frequency-domain results. As can be seen from Fig. 1, the obtained decay is faster than exponential which is in qualitative agreement with our time-resolved results. Detailed analysis [4] shows that the characteristic decay time (70 $\pm$ 15 ps) is in good agreement with our TR-CARS results (66 $\pm$ 5 ps). We conclude that homodyne time-

Springer Proceedings in Physics, Vol. 74 Time-Resolved Vibrational Spectroscopy VI
Editors: A. Lau, F. Siebert, W. Werncke

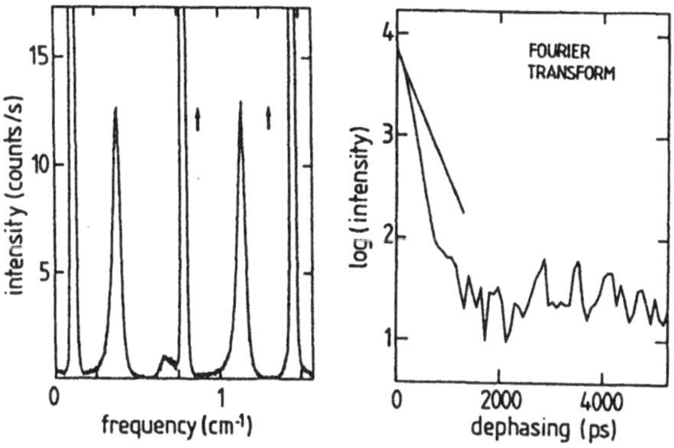

Fig. 1 High-resolution Raman spectrum of the $^{14}N_2$ vibrons in a mixed crystal of α-$(^{15}N_2)_{0.013}(^{14}N_2)_{0.937}Ar_{0.05}$ and the corresponding Fourier tranform of the A_g vibron mode (interval indicated by the two arrows), plotted as $|\chi^{(3)}(t)|$. The intense features in the frequency domain results are due to elastically scattered light. A tangent to the curve was drawn at short dephasing times, showing the faster-than-exponential decay.

domain techniques have one particular substantial drawback: They are unable to distinguish between asymmetrical and symmetrical lineshapes.

The vibrational transition of localized transitions of isolated $^{15}N_2$ molecules provides evidence [4] for the fact that the asymmetry is due to differences in local configurations of the N_2 molecules and the Ar atoms. Due to substantial concentration of Ar dopant (5 %), the probability for dimers and higher aggregates of Ar atoms is substantial.

JDK acknowledges financial support from the National Fund for Scientific Research Belgium (NFWO). This workd was financed by the Inter-University Institute for Nuclear Sciences (IIKW).

References

[1] J. De Kinder, E. Goovaerts, A. Bouwen, and D. Schoemaker, Phys. Rev. **B 42**, 5953 (1990).

[2] J. De Kinder, E. Goovaerts, A. Bouwen, and D. Schoemaker, J. Chem. Phys. **95**, 2269 (1991).

[3] R.W. Hellwart, Prog. Quant. Electr. **5**, 1 (1977).

[4] J. De Kinder, A. Bouwen, and D. Schoemaker, accepted for publication in Phys. Rev. **B** (1993).

Dephasing of the Internal Vibrations in Ortho/Para H_2 Mixed Crystals

J. De Kinder, A. Bouwen, and D. Schoemaker

Physics Department, University of Antwerp (U.I.A.), B-2610 Wilrijk (Antwerp), Belgium

For the dephasing of the vibron in α-N_2, Abram *et al.* [1] found a stretched exponential decay on the ns timescale, while our results in the ps domain [2] show a faster-than-exponential decay. The latter results were found to be in good agreement with frequency domain experiments [3]. This faster-than-exponential decay was interpreted in terms of homogeneous and inhomogeneous line broadening [2]. For the dephasing of the p-H_2 vibron in mixed crystals with a varying concentration of o-H_2, similar results were obtained on the ns timescale [4].

We have studied the linewidth of coupled p-H_2 vibron states ($\Omega = 4150$ cm^{-1}) and isolated vibrational states in $(o\text{-}H_2)_x(p\text{-}H_2)_{1-x}$ mixed crystals. Hereto, single crystals of p-H_2 with a varying concentration of o-H_2 were grown by slowly cooling the liquid. Isolated vibrational states were studied, either through the presence of a low concentration ($< 3\%$) of o-H_2 molecules ($\Omega = 4146$ cm^{-1}), or intentionally admixed o-D_2 molecules ($\Omega = 2994$ cm^{-1}). The light of a single mode Ar laser, scattered by these crystals, is collected in a back scattering geometry and sent through the tandem combination of a Fabry-Perot interferometer and a single monochromator. Linewidths down to 0.002 cm^{-1} can be obtained by deconvolution techniques.

For o-H_2 concentrations lower than 2 %, the linewidth of the p-H_2 vibron is narrower than the resolution of our setup (0.002 cm^{-1}). This indicates that the linewidth observed in a recent stimulated Raman gain experiment [6] 0.0014 cm^{-1}, for an ortho concentration of 0.06 %, is still limited by their resolution. For higher o-H_2 concentrations, we observe a lineshape which is asymmetric towards higher frequencies (see Fig. 1). The global width and the asymmetry of the line increase with increasing o-H_2 dopant level. For an o-H_2 concentration $> 30\%$, i.e., well within the mixed crystal regime, the lineshapes of both the o-H_2 and the p-H_2 transition no longer show this asymmetry.

After Fourier transformation, the dephasing signals in the ps domain have the same characteristics as in the ns domain [4], in contrast to the N_2 results. Our frequency domain results provide more information [5] than the time-resolved measurements, by showing the clear asymmetry in the lineshape.

For the uncoupled vibrational states, we obtain a lineshape which is broader than for the coupled states, showing the line narrowing effect of the coupling process. For concentrations up to 6.85 % o-H_2, we cannot resolve any clear asymmetry in the lineshape. Since the spectrum can be well fitted by a Lorentzian lineshape, we conclude that inhomogeneous line broadening, caused

Springer Proceedings in Physics, Vol. 74 Time-Resolved Vibrational Spectroscopy VI
Editors: A. Lau, F. Siebert, W. Werncke

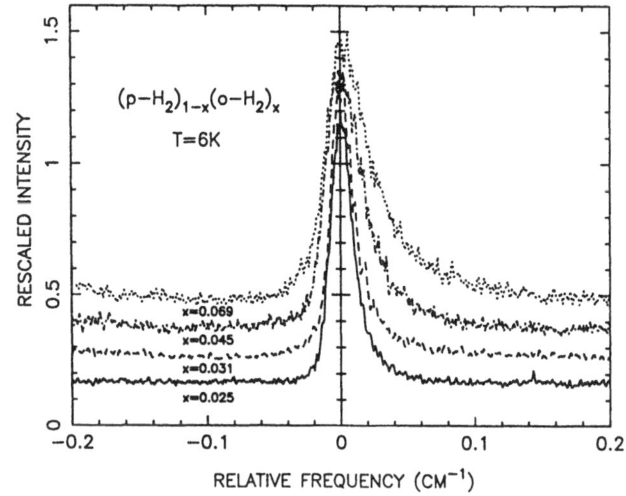

Fig. 1 Lineshapes of the p-H_2 vibron as a function of o-H_2 concentration in $(o\text{-}H_2)_x(p\text{-}H_2)_{1-x}$ mixed crystals at 6 K. The frequency indication is only a relative measure.

by a difference in the energy of different configurations of o-H_2 and p-H_2 molecules is small.

JDK acknowledges financial support from the National Fund for Scientific Research Belgium (NFWO). This work was financed by the Inter-University Institute for Nuclear Science (IIKW).

References

[1] I.I. Abram, R.M. Hochstrasser, J.E. Kohl, M.G. Semack, and D. White, J. Chem. Phys. **71**, 153 (1979).

[2] J. De Kinder, E. Goovaerts, A. Bouwen, and D. Schoemaker, Phys. Rev. **B 42**, 5953 (1990).

[3] R. Ouillon, C. Turc, J.P. Lemaistre, and P. Ranson, J. Chem. Phys. **93**, 3005 (1990).

[4] I.I. Abram, R.M. Hochstrasser, J.E. Kohl, M.G. Semack, and D. White, Chem. Phys. Lett. **71**, 405 (1980).

[5] J. De Kinder, A. Bouwen, and D. Schoemaker, contribution to this volume.

[6] T. Momose, D.P. Weliky, and T. Oka, J. Mol. Spectr. **153**, 760 (1992).

CARS with Femtosecond Time Resolution Using Broadband Dye Lasers

A. Kummrow, A. Lau, M. Pfeiffer, W. Werncke, and H.G. Ludewig

Max-Born-Institut für Nichtlineare Optik und Kurzzeitspektroskopie, Rudower Chaussee 6, D-12474 Berlin, Germany

Abstract. Radiation-matter detuning oscillations in CARS with incoherent light are applied to measure vibrational frequencies and dephasing times of neat solvents and crystals. A coherence peak found for some liquids is attributed to the non-Raman resonant susceptibility.

1. Introduction

In four-wave-mixing the time resolution is in principle determined by the coherence time of the laser pulses rather than their duration [1]. Using the incoherent light of broadband dye lasers femtosecond time resolution was achieved [2-5]. We use the method developed by the group of Dugan and Albrecht [3-6] for different solvents and crystals. A peculiarity shows up for toluene where in addition to the radiation-matter detuning oscillations a coherence peak is observed.

2. Experiment and discussion

The experiment uses a frequency doubled Q-switched Nd:YAG laser as a narrowband Stokes laser. A non-transform-limited broadband coumarin 307 dye laser is used as pump radiation. The pump pulse is split into two twin pulses in a Michelson interferometer. The three beams are overlapped in a folded BOXCARS arrangement. A spectrometer permits the narrowband detection of the CARS signal, which is necessary to obtain the detuning oscillation [2,3]. A time resolution of about 30fs is achieved using the inherent short pulse fluctuations of the broadband laser.

Liquids like toluene or nitrobenzene and crystals like diamond give intense signals with damped periodic modulations as function of delay between the two pump beams. These modulations give vibrational frequencies of 1003 cm^{-1}, 1349 cm^{-1} and 1332cm^{-1} and dephasing times of 4.7ps, 2.0ps and 5.2ps respectively (Fig.1). The dephasing times listed here are obtained using a correction formula for finite Stokes laser linewidth [6]. Dephasing time and vibrational frequency derived from these modulations agree with linewidth data of spontaneous Raman scattering spectra [7]. This fact demonstrates a predominantly homogeneous broadening of the vibrational transition.

Detecting toluene with higher time resolution an additional peak can be resolved. Peaks with an equal half width of the coherence peak are found for methanol and ethanol, but without radiation-matter detuning oscillations (Fig.2). The

Springer Proceedings in Physics, Vol. 74 Time-Resolved Vibrational Spectroscopy VI
Editors: A. Lau, F. Siebert, W. Werncke

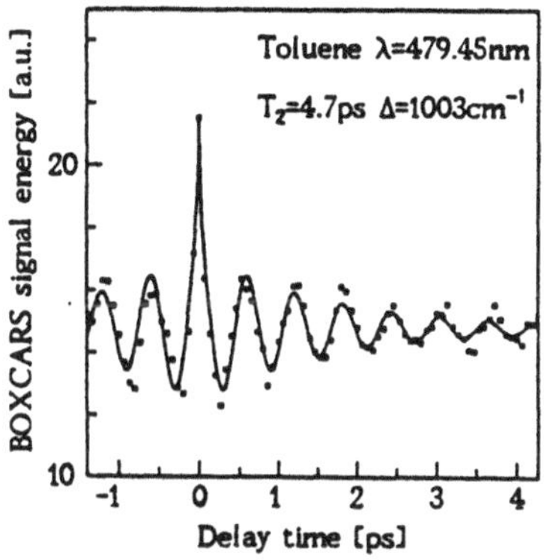

Fig.1: CARS signal as function of delay time for toluene. The monochromator is detuned by 55cm^{-1} from the two times Raman transformed Stokes laser wavelength. This leads to the oscillations with 600fs period. The fit includes an additional exponentially damped term with a 33fs time constant.

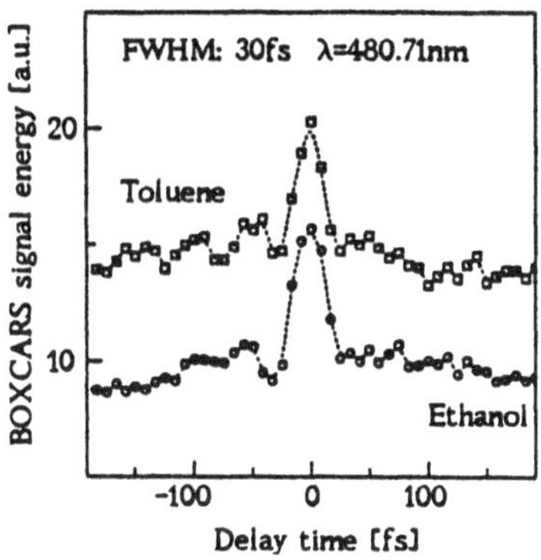

Fig.2: CARS signal as function of delay time for toluene and ethanol. The spectral position of the monochromator slit was chosen to coincide with the two times Raman transformed Stokes laser wavelength (i.e. 480.71nm) for the 1003cm^{-1} vibration of toluene (pump energy 20μJ and 27μJ, Stokes energy 20μJ for toluene and 60μJ for ethanol).

lack of these oscillations is due to the large Raman shift of their lines (2928cm^{-1} and 2835cm^{-1} respectively) with respect to the frequency position of the slit. The coherence peak can be explained by non-Raman resonant scattering which was not included in the calculations of Dugan and Albrecht [6]. A quantitative analysis for non-Raman resonant contributions is under way (see the paper of A.C. Albrecht in these proceedings). This will allow to obtain the ratio of the Raman resonant and the non-Raman resonant susceptibility.

3. References

1. N. Morita, T. Yajima, Phys. Rev. A **30**, 2525-2536 (1984)
2. T. Hattori, A. Terasaki, T. Kobayashi, Phys. Rev. A **35**, 715-724 (1987)
3. M.A. Dugan, J.S. Melinger, A.C. Albrecht, Chem. Phys. Lett. **147**, 411-419 (1988)
4. K. Wynne, M. Müller, J.D.W.VanVoorst, Phys. Rev. A **41**, 6361-6375 (1990)
5. A. Lau, M. Pfeiffer, W. Werncke, Nuovo Cimento **14D**, 1023-1032 (1992)
6. M.A. Dugan, A.C. Albrecht, Phys. Rev. **43**, 3877-3921, 3922-3933 (1991)
7. A. Penzkofer, A. Laubereau, W. Kaiser, Prog. Quant. Electr. **6**, 55-140 (1979)

Part II

Progress in Instrumentation and Measurements, NLO-Techniques, Surface Phenomena

Phonon-Polariton Transmission Through a Vacuum Layer

F. Vallée and Ch. Flytzanis

Laboratoire d'Optique Quantique du C.N.R.S.,
Ecole Polytechnique, F-91128 Palaiseau Cédex, France

Abstract. the transmission of a short phonon-polariton pulse through a vacuum layer separating two identical noncentrosymmetric crystals (NH_4Cl) is investigated in real time and space by use of the nonlocal time-resolved CARS technique. The results yield selective information on the impact of an interface on the polariton wave-packet coherence.

The propagation of a resonantly driven dipolar excitation in a crystal, namely the polariton mode, is frequently encountered in physics and also expected to have important repercussions in the operation of optoelectronic devices close to material resonances. Besides being connected with the absorption of electromagnetic waves it also has a bearing on the storage, redistribution and propagation of electromagnetic energy in condensed matter.

A conspicuous feature of the polariton mode is the wide range of values over which its group and phase velocities can vary namely from close to that of the light velocity in vacuum down to those of the sound velocity. This results from the hybrid character of the polariton: partly electromagnetic partly material excitation, with their relative strength monotonically varying along the polariton dispersion curve [1]. Consequently the polariton behavior can be drastically affected as its material support is altered for instance in the course of its propagation in inhomogeneous media [2]. In this context the polariton transmission through an interface or a layer is a key problem.

We have extended and applied to this problem the nonlocal time resolved nonlinear technique we previously used for the study of polariton in bulk crystals [2,3]. We briefly recall that this technique gives access to the wave packet envelope and has the advantage of being frequency selective and is particularly suited to high frequency polaritons (typically $\geq 100\ cm^{-1}$). It is in essence a "time of flight" experiment and consists in separating in space the coherent excitation and probing stages of the time resolved CARS, the excited polariton wave-packet being tracked in real time and space during its propagation. In the present case we have accessed phonon-polariton transmission by performing the stages in two identical halves of an ammonium chloride (NH_4Cl) crystal separated by a vacuum layer.

The coherent excitation of a phonon-polariton packet is realized at time t=0 and position y=0 in the one half with two picosecond copropagating pulses with frequencies ω_L and ω_S and with wave vectors $\underline{k}_L$ and $\underline{k}_S$ respectively (Fig. 1). The frequencies and the experimental geometry are chosen to resonantly excite a polariton wave packet with central frequency ω_π and wave vector $\underline{k}_\pi$

$$\omega_L - \omega_S = \omega_\pi \ ; \ \underline{k}_L(\omega_L) - \underline{k}_S(\omega_S) = \underline{k}_\pi(\omega_\pi).$$

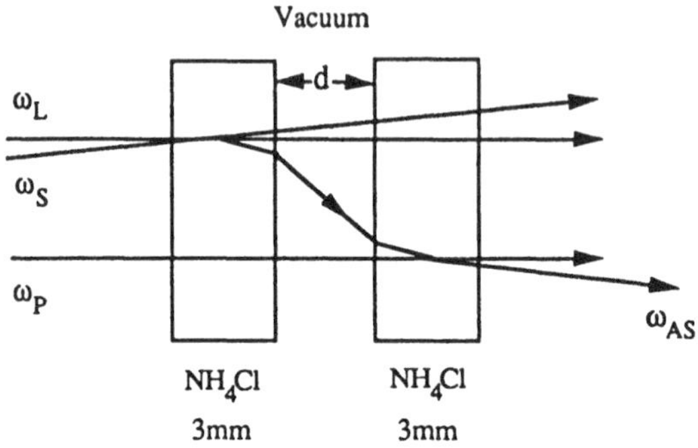

Fig 1 : Principle of the experiment. The polariton wave packet created in the first NH_4Cl crystal is transmitted through vacuum and detected in the second NH_4Cl crystal by anti-Stokes Raman scattering of the probe pulse (ω_p).

This wave packet freely propagates with its group velocity $V_g(\omega_\pi)$ along a direction fixed by its wave vector k_π, which makes an angle with the direction of $\underline{k}_L$ and $\underline{k}_S$ (typically of the order of 20°). Its temporal and spatial evolution in the first half and after, transmission through the vacuum layer, in the second half is monitored by phase matched coherent anti-Stokes Raman scattering at $\omega_{AS} = \omega_P + \omega_\pi$ of a third picosecond pulse of frequency ω_P. The probe pulse is displaced with respect to the excitation in time by t_D and in space by x_D. The nonlocal probing allows a separation of the temporal and spatial features and gives a direct and independent access to the influence of the interface on the polariton envelope.

The three independent picosecond pulses for the CARS-type experiment are created from a single 5ps pulse delivered by a passively mode-locked cavity dumped Nd/glass oscillator. After amplification to 2GW the infrared pulse at 1.054μm is doubled in frequency in a KDP crystal. The visible pulse is divided into three parts. The first one is shifted by stimulated Raman scattering in deuterated methanol to create the ω_s beam while the second one is used as the probe beam (ω_P). The last part is tightly focussed into a carbon tetrachloride cell to generate a picosecond quasi-continuum from which a narrow frequency band is selected by a grating and amplified in a dye amplifier. This creates the second excitation beam ω_L which is tunable. The temporal resolution of the system is routinely of 0.4ps. After polarization adjustment the beams are focussed with 20cm lenses into the sample with a calculated non-collinear geometry. The anti-Stokes signal is detected by a photomultiplier after spatial spectral and polarization selections allowing a very high signal over noise discrimination.

The transmission was studied for two layer thicknesses (d=3mm and 1mm) and two different polariton modes $\omega_\pi = 1270cm^{-1}$ and $\omega_\pi = 1170cm^{-1}$ of the same lower branch of the ν_4 polariton in NH_4Cl. Only the lower branch is accessible in a cubic crystal by the coherent excitation stage. The polariton group velocity has been previously measured as a function of frequency in a 1cm thick NH_4Cl crystal and is V_g= 0.35c and V_g= 0.47c for, respectively, the $1270cm^{-1}$ and $1170cm^{-1}$ polariton [2].

In Fig. 2 we show the intensity of the CARS signal on a logarithmic scale as a function of the probing point for the $1270cm^{-1}$ polariton mode and a layer

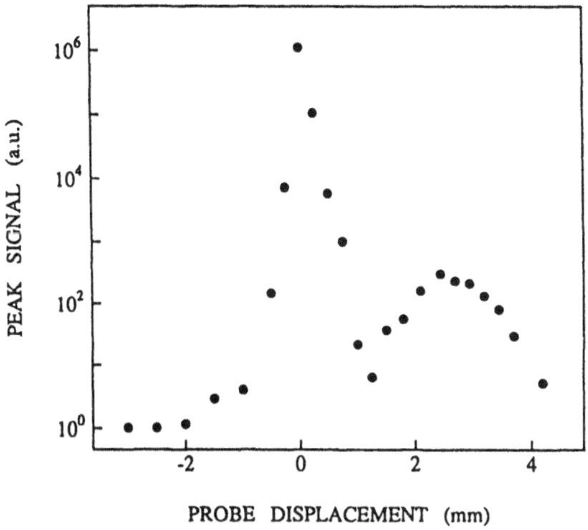

Fig 2 : Measured peak CARS signal as a function of probing position for the $1270cm^{-1}$ polariton in NH_4Cl. The vacuum layer is 3mm thick and the crystal temperature is 77K.

thickness d = 3mm at nitrogen temperature. The first maximum corresponds to probing of the polariton wave packet inside the first crystal, with an asymmetric shape due to transverse propagation along the positive X direction. The residual signal for negative displacement reflect the spatial shape of the wave-packet, essentially the one of the focal spot. The second maximum corresponds to probing in the second crystal of the phonon-polariton wave packet excited in the first one and transmitted through the vacuum layer. The position of the maximum signal is concomitant with a transmission through a 3mm vacuum layer and with the excitation geometry. For a smaller spacing between the two crystals the second maximum occurs for a smaller displacement. A correlated dependence is observed for the $1170cm^{-1}$ polariton.

In Fig.3 we show the probe delay as a functions of the probing position. The slope of the two lines corresponds to polariton propagation inside the first and the second crystal with the group velocity $V_g(1270cm^{-1})$. In the intermediate region (between 1 and 3 mm probing position) most of the excitation propagate into the vacuum at the speed of light.

It is evident from the above results that the polariton is weakly perturbed by its environnement and can propagate coherently without distortion over macroscopic distances. The spatial quality of the interface apparently has a minor influence as long as the possible surface irregularities are much less than the polariton wavelength as was in our case.

The above technique can be extended to measure polariton transmission and reflection coefficients on interfaces or dielectric layers that separate different crystalline media or other inhomogeneous configurations. Such information is very important when envisaging polariton pulse processing and interaction with other injected pulses.

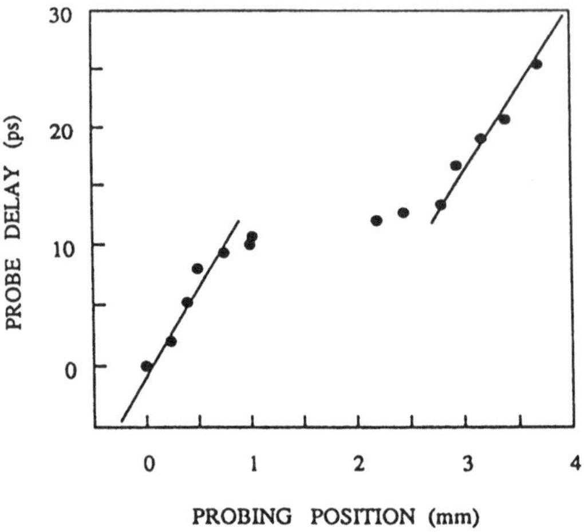

Fig 3 : Measured probe time delay where the signal is maximum as a function of probing position for the 1270cm^{-1} polariton in NH_4Cl. The slope of the lines corresponds to the polariton group velocity V_g(1270cm^{-1}).

References

[1] D.L.mills and E.Burstein, Rep. Prog. Phys. 37, 817 (1974)
[2] G.Gale, F.Vallee and Ch.Flytzanis, Phys. Rev. Lett 57, 1867 (1986)
[3] F.Vallee and Ch.Flytzanis, Phys. Rev. B 46, 13799 (1992)

Picosecond Time-Resolved Raman/Coherent-Raman Spectroscopy Using the "Pico-Pump and Quasi-CW Probe" Method with a Streak Camera

T. Tahara and H. Hamaguchi

Molecular Spectroscopy Laboratory, The Kanagawa Academy of Science and Technology (KAST), KSP East 301, 3-2-1 Sakato, Takatsu-ku, Kawasaki-shi 213, Japan

Abstract A Picosecond time-resolved spontaneous Raman measuring system using a streak camera was constructed. Nanosecond probing laser pulses were used with picosecond UV pumping. Raman scattering at each delay time was time-resolved and detected by a streak camera. Two dimensional (the time and the frequency domain) simultaneous detection of Raman scattering has been carried out. Time-resolved Raman spectra of benzophenone at delay times ranging from -1.25 ns to 1.5 ns were obtained with one single measurement. A picosecond time-resolved multiplex CARS measuring system based on the same principle was also constructed. Nanosecond narrow band laser pulses and broad band dye laser pulses were used as CARS probing ω_1 and ω_2, respectively, with the picosecond UV pumping pulses. The cis-trans photoisomerization dynamics of all-trans and 9-cis retinal was investigated. It was clarified that the conformational change from the 9-cis form to the all-trans takes about 880 ps in the triplet manifold.

1. Introduction

In ordinary picosecond time-resolved Raman spectroscopy, we use two picosecond laser pulses, i.e. the pumping pulse and the probing pulse. The pumping laser pulse first irradiates the sample to produce transient species and then the probing laser pulse comes in after a certain delay time to monitor Raman scattering. Thus, each measurement gives one Raman spectrum at one fixed delay time. Time-resolution is determined by the cross correlation between the pump and probe laser pulses, and no time-resolving capability is required for the detector.

If we use a streak camera as a detector, we can carry out another type of time-resolved measurement. In this case, we use a picosecond laser pulse for the photoexcitation and a quasi-cw laser radiation for probing. Raman scattering at each delay time is time-resolved and detected by a streak camera. Consequently, Raman spectra at various delay times are obtainable with one single measurement. Time-resolution in this scheme is determined by the duration of the pump pulse and also by the time-resolution of the streak camera, but it does not depend on the duration of the probing laser pulse. In this case, the two-dimensional feature of a streak image allows us multichannel detection both in the time and frequency domains.

In this paper, we report the construction and the application of picosecond time-resolved spontaneous Raman and CARS measuring systems based on this "pico pump and quasi-cw probe" method.

Springer Proceedings in Physics, Vol. 74 Time-Resolved Vibrational Spectroscopy VI
Editors: A. Lau, F. Siebert, W. Werncke

2. Picosecond time-resolved spectroscopy by using a streak camera

The experimental set-up has already been described elsewhere[1]. The fourth harmonic of the output from a Nd:YAG cw regenerative amplifier (266 nm, 50 ps) was used as the pumping pulse. The second harmonic of a diode pumped cw Q-switched Nd:YLF laser (523.5 nm, 7 ns) was used to probe Raman scattering. Raman scattering was analyzed by a spectrograph and detected with a streak camera (10 ps time resolution at the highest sweeping speed). The 7-ns probing laser pulse can be regarded as "quasi-cw" in the time window covered by a streak image.

We measured the time-resolved Raman spectra of benzophenone. Figure 1 shows the time-resolved Raman spectra obtained from a 8×10^{-3} M hexane solution of benzophenone. All spectra shown in this figure were obtained from one streak image, i.e. one single measurement. In the spectra before the photoexcitation, only solvent Raman bands are observed. With the photoexcitation, the fluorescence emitted from S_1 benzophenone covers all the measured wavenumber region, and next, transient Raman band due to the T_1 state appears around 970 cm^{-1} (marked by

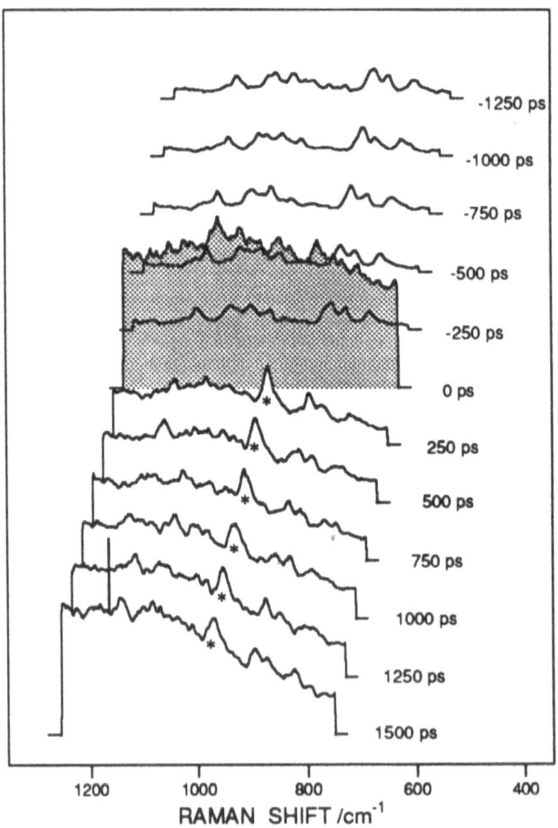

Fig. 1. Time-resolved Raman spectra of benzophenone in hexane measured with the streak camera system.

asterisks) [2]. Finally, the fluorescence background due to the ketyl radical increases in accordance with the decay of the T_1 state. It is known that T_1 benzophenone abstracts hydrogen from the solvent to form the ketyl radical. In this experiment, the temporal behavior of all the three transient species appearing in the early stage of benzophenone photochemistry, the S_1 and T_1 states, and the ketyl radical were monitored through either fluorescence or Raman scattering.

3. Picosecond time-resolved multiplex CARS spectroscopy by using a streak camera

The "pico pump and quasi-cw probe" method is also applicable to coherent Raman spectroscopy. We constructed a picosecond time-resolved multiplex CARS measuring system. The third harmonic of the output from a Nd:YAG cw regenerative amplifier (355 nm, 50 ps) was used as the pumping pulse. The second harmonic of the output from the Nd:YLF laser (523.5 nm, 7ns) was used as ω_1 and the output of a broad band dye laser (ω_1-ω_2 = 900 cm^{-1}- 1700 cm^{-1}) was used as ω_2. Multiplex CARS beam was analyzed by a spectrograph and detected with the streak camera.

The cis-trans isomerization dynamics of retinal in the lowest excited triplet (T_1) state was investigated. In the case of the photoexcitation of all-trans retinal, three transient CARS signals appeared around 1555 cm^{-1}, 1255 cm^{-1} and 1190 cm^{-1} immediately after the photoexcitation. The observed vibrational frequencies agreed well with those of the three strongest spontaneous Raman bands of "all-trans-like" T_1 state which was observed with nanosecond time-resolved Raman spectroscopy[3]. These three transient CARS bands were observed also in the case of the photoexcitation of 9-cis retinal. However, their appearance was markedly delayed, in contrast to the case of all-trans retinal. This indicates the existence of a precursor of the "all-trans-like" T_1 state in the photoexcitation of 9-cis retinal. The precursor is most likely to be the T_1 state having a "9-cis-like" conformation (the "9-cis-like" T_1 state).

Figure 2 shows the population change of "all-trans-like" T_1 retinal obtained by the fitting analysis. The rise time agrees well with the fluorescence lifetime (34 ps)

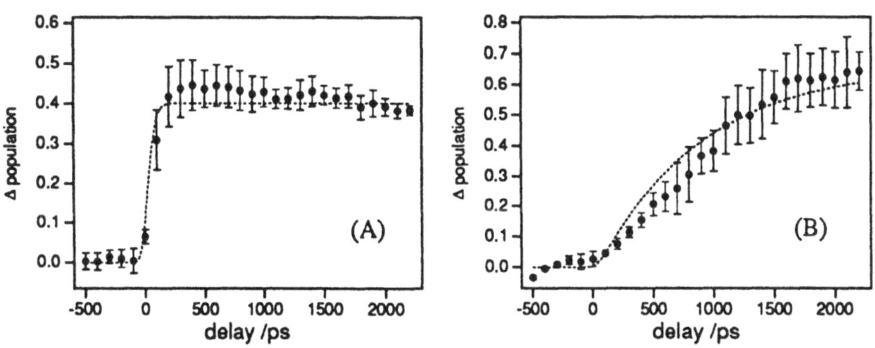

Fig. 2. The temporal change of the fractional population of the "all-trans-like" T_1 state. (A) Photoexcitation of all-trans retinal, (B) photoexcitation of 9-cis retinal.

in the case of the photoexcitation of the all-trans isomer, indicating that the "all-trans-like" T_1 state is directly produced from the S_1 state. In the case of the 9-cis isomer, the rise time of the "all-trans-like" T_1 state is about 880 ps. It implies that the conformational change from the 9-cis form to the all-trans takes about 880 ps in the triplet manifold.

References

1. T. Tahara and H. Hamaguchi, Appl. Spectrosc., **47**, 391(1993).
2. T. Tahara, H. Hamaguchi and M. Tasumi, J. Phys. Chem. **91**, 5875 (1987).
3. H. Hamaguchi, H. Okamoto, M. Tasumi, Y. Mukai and Y. Koyama, Chem. Phys. Lett. **107**, 355 (1984).

Vibrational Dynamics of Condensed-Phase Molecules Studied by Ultrafast Infrared Spectroscopy

J.C. Owrutsky, M. Li, J.P. Culver, M.J. Sarisky, A.G. Yodh, and R.H. Hochstrasser

Department of Chemistry and Department of Physics, University of Pennsylvania, Philadelphia, PA 19104, USA

Abstract. Infrared pump/probe measurements of T_1 relaxation and oritentational relaxation of ions in hydrogen bonding solvents are discussed in terms of solvent induced anharmonic coupling.

Introduction

Transient IR spectroscopy was used to measure T_1 vibrational energy relaxation times and orientational (P_2) relaxation times τ_2. The results of such measurements on simple ions and on FeCO groups in proteins provide insight into solvent dynamics.

Experimental Methods

The experiments involve IR spectroscopy of molecules in excited vibrational states. An intense, tunable IR pulse in 2000 cm^{-1} range was used to excite the samples. Probing was effected by a tunable IR pulse (duration ca. 20 ps) which was gated before the detector by a 300 fs optical pulse. By this technique sub-picosecond IR pump/ IR probe experiments could be carried out [1].

Pseudohalide Ions: Results Summary

The observation that high frequency vibrational states of diatomic [2] and triatomic ions [1, 3] relax on the few picosecond timescale in hydroxyllic solvents demonstrates the importance of Coulomb forces in the coupling to the solvent. Infrared pump/probe experiments with tunable pulses have enabled measurements of T_1 relaxation and orientational relaxation (τ_2) of a number of simple pseudo-halogen ions, N_3^-, NCO^- and NCS^-. These ions have an asymmetric stretch mode near 2000 cm^{-1} which is studied in this work. The principal results were as follows and in Table I:

An inverse relation between T_1 and τ_2 was found: For $T_1 << \tau_2$ Vibrational relaxation occurs within a well-defined solvent structure - i.e. supermolecule relaxation. For $T_1 > \tau_2$ Vibrational relaxation averaged over a broad range of structures.

· The T_1 relaxation is charge related indicating importance of Coulombic forces. This may mean that not only the nearest neighbor solvent atoms are involved. Relaxation rates are roughly consistent with being proportional to (charge)^2.

· The T_1 relaxation depends on the extent of motion of the end atoms in the asymmetric stretch coordinate.

· Hydrogen bonding is a crucial factor in ultrafast relaxation. In nonhydroxyllic solvents the rates are much slower.

· A possible role seen for internal modes of water is inferred from isotope effects (H_2O/D_2O) in the T_1 relaxation of N_3^-.

Table I: T_1 Values for Pseudohalide Ions [1]

Ion	Solvent	T_1 (ps)	τ_2 (ps)
N_3^-	H_2O	1.2	...
	D_2O	2.4	7.1
	MeOH	2.4	12.7
NCO^-	MeOH	3.0	6.7
SCN^-	D_2O	18.3	4.7
	MeOH	11.0	8.7

Pseudohalide Ions: Discussion

A conceptually useful approach to describing vibrational relaxation processes involves the Landau-Teller method which incorporates the Zwanzig [4] notion that the fluctuations in forces exerted by the solvent on the mode of vibration with characteristic frequency ω is what causes the dissipation of vibrational energy. According to this idea which has been taken much further by a number of different groups [5], the T_1 relaxation is given by:

$$\frac{1}{T_1} = \frac{G(\omega)}{\mu k_B T}$$

where μ is a reduced mass for the vibration and $G(\omega)$ is the power spectrum (i.e. the cosine Fourier transform of the autocorrelation function) of the forces acting on the vibrational coordinates. The power spectrum will incorporate forces from all types of solvent motion. When translational, rotational or librational fluctuations determine $G(\omega)$, the dissipation mechanism is $V \rightarrow T$, R and L: when internal modes of solvent molecules determine $G(\omega)$ the relaxation is mainly of $V \rightarrow V$ type. In general, the molecular (solute) vibrational energy relaxes into a distribution of solvent motions. It is important to note that $G(\omega)$ contains the forces on the oscillator. This means specifically that to be effective, the collisions should involve the solute atoms that are involved in the vibrational motion. Then if this mode has the form

$Q_\alpha = \sum C_{k_r\alpha} q_{k_r}$, the effective force should take the form:

$$F = \mu^{1/2} \sum_{k_r} F_{k_r} C_{\alpha k_r} M_k^{-1/2} ,$$

where $\vec{F}_{k_r}$ is the force in the kth atom in the direction r (= x_k, y_k, z_k). In principle the forces F_{k_r} can be obtained from a dynamics simulation [5, 6] and if the forces on different atoms are uncorrelated the relevant spectral density reduces to:

$$\frac{1}{T_1} = \left(\frac{1}{k_B T}\right) \sum_{k_r} (C^2_{\alpha k_r}/M_k)\ G_{k_r}(\omega) ,$$

where $G_{k_r}(\omega)$ is the spectral density at ω of solvent forces on atom k in direction r.

If the forces are dominated by Coulomb interactions the $G_{k_r}(\omega)$ will vary as ρ_k^2, with ρ_k the charge on the kth atom, and the relaxation rate becomes proportional to $\sum_{k_r} C^2_{\alpha k_r} \rho_k^2/M_k$. This notion was introduced in our earlier paper [1]. Of course if the Coulomb interactions become less important than the forces from other sources this simple rule will not hold. This relation predicts N_3^- will relax 6.7 times faster than SCN^- when the parameters in Table III of reference [1] are utilized. The observed ratios are 7.6 for D_2O and 4.6 for MeOH. Although this suggests that the approach has merit, the significant difference in relaxation predicted from these same data between N_3^- and NCO^- is not observed. Needless to say the results are very sensitive to the charges ρ_k which must be subject to considerable uncertainty as to their values in solution.

The effect of anharmonicity needs careful consideration. In our first paper on N_3^- [3] we suggested that the coupling between the asymmetric stretch and combination modes is sufficiently small in the isolated ion that its contribution to the relaxation likely would be small. From another standpoint, the existence of anharmonic coupling large enough to generate a T_1 in the range 1 ps would be likely to have spectral manifestations (e.g. a splitting) even in solution phase. However we observe a single Lorenzian lineshape in the 2000 cm^{-1} spectral region with no trace of combination bands. However we wish to emphasize here that the role of G(ω) should be crucial in determining the importance of weak anharmonicity effects. As an illustration of the effect of the energy gap, according to the simulation by Klein, Ferrario and McDonald [6] the value of $G(k_BT/hc)$ / G (2000 cm^{-1}) for N_3^- in H_2O is approximately 10^4 and $G(k_BT/hc)$ / G (1000 cm^{-1}) is ca. 10^2 at 300 k. Thus the spectral density for the forces of H_20 molecules on N_3^- is about 10^4 times larger at the combination difference frequencies $\nu_3 - 3\nu_2 = 133$ cm^{-1} and $\nu_3 - (\nu_1 + \nu_2) = 48$ cm^{-1} than at $\nu_3 \sim 2000$ cm^{-1}. Whenever there exists strong interactions between solute and solvent there will be frequency regimes where G(ω) is very large. In order to describe a <u>solvent induced anharmonic pathway</u> in the Landau-Teller picture it seems necessary to calculate the forces on the ν_3 mode resulting from the motions of solvent atoms as before but with the solute atoms moving according to modes ν_2 and ν_1. In other words the very high frequency part of the fluctuating force would be needed. Any intrinsic anharmonicity of the solute ion potential function might be dominated by this solvent effect in polar and ionic systems.

For example, the ion water hydrogen bonded structure that must persist throughout the decay of ν_3 in N_3^- may involve forces that are sufficiently strong to mix the intramolecular modes. In the partial structure:

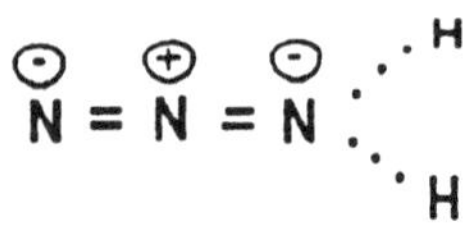

the ν_3 normal mode of N_3^- is coupled to the ν_1 mode (symmetric stretch) by symmetric stretches and bends of the NH_2 unit while the ν_3 mode couples to ν_2 as a result of asymmetric stretches of the NH_2 group.

The isotope effect showing N_3^- relaxing twice as fast in H_2O compared with D_2O suggests an internal mode of water is involved in the relaxation, arguing against solvent modification of the surfaces. However, the correlation observed between the relaxation times and the solvent shifts of the vibrational spectra do evidence the involvement of solvent forces in the relaxation mechanism. The idea of the persistence of a supermolecule structure throughout the T_1 relaxation of N_3^- and NCO^- in water does argue in favor of "polyatomic" behavior, incorporating mode mixing as suggested above.

Proteins

The results on the T_1 relaxation of the carbonyl v=1 state in myoglobin, hemoglobin and protoporphyrin in D_2O using the same approach showed much slower relaxation times of 18, 18 and 31 ps respectively. These relaxations are nevertheless much faster than normally found for metallocarbonyls and are proposed to involve anharmonic coupling to the modes of the heme-CO group, mediated by the iron atom.

Further development of these ideas will require both new experiments and theory.

Acknowledgements

This work was supported by NIH and NSF.

References

1. M. Li, J. Owrutskv, J.P. Culver, A.G. Yodh, R.M. Hochstrasser, J. Chem. Phys. 98, 7 (1993).
2. E.J. Heilweil, F.E. Doany, R. Moore, R.M. Hochstrasser, J. Chem. Phys.76, 5632 (1982).
3. J. C. Owrutsky, Y.-R. Kim, M. Li, M.J. Sarisky and R.M. Hochstrasser, Chem. Phys. Lett. 184, 368 (1991).

4. R. Zwanzig, J. Chem. Phys. 34, 1931 (1961).
5. See, for example: (a) D.W. Oxtoby, Adv. Chem. Phys. 47, 482 (1981); (b) S.A. Adelman and R.H. Stote, J.Chem. Phys. 88, 4397 (1988); (c) R.M. Whitnell, K.R. Wilson and J.T. Hynes, J.Chem. Phys. 94, 8625 (1990).
6. M. Ferrario, I.R. McDonald and M.L. Klein (unpublished).

Optical Stark Shift of the Raman-Gain Spectrum in a Polydiacetylene PDA-DFMP Single Crystal

T. Kobayashi and S. Halle

Department of Physics, University of Tokyo, Hongo 7-3-1, Bunkyo, Tokyo 113, Japan

Optical Stark shift in the Raman gain spectrum was observed for a single crystal of polydiacetylene (PDA), PDA-DFMP (2,5-bis (trifluoromethylphenyl) PDA) by using pump-probe spectroscopy apparatus based on a femtosecond CPM laser.

The excitation intensity dependence of the Raman gain spectra of PDA-DFMP was studied between 19 GW/cm^2 and 58 GW/cm^2. The Stokes Raman gain spectra corresponding to 2100 cm^{-1} and 1500 cm^{-1} frequency shifts due to C=C and C=C bond streching, respectively, shift to red by 50 cm^{-1} at the highest intensity from the weakest intensity. The shift can be explained in terms of the optical Stark effect due to near resonance between the vibrationally excited level in the ground state and the zero-phonon exciton state.

I. Introduction

Conjugated polymers have been studied extensively since they have several unique properties such as one-dimensionality and strong correlation and electron-phonon coupling. Ultrafast geometrical relaxation [1-10] and short exciton lifetime [3,4,7,9,10,13,14] are of interest from the basic physical point of view and also from the stand point of application as ultrafast nonlinear materials.[3]

Among various optical nonlinear processes, which have been reported and analyzed, characteristic nonlinear processes which cannot be found in semiconductors either in bulk or in quantum confined structure are those induced by exciton-phonon coupling [10,12,15-17]. The processes are phonon side band hole [12], induced Raman gain [10,12], inverse Raman [12,15-17], phnon mediated optical Stark effect [12,15-17], and phonon mediated pump polarization coupling [12].

The optical Stark effect was reported for exciton transitions in semiconductor quantum wells [18,19]. The time response of the effect is only limited by the pulse width and the effect was considered to be usefull for ultrafast opitcal switch. However other optical nonlinear effects such as two photon absorption takes place at the same time, it results in real excitation of excitons and hence the response is limited by the exciton lifetime, which ranges from several hundred picoseconds to a few nanoseconds.

In case of the optical Stark effect in polymers, [10] since the lifetime of exciton is as short as a few picosecond [9,10], the response is still fast enough even when two-photon absorption takes place. In case of phonon-meidated Stark effect, which appears in polymers, is even more advantageous, since the probe light can be far from the strong exciton absorption and can be still in resonance with the transition between two unpopulated levels, i.e. a vibrationally excited level in the ground state and a zero-phonon exciton state.

Springer Proceedings in Physics, Vol. 74 Time-Resolved Vibrational Spectroscopy VI
Editors: A. Lau, F. Siebert, W. Werncke

In the present paper we studied the phonon-mediated optical Stark effect, which changes the Raman gain spectrum of a polydiacetylene PDA-DFMP.

II. Experimental

Femtosecond pump-probe spectroscopy is performed with a 10 Hz amplified colliding-pulse mode-locked dye laser yielding 100 fs pulses. White-light continuum generation of probe pulses is employed, where the linearly polarized pump and probe beams are oriented parallel to the conjugated chain direction in the polymer and are overlapped on the free-standing crystal sample in air at 297 K. The relative intensity of the pump beam, with a beam diameter of 0.04 cm, is adjusted with a variable neutral density filter.

For the study of phonon-mediated optical Stark effect was done for the single crystal of polydiacetylene with a backbone structure $\{RC-C\equiv C-CR\}_n$ where R=R'=2,5-bis trifluoromethyl-phenyl (PDA-DFMP). All the experiment was performed at room temperature.

III. Optical Stark effect in Raman gain spectrum

The experiment of the optical Stark effect in the Raman gain spectrum was performed with a pump photon energy, ω_p=16010 cm^{-1} (1.99 eV), which is lower in energy than the peak of the absorption band, at 18270 cm^{-1} (2.27 eV), due to the ^1B_u exciton transition. In the photoinduced absorption spectra of PDA-DFMP at delay times from t=-170 to +200 fs, two peak minima can be observed and are attributed to Raman gain signals [20]. In Fig. 1, the Raman gain spectra at 0 ps for both v_1 and v_2 are shown for six different pump intensities varying from 19 to 58 GW/cm^2. At the weakest pump intensity the gain signals occur at probe photon energies $\omega_{pr}(I_{min})$ 14510 cm^{-1} (1.78 eV) and 13845 cm^{-1} (1.72 eV). The measured RG Stokes shifts of 2165 and 1500 cm^{-1}, are in approximate agreement with the ground state stretching mode frequencies of 2135 cm^{-1} (0.265 eV) for v_1 and 1500 cm^{-1} (0.186 eV) for v_2, respectively, obtained by the resonant Raman scattering in PDA-DFMP. At the highest pump intensity, the peak probe photon energies for the RF signal are $\omega_{pr}(I_{max})$ 13900 cm^{-1} (1.72 eV) for v_1 and 14550 cm^{-1} (1.80 eV) for v_2, corresponding to a Stark shift of the peak probe photon energy of -55 and -40 cm^{-1}, respectively, where the experimental Stark shift is defined as $\omega_{pr}(I_{min}) - \omega_{pr}(I_{max})$.

In the model calculations presented below, the Raman gain process in PDA in the presence of intense pump laser excitation is modeled by choosing the appropriate coupling of the near-resonant optical fields to a 3-level system. In the 3-level system considered here, the time-independent (transformation) matrix representation of the Liouville super-operator X is a 9x9 matrix which includes the off-diagonal optical coupling Rabi frequency terms R_1, R_2, and R_3. By assuming Markovian relaxation dynamics, damping parameters such as electronic dephasing rates, (Γ_{ca} and Γ_{cb}), a vibrational relaxation rate, (Γ_{ba}) and the population decay rates (Γ_{cc} and Γ_{bb}) are introduced as off-diagonal elements in X. The rotating frame for this 3-level system assumes that the transition energies ω_{ca} and ω_{bc} (i.e. the Stokes emission) are considered to be nearly resonant with the pump energy ω_p, and the probe energy ω_{pr}, respectively. For a given set of Rabi freuencies, the solution of the steady state equation of motion is found by numerically calculating the inverse of the full 9x9 Liouville superoperator, over a range of probe energies. The calculated spectrum

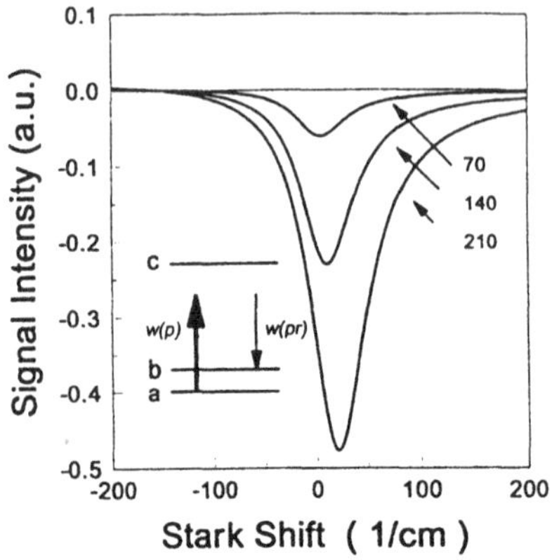

Fig.1. The experimental RG spectra plotted as the normalized absorbance change versus the probe photon energy ω_{pr} at six different pump intensities, where I=58 GW/cm^2, for (A) v_1 and (B) v_2.

$q(\omega_{pr})$ is obtained by evaluating the correlation functions [21] upon the inverse matrix elements of the Liouville superoperator X' in the rotating frame. For a given set of Rabi frequencies, the spectrum, $q(\omega_{pr})$, at a particular probe energy is evaluated as [21]

$$q(\omega_{pr}) = Re\{X'_{cbab}\rho_{ac} + X'_{cbbb}\rho_{bc} - X'_{cbcc}\rho_{bc}\}, \quad (1)$$

where ρ_{ac} and ρ_{bc} are the steady state density operators associated with the (Stokes) Raman process and the relevant coherence terms, respectively.

Over the range of experimental pump intensities, R_3 is varied from 93 to 280 cm^{-1} and 80 to 240 cm^{-1} for v_2 and v_1, respectively. The values of the transition dipole moments are taken to be μ_{bc}=2.14 and 2.5 D for v_1 and v_2, respectively. The ratio of μ_{bc} transition moments for v_1 and v_2 are deduced from the experimental Raman cross sections of the related polydiacetylene, PDA-pTS [16,17]. For v_2, where R_3=280 cm^{-1} > Δ_{pr} and where $\Delta_{ph} = \omega_{ac} - \omega_p - \omega_{ab}$ = 760 cm^{-1}, Stark shifting of b> and c> states is expected to be significant.

The same pump laser, which optically couples the c> - b> transition, is detuned Δ=125 and 760 cm^{-1} for v_1 and v_2, respectively. The 9x9 density matrix is separately evaluated with the b> as either v_1 and v_2 using the parameters listed in Fig. 2 and the following range of values for the Rabi frequency : R_2 is varied from 0 to 210 cm^{-1} (for either v_1 or v_2) and R_3 is varied from 0-240 cm^{-1} and 0 to 280 cm^{-1} fir v_1 and v_2, respectively.

Over the range of the experimental pump intensities, with increasing Rabi frequency the peak of the Raman gain signal shift to the higher probe photon energies by 53 and 58 cm^{-1}, for v_1 and v_2, respectively. Calculated spectra for v_1 and v_2, are shown in Fig. 2. The direction of the resulting blue shift is consistent with the experiment, and the magnitude of the calculated shift is in semi-quantitative agreement with the data.

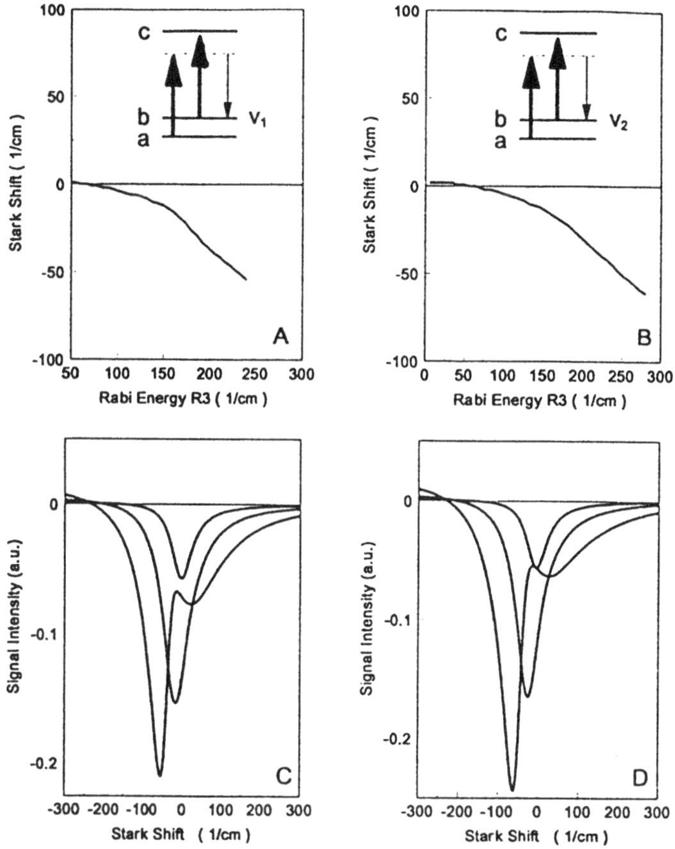

Fig. 2. Maximum Stark shift as a function of Rabi energy R_3 for v_1 (A) and v_2 (B) calculated with the following parameters: ω_{ca}=18270 cm^{-1}; ω_{cb}=2135 and 1500 cm^{-1} for v_1 and v_2, respectively; ω_p=16010 cm^{-1}; Γ_{ca}=300 cm^{-1} and Γ_{cb}=300 cm^{-1}; Γ_{ba}=30 cm^{-1}; $\Gamma_{cc}=\Gamma_{bb}$=0 cm^{-1}. The calculated spectrum versus the Stark shift for (C) v_1 at R_3=80, 160, and 240 cm^{-1} and (D) v_2 at R_3=93, 186, and 280 cm^{-1}..pa

VI. Conclusion

We have observed the optical Stark shift of the Raman gain spectra in polydiacetylene PDA-DFMP under resonant condition between unpopulated levels.

References

[1] W.P. Su, J.R. Schrieffer, and A.J. Heeger, Phys. Rev. Lett. 42, 1698 (1979)

[2] W.P. Su and J.R. Schrieffer, Proc. Natl. Acad. Sci. USA 77, 5626 (1980)

[3] T. Kobayashi, IEICE Trans. Fundamentals E-75A, 38 (1992)
[4] M. Yoshizawa, Y. Hattori, and T. Kobayashi, Phys. Rev. B47, 3882 (1993)
[5] C.V. Shank, R. Yen, R.L. Fork, J. Orenstein, and G.L. Baker, Phys. Rev. Lett. 498, 1660 (1982)
[6] L. Rothberg, T.M. Jedju, S. Etemad, and G.L. Baker, Phys. Rev. Lett. 57, 3229 (1986)
[7] M. Yoshizawa, T. Kobayashi, K. Akagi, and H. Shirakawa, Phys. Rev. B37, 10301 (1988)
[8] S. Takeuchi, M. Yoshizawa, T. Masuda, T. Higashimura, and T. Kobayashi, IEEE J. Quantum electron. QE-28, 2508 (1992)
[9] M. Yoshizawa, M. Taiji, and T. Kobayashi, IEEE J. Quantum electron. QE-25, 2532 (1989)
[10] T. Kobayashi, M. Yoshizawa, M. Hasegawa, na dM. Taiji, J. Opt. Soc. Am. B7, 1558 (1990)
[11] J.M. Huxley, P. Mataloni, R.W. Schoenlein, J.G. Fujimoto, E.P. Ippen, and G.M. Carter, Appl. Phys. Lett. 56, 1600 (1990)
[12] M. Yoshizawa, A. Yasuda, and T. Kobayashi, Appl. Phys. B53, 296 (1991)
[13] M. Yoshizawa, K. Nishiyama, M. Fujihira, and T. Kobayshi, Chem. Phys. Lett., in press
[14] A. Yasuda, M. Yoshizawa, and T. Kobaysahi, Chem. Phys. Lett., in press.
[15] B.I. Greene, J. Orenstein, D.H.Rapkine, S.Schmitt-Rink, and M. Thakur, Phys. Rev. Lett. 61, 325 (1988)
[16] G.J. Blanchard, J.P. Heriage, A.C. von Lehmen, M.K. Kelly, G.L. Berker, and S. Etemad, Phys. REv. Lett. 63, 887 (1989)
[17] G.J. Blanchard and J.P. Heritage, J. Chem. Phys. 93, 4377 (1990)
[18] A. Mysyrowicz, D. Hulin, A. Antonetti, A. Migus, W.T. Masselink, and H. Morkoc, Phys. Rev. Lett. 55, 1335 (1985)
[19] A.von Lehmen, J.E. Zucher, J.P. Hritage, and D.S. chemla, Optics Lett. 11, 609 (1986)
[20] S.D. Halle, M. Yoshizawa, H. Matsuda, S. Okada, H. Nakanishi, and T. Kobayashi, in Time-Resolved Vibrational spectroscopy, ed. H. Takahashi (springer-Verlag, Berlin 1992) pp.260-261.
[21] B. Dick and R.M. Hochstrasse, J. Chem. Phys. 81, 2897 (1984)

Femtosecond Time-Resolved Raman-Gain Spectrum in a Polydiacetylene PDA-3BCMU

T. Kobayashi and M. Yoshizawa

Department of Physics, University of Tokyo, Hongo 7-3-1, Bunkyo, Tokyo 113, Japan

Raman gain spectrum was observed for a polydiacetylene (PDA), epitaxially grown PDA-3BCMU (3-butoxycarbonylmethylurethane) microcrystalite on a KCl substrate by using pump-probe spectroscopy apparatus based on a femtosecond CPM laser.

The femtosecond time-resolved Raman gain spectra of PDA-3BCMU were measured using two pump pulses at 1.97 eV and 1.78 eV. The spectra at various delay times between -500 fs and 1 ps are explained in terms of the conformational change due to the geometrical relaxation taking place from acetylene-type to butatriene-type configuration associated with the self-trapping of free exciton followed by quasi-thermalization with time constant shorter than 200 fs, which is consistent with the formation time constant of the quasithermal self-trapped exciton being 140 fs determined by the absorption spectral change.

I. Introduction

Various optical nonlinear processes in conjugated polymers have been extensively studied.[1-14] Among them characteristic nonlinear processes which cannot be found in semiconductors either in bulk or in quantum confined structure are those induced by exciton-phonon coupling [10,12,15-17]. The processes are phonon side band hole [12], induced Raman gain [10,12], inverse Raman [12,15-17], phnon mediated optical Stark effect [12,15-17], and phonon mediated pump polarization coupling [12]. The exciton-phonon coupling is also of key importance to determine the dynamics of the nonlinear response induced by free exciton and self-trapped exciton. The disappearance and formation of these two excitons are triggerd by the exciton-phonon coupling. The relaxation of the latter to the ground state is due to the tunneling from the potential curve of the self-trapped exciton (STE) to that of the ground state.

The speed of self-trapping was estimated to be 10-20 fs and that of quasi-thermalization of STE is determined between 70 fs and 200 fs depending on the structures of the main chain and side groups. The self-trapping is considered to be associated with the geometrical relaxation of the conjugated main chain in the case of polydiacetylene and that from thiophene-type to quinone-type configuration in the case of polythiophene.

Time-resolved Raman spectroscoy is a powerful method for studying structures of transient species and excited states with short life. Theoretical study in polyacetylene shows that soliton and polaron have different phonon modes and they can be distinguished by Raman spectroscopy [18]. However, only a few time-resolved Raman experiments have been done in conjugated polymers, because of the difficulty due to very short lifetime of the excited states [16,17]. However to the best of our knowledge transient Raman signal of excited states has not been observed in conjugated polymers.

II. Experimental

The femtosecond-resolved Raman gain spectroscopy was performed using three pulses of femtosecond durations. The 1.97 eV femtosecond pulse was generated and amplified by a colliding-pulse mode-locked dye laser and a four-stage dye amplifier pumped by the second harmonic of a 10 Hz Q-switched Nd:YAG laser, respectively [9]. The duration and energy of the amplified pulse were 100 fs and 200 μJ, respectively. The amplified pulse was split into three beams. The second one was focused in a 3-mm cell containing CCl_4 and femtosecond white continuum was generated by self-phase modulation (SPM). a part of the white continuum was selected with a set of prism pairs and a slit and amplified by a two-stage dye amplifier pumped by the second-harmonic of the Q-switched Nd:YAG laser. The amplified pulse has the center photon energy of 1.78 eV and the duration of 200 fs and was used for the pump pulse of the Raman gain spectroscopy (pump 2). The probe pulse was detected with a polychromator/CCD detection system and the Raman gain signal was obtained from the time-resolved transmittance change spectra. Polarizations of the three beams were parallel to the oriented polymer chains of PDA-3BCMU (3-butoxycarbonylmethylurethane) deposited on a KCl crystal [2].

III. Time-resolved Raman gain spectrum

Figure 1 shows transient absorption spectra induced by the 1.97 eV pulse which is just resonant with an excitonic absorption peak of PDA-3BCMU. Two negative peaks at 1.79 and 1.71 eV at the delay time of 0.0 ps are due to Raman gain of the stretching vibrations of the C=C and C=C bonds, respectively. This spectral change has been explained by the relaxation model shown in Fig. 2.[12,13] The free exciton (FE, state 1) photoexcited by the pump 1 pulse is coupled with the C-C stretching modes within 10-20 fs and becomes nonthermal self-trapped exciton (STE, state 2). Then the nonthermal STE relaxes to the bottom of the potential surfaces (quasithermal STE, state 3) by quasithermalization process. Therefore, the transient absorption is

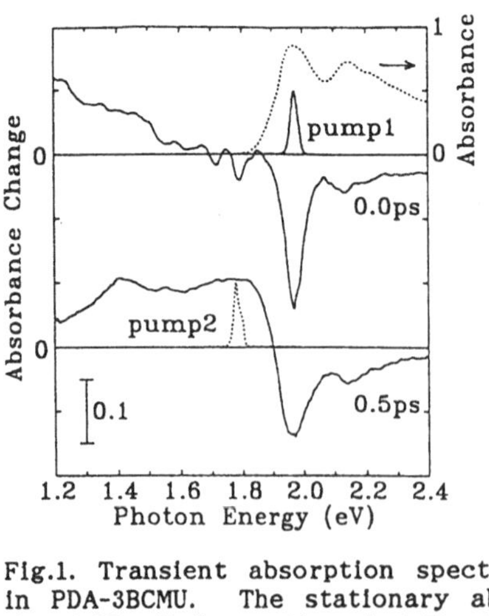

Fig.1. Transient absorption spectra induced by the 1.97 eV pump pulse in PDA-3BCMU. The stationary absorption spectrum and the spectra of the 1.97 eV (pump 1) and 1.78 eV (pump 2) pulses are shown together. The excitation photon density is 1.7×10^{15} photons/cm^2.

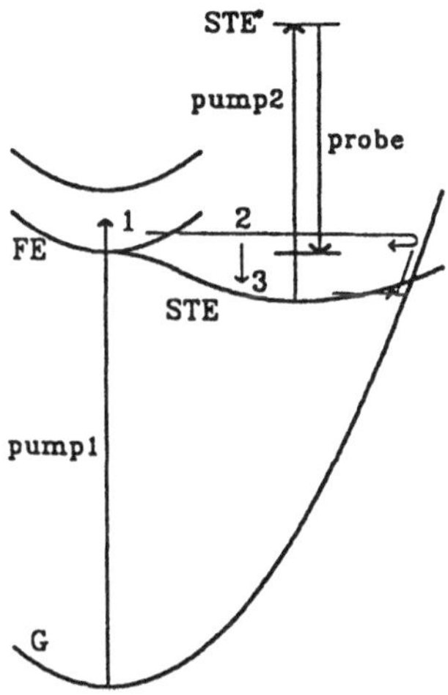

Fig. 2. Relaxation model in polydiacetylene shown in potential surfaces of excitons. States 1,2,3, and G are free exciton (FE), nonthermal self-trapped exciton (STE), quasithermal STE, and the ground state, respectively. STE* is a biexciton state with the ^1A_g symmetry.

shifted to higher energy and the transient fluorescence of the FE decays very rapidly as observed by the probe-saturation spectroscopy developped recently by our group. [14] The STE relaxes to the ground state (G) by tunneling in the configuration space. Since the time constant of the tunneling in a blue-phase PDA is 1.5 ps at room temperature, the STE disappears before thermalization. The 1.78 eV pulse (pump 2) is resonant with the transition observed at 1.8 eV, which was assigned to the transition from the STE to a biexciton with an ^1A_g symmetry [12].

Figure 3 shows the Raman gain spectra obtained from the transient absorption spectra induced by the 1.78 eV pulse after the photoexcitation of the 1.97 eV pulse. At -0.5 ps, two Raman gain peaks are observed at 1440 and 2060 cm^{-1}. They are assigned to the stretching vibrations of the carbon-carbon double and triple bonds in the acetylene-like structure of the ground state. The spectrum at 0.0 ps has broad signal below the 1440 cm^{-1} Raman peak down to 1000 cm^{-1}. At delay time longer than 0.2 ps the Raman signal has a clear peak at 1200 cm^{-1}. The width of the 2060 cm^{-1} Raman signal becomes slightly broader after the photoexcitation but no new Raman peak is observed around 2000 cm^{-1}.

Resonance Raman spectroscopy of trans-1,3-butadiene in the ground state shows a weak Raman signal around 1200 cm^{-1}.[19] It is assigned to the bending mode of the C=C bond.[20] Therefore, enhancement of the Raman scattering of the bending mode can explain the observed new peak. However, the broad signal at 0.0 ps cannot be explained by the enhancement. The spectral change of the Raman signal indicates that the geometircal relaxation of main chain occurs within 200 fs after the 1.97 eV photoexcitation.

The 1200 cm^{-1} Raman peak suggests that the C=C bond in the ground state changes to C-C bond by the formation of STE. In other words, the acetylene-like structure is expected to relax geometrically to the butatriene-like structure $(-CR=C=C=CR-)_x$. The stretching modes of center bonds in linear molecules with four carbon atoms [21] are compared with the observed Raman signal. The frequency of the C-C single

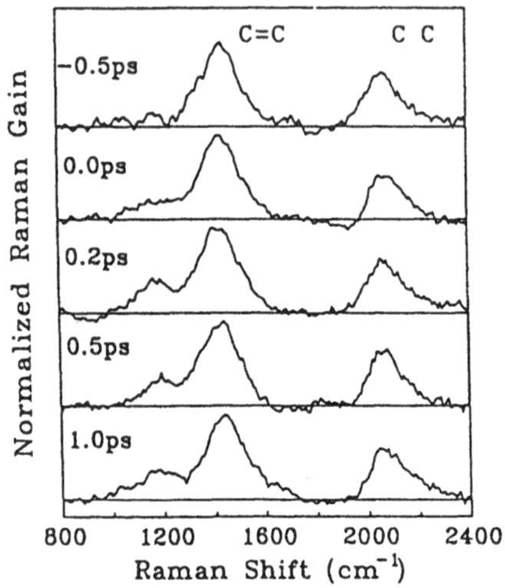

Fig. 3. Normalized stimulated Raman gain spectra of PDA-3BCMU at several delay times after the excitation of 1.97 eV pump pulse. The excitation photon densities of the 1.97 and 1.78 eV pulses are 2.9×10^{15} and 9.0×10^{15} photons/cm^2, respectively.

bond in trans-1,3-butadiene (CH_2=CH-CH=CH_2) is 1202 cm^{-1}. It is consistent with the observed new Raman peak at 1200 cm^{-1} which has smaller Raman shift than the peak of the C=C bond.

IV. Conclusion

We observed the femtosecond-resolved resonance Raman gain spectrum in polydiacetylene PDA-3BCMU and found a few Raman mode appearing within 200 fs. The spectrum corresponds to the self-trapped exciton.

References

[1] W.P. Su, J.R. Schrieffer, and A.J. Heeger, Phys. Rev. Lett. 42, 1698 (1979)
[2] W.P. Su and J.R. Schrieffer, Proc. Natl. Acad. Sci. USA 77, 5626 (1980)
[3] T. Kobayashi, IEICE Trans. Fundamentals E-75A, 38 (1992)
[4] M. Yoshizawa, Y. Hattori, and T. Kobayashi, Phys. Rev. B47, 3882 (1993)
[5] C.V. Shank, R. Yen, R.L. Fork, J. Orenstein, and G.L. Baker, Phys. Rev. Lett. 498, 1660 (1982)
[6] L. Rothberg, T.M. Jedju, S. Etemad, and G.L. Baker, Phys. Rev. Lett. 57, 3229 (1986)
[7] M. Yoshizawa, T. Kobayashi, K. Akagi, and H. Shirakawa, Phys. Rev. B37, 10301 (1988)
[8] S. Takeuchi, M. Yoshizawa, T. Masuda, T. Higashimura, and T. Kobayashi, IEEE J. Quantum electron. QE-28, 2508 (1992)

[9] M. Yoshizawa, M. Taiji, and T. Kobayashi, IEEE J. Quantum Electron. QE-25, 2532 (1989)
[10] T. Kobayashi, M. Yoshizawa, M. Hasegawa, na dM. Taiji, J. Opt. Soc. Am. B7, 1558 (1990)
[11] J.M. Huxley, P. Mataloni, R.W. Schoenlein, J.G. Fujimoto, E.P. Ippen, and G.M. Carter, Appl. Phys. Lett. 56, 1600 (1990)
[12] M. Yoshizawa, A. Yasuda, and T. Kobayashi, Appl. Phys. B53, 296 (1991)
[13] M. Yoshizawa, K. Nishiyama, M. Fujihira, and T. Kobayshi, Chem. Phys. Lett., in press
[14] A. Yasuda, M. Yoshizawa, and T. Kobaysahi, Chem. Phys. Lett., in press.
[15] B.I. Greene, J. Orenstein, D.H.Rapkine, S.Schmitt-Rink, and M. Thakur, Phys. Rev. Lett. 61, 325 (1988)
[16] G.J. Blanchard, J.P. Heriage, A.C. von Lehmen, M.K. Kelly, G.L. Baker, and S. Etemad, Phys. REv. Lett. 63, 887 (1989)
[17] G.J. Blanchard and J.P. Heritage, J. Chem. Phys. 93, 4377 (1990)
[18] A. Terai, Y. Ono, and Y. Wada, J. Phys. Soc. Jpn. 58, 3798 (1989)
[19] D.N. Batchelder and D. Bloor, J. Phys. C : Solid State Phys. 15, 3005 (1982)
[20] W.F. Lewis and D.N. Batchelder, Chem. Phys. Lett. 60, 232 (1979)
[21] L.M. Sverdlov, M.A. Kovner, and E.P. Krainov, Vibrational Spectra of Polyatomic Molecules, (John Wiley & Sons, New York, Toronto, 1970), pp.282-323

Construction of a Nanosecond Time-Resolved IR Spectrometer Using a Photovoltaic MCT Detector

C. Kato[1,2], *M.W. George*[1,3], *and H. Hamaguchi*[1]

[1]Molecular Spectroscopy Laboratory,
Kanagawa Academy of Science and Technology (KAST),
KSP East 301, 3-2-1 Sakato, Takatsu-ku, Kawasaki-shi 213, Japan
[2]R D Department, Yamazaki Works, Hitachi Chemical Co., Ltd.,
4-13-1 Higashi-cho, Hitachi-shi 317, Japan
[3]Department of Chemistry, University of Nottingham,
Nottingham NG7 2RD, UK

We have previously reported the development of a scanning (4000 - 700 cm^{-1}) microsecond time-resolved IR (TRIR) spectrometer with very high sensitivity (ΔAbs $< 10^{-6}$) [1]. This consisted of a dispersive IR monochromator and a photoconductive MCT detector. We found that the measurement of TRIR spectra in solution was hampered by the strong thermal artifact which occur at IR frequencies corresponding to solvent absorptions [2]. The risetime of the thermal artifact was found to have a microsecond delay with respect to the transient signal. With our microsecond TRIR spectrometer, however, we were not able to remove totally the thermal artifact. To solve this problem, we have developed a new spectrometer using a photovoltaic MCT detector with a 30 ns risetime.

Figure 1a shows temporal behaviour of the transient signal at 1325 cm^{-1} and the thermal artifact at 1440 cm^{-1} of 4-phenylbenzophenone (4-PhBzP) in acetonitrile. The observed laser profile is also shown in figure 1a. The transient signal at 1325 cm^{-1}, which has been assigned to the T_1 state of 4-PhBzP [3], rises instantaneously upon laser excitation. On the other hand, the thermal artifact at 1440 cm^{-1} shows several hundred nanoseconds of delay. Figure 1b shows TRIR spectra obtained with photovoltaic (full line) and photoconductive (broken line) MCT detectors. The former spectrum, which was obtained with the first 100 ns gating, has no contribution of the thermal artifact at 1440 cm^{-1}, in contrast to the latter spectrum with microsecond time resolution.

We also applied this apparatus to another simple and interesting molecule. Figure 1c shows TRIR spectra of biphenyl in deuterated acetonitrile with different gate positions. The observed TRIR band at 1560 cm-1 shows a few hundred nanoseconds of lifetime, and is not quenched by adding an ammonia water, indicating that this band is due to the T_1 state of biphenyl.

Springer Proceedings in Physics, Vol. 74 Time-Resolved Vibrational Spectroscopy VI
Editors: A. Lau, F. Siebert, W. Werncke

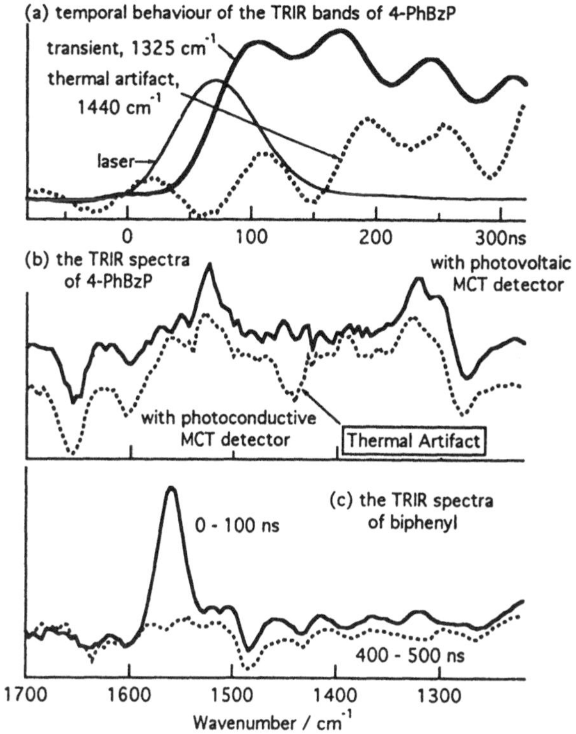

Figure 1: (a) temporal behaviour of the transient signal and the thermal artifact of 4-PhBzP in acetonitrile; and (b) the TRIR spectra of 4-PhBzP obtained with photovoltaic and photoconductive MCT detectors; and (c) the TRIR spectra of biphenyl obtained with the first (0 - 100 ns) and delayed (400 - 500 ns) gate positions.

References

1) C. Kato, K. Iwata and H. Hamaguchi, J. Spectrosc. Soc. Jpn., 40, 255 (1991).

2) C. Kato, M. W. George and H. Hamaguchi, Appl. Spectrosc., to be published.

3) M. W. George, C. Kato and H. Hamaguchi, Chem. Lett., 873 (1993).

Versatile Broadband CARS Spectrometer Based on a Diode-Pumped Nd:YLF Laser: Polarization-Sensitive and Time-Resolved Measurements

B. Toleutaev, T. Tahara, and H. Hamaguchi*

Kanagawa Academy of Science and Technology (KAST),
KSP East 301, 3-2-1 Sakato, Takatsu-ku, Kawasaki-shi 213, Japan
*Present address: Physics Department, Moscow State University, 119899 Moscow, Russia

Abstract. Construction and performance of a multiplex CARS spectrometer with an expanded (1000 cm^{-1}) spectral range and good signal-to-noise ratio are reported. The potentials of the setup in the polarization sensitive CARS measurements were demonstrated for accurate depolarization ratio determination and for detection of weak (overlapped) Raman bands. The nanosecond time resolved resonance CARS spectra of the lowest excited triplet state of all-trans retinal were measured.

1. Introduction

Coherent anti-Stokes Raman spectroscopy (CARS) is now widely used for studying the vibrational structure and time-resolved dynamics of molecules [1,2]. A high intensity of coherent signal in the anti-Stokes spectral region determines the feasibility of this technique in excited state experiments, when short accumulation times, low probe laser power and rejection of strong stray and luminescence lights are required. CARS also offers a variety of polarization measurements that increase the amount and accuracy of information obtained.

The purpose of the present work was to build a spectrometer with the improved performance for polarization-sensitive and time-resolved CARS applications.

2. Results and Discussion

The versatility of the setup has been achieved thanks to the use of a cw diode-pumped, 2 kHz repetition rate Q-switched Nd:YLF laser (Spectra Physics, TFR), which enhanced the stability of operation and easiness of alignment.

To minimize the signal fluctuations we have implemented the multiplex CARS scheme with multichannel spectrum detection system. The broadband dye-laser (Spectra Physics, 375B) was optimized for pulsed pumping by Nd:YLF second harmonic. Excellent mode characteristics of both lasers have resulted in significantly low noise in the spectra. The fluctuations in the dye laser emission profile were averaged out due to the high pulse repetition rate at accumulation times as short as few seconds.

In order to expand an emission bandwidth of the dye laser we used a mixture of fluorescein 548 and rhodamine 6G dyes. This enabled us to cover a 1000 cm^{-1}

Springer Proceedings in Physics, Vol. 74 Time-Resolved Vibrational Spectroscopy VI
Editors: A. Lau, F. Siebert, W. Werncke

range in a single CARS spectrum, which is, to the best of our knowledge, about twice as wide as ever reported. Simple calculations have also shown the way to avoid the spectral range limitation brought by the phase-matching condition. For this purpose we used a jet sampling method. To compensate for signal loss due to the short sample length of about 300 μm we used a tight focusing of laser beams.

The CARS signal from carbon tetrachloride, showing no prominent Raman bands in the region of interest, was used for spectral intensity normalization. When measured under the same experimental conditions as a sample spectrum, it serves as a reference which accounts for all the factors such as dye-laser emission profile, detector spectral response and residual phase-matching distortions.

A series of polarization-sensitive multiplex CARS spectra of neat cyclohexane were measured. The efficiency of polarization suppression of strong Raman bands was demonstrated for resolving weak overlapped bands. Good signal-to-noise ratio achieved and least-squares fitting of the measured spectra have resulted in accurate (within ±5%) determination of the Raman depolarization ratios.

In the time-resolved CARS experiment, the third harmonic pulses from regeneratively amplified mode-locked Nd:YAG laser (355 nm, 50 ps, 2 kHz) [3] were used to produce the transient species. Multiplex resonance CARS spectra of the T_1 state of all-trans retinal in cyclohexane were measured at a series of delay times. The temporal resolution was determined by the laser pulse duration of 6 ns. Low energy of CARS probing pulses (1..5 μJ) ensured minimal disturbance of the excited state population. The transient CARS spectrum detected at early delay time in the presence of fluorescence, gives an unique possibility of studying the excited states inaccessible by spontaneous Raman spectroscopy [4].

References

1. V. F. Kamalov, N. I. Koroteev and B. N. Toleutaev. In: *Time Resolved Spectroscopy*, Eds. R. J. H. Clark and R. E. Hester (Wiley, Chichester, 1989) and references therein.
2. G. H. Atkinson, in Proc. XIII Int. Conf. on Raman Spectr. (ICORS), Würzburg, Germany, 1992. Eds. W. Kiefer et al. (J. Wiley, Chichester, 1992), p. 192.
3. K. Iwata, H. Hamaguchi, Chem. Phys. Lett. **196**, 462 (1992).
4. H. Hamaguchi, in *Vibrational Spectra and Structure*, J. R. Durig Ed. (Elsevier, 1987), Vol. 16, p. 286.

Part III

Excited States

Picosecond Time-Resolved Raman Observation of the Mode-Dependent Vibrational Dynamics in S_1 Trans-Stilbene

K. Iwata and H. Hamaguchi

Molecular Spectroscopy Laboratory,
Kanagawa Academy of Science and Technology (KAST),
KSP East 301, 3-2-1 Sakato, Takatsu-ku, Kawasaki-shi 213, Japan

Abstract. The time and solvent dependent changes of the picosecond time-resolved Raman spectra of S_1 trans-stilbene (tSB) are recorded. The four Raman bands at 1570, 1240, 1180, and 285 cm^{-1} show different time dependence. The position and width of the 1570 cm^{-1} band change systematically in the alcohol solutions from butanol to decanol. Our model assuming an exchange mechanism between two electronic states can explain both the time and solvent dependence of the 1570 cm^{-1} band successfully.

1. Introduction

In the photoisomerization reaction of tSB in the first excited singlet (S_1) state, a typical potential barrier for the rotation around the C=C bond is 1200 cm^{-1} [1]. Thus, it is expected that the vibrational excitation and the subsequent energy redistribution in the S_1 state play an important role in the isomerization reaction. With the use of a picosecond time-resolved Raman spectrometer, we are able to observe the dynamic change following the photoexcitation in a direct manner. The time dependent change in the position and width of the 1570 cm^{-1} (central C=C stretching) and 1180 cm^{-1} (C-Ph stretching) bands of S_1 tSB have been reported [2-4]. In this paper, we report the time dependent change of four S_1 tSB Raman bands in several solvents. The time and solvent dependent changes of the 1570 cm^{-1} band are well explained by a simple model based on an exchange mechanism between two electronic excited states [5].

2. Experimental

We use a standard pump-probe technique for our transform-limited picosecond time-resolved Raman spectrometer. The details of the system have been published elsewhere [6]. In short, a synchronously pumped dye laser output (588 nm, 82 MHz 110 mW) is amplified with the use of the second harmonic of a cw Nd:YAG regenerative amplifier (532 nm, 2 kHz, 700 mW). The second harmonic of the amplified pulse (294 nm, 2 kHz) is used for the pump light while the remaining visible pulse (588 nm) is used for the probe light. The time resolution and the wavenumber resolution of the system are 2.2 ps and 3.5 cm^{-1} [6]. The Raman scattering light is analyzed by a single spectrograph and detected with a CCD detector.

3. Results and Discussion

We fitted Lorentzian functions to the observed Raman bands and obtained the band parameters. The position and the width of each S_1 tSB Raman band show characteristic temporal changes. Figure 1 shows the time dependence of the peak positions of four different Raman bands, 1570, 1240, 1180, and 285 cm^{-1}. The changes for the 1570 and 1180 cm^{-1} bands are essentially the same as those reported earlier [2-4]. In the figure, the 285 cm^{-1} band shifts to the lower wavenumber direction as opposed to other three bands which show upshifts as the time delay increases. We suggest, however, that the changes for the time region of -2 to 5 ps is ascribable to the apparent shift of the energy levels caused by the presence of the strong electromagnetic field of the pump light. The details for this effect will be published separately.

In the following part of this paper, we focus our attention on the 1570 cm^{-1} band which shows the biggest time and solvent dependent changes for the position and width. It should be noted that all the observed 1570 cm^{-1} bands are fitted by Lorentzian functions well regardless of the time delay or the solvent. In Figure 2, the time dependent change of the peak position of the 1570 cm^{-1} band is plotted for three different solvents, octanol, butanol, and heptane. Although the absolute peak positions are different form each other, the differences of the peak positions between 0 ps and 50 ps or the ways they reach the plateaus for the three solvents are quite similar. These results suggest that the general trend of the temporal change for the 1570 cm^{-1} band does not depend on the polarity nor viscosity of the solvent. It is noteworthy that the lifetimes of the S_1 tSB in these solutions range from 57 ps (butanol) to 80 ps (heptane) or 98 ps (octanol). The relative peak position change seems to have little connection with the S_1 lifetime.

The time and solvent dependent changes of the position and width of the 1570 cm^{-1} band are well explained by a simple model based on an exchange mechanism [5]. In this model, we assume that there is an exchange between the S_1 and some other upper state, possibly the S_2 state. Hereafter we call the second state S_2 for simplicity. The vibrational frequencies for the S_1 and S_2 states are ω_1 and ω_2, re-

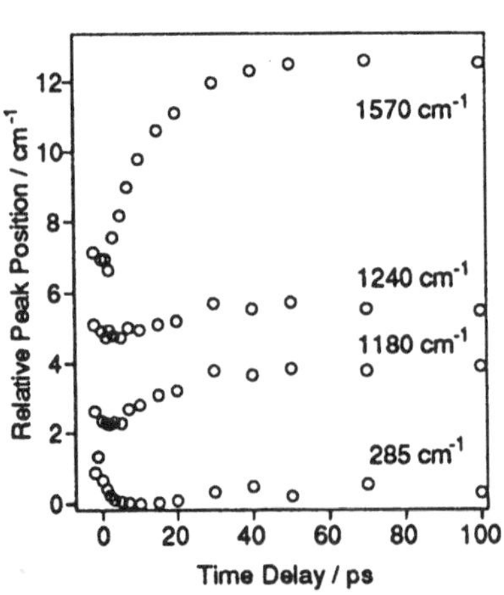

Fig.1 Time dependence of the Raman band positions of S_1 trans-stilbene in heptane.

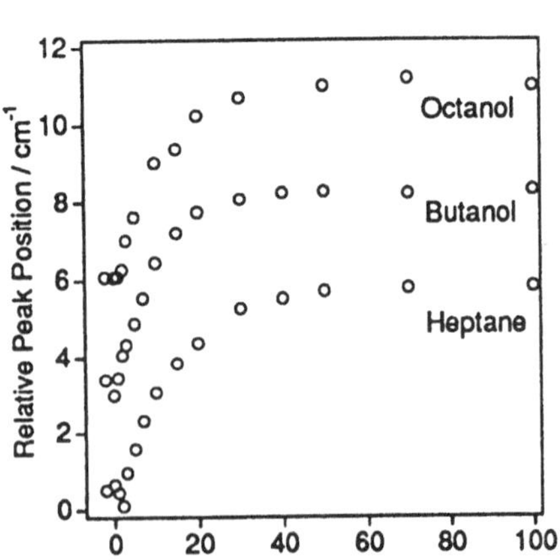

Fig.2 Time dependence of the 1570 cm^{-1} band position in three different solvents.

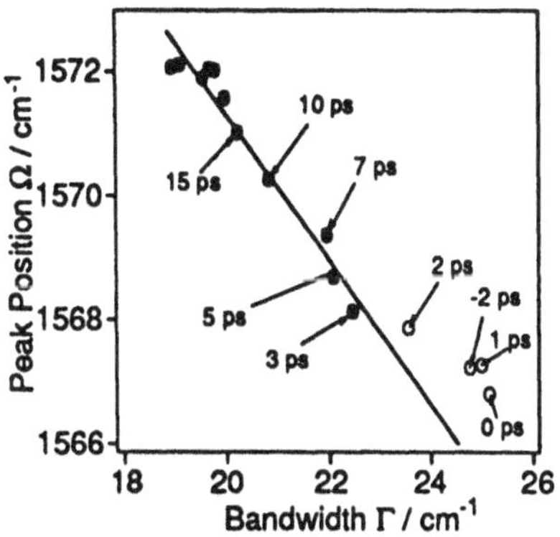

Fig.3 Peak position vs bandwidth of the 1570 cm^{-1} band in the butanol solution for time delays from -2 ps to 100 ps.

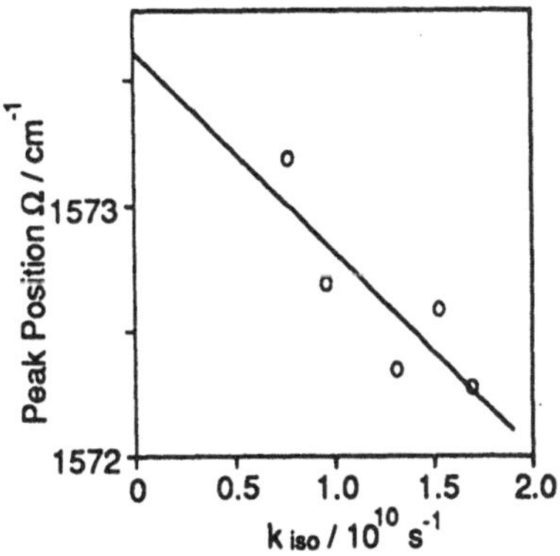

Fig.4 Position of the 1570 cm^{-1} Raman band vs trans-cis isomerization rate in alcohol solutions.

spectively, with the frequency difference $d\omega=\omega_1-\omega_2$. The forward hopping rate from ω_1 to ω_2 is W_1 while the backward hopping rate from ω_2 to ω_1 is W_2. It is probable that these hopping processes have large contributions to the whole dephasing process in the S_1 vibrational level. The original Lorentzian band shapes are changed as the result of the exchange motion. In the slow modulation case (W_1, $W_2 \leq d\omega$), the changed band shape can be expressed with a new Lorentzian function whose position and width are different from those of the original Lorentzian by $\Delta\Omega$ and $\Delta\Gamma$ [7],

$$\Delta\Omega = W_1\, W_2\, d\omega \,/\, (d\omega^2 + W_2{}^2), \tag{1}$$

$$\Delta\Gamma = 2W_1\, d\omega^2 \,/\, (d\omega^2 + W_2{}^2). \tag{2}$$

From equations (1) and (2), the ratio $\Delta\Omega / \Delta\Gamma$ is calculated as

$$\Delta\Omega / \Delta\Gamma = W_2 / 2d\omega. \tag{3}$$

Because $d\omega$ is constant, if W_2 is constant for different dime delays, the ratio $\Delta\Omega/\Delta\Gamma$ is constant. Thus, the plot $\Delta\Omega$ vs $\Delta\Gamma$ for various time delays should give a straight line. The observed results are shown in Figure 3. After t = 3 ps, the observed points in the figure can be well fitted with a single line, as expected. Similar results have been obtained for a heptane solution [5]. In either case, there is a deviation from the straight line for t = -2 to 5 ps, which we attribute to the effect of the pump light electromagnetic field at the present stage. The different locations on the straight line correspond to different values for W_1, suggesting that the forward hopping rate W_1 changes depending on the time delay. We think that this change in W_1 corresponds to the cooling process of the S_1 tSB and its environment after the photoexcitation.

The model presented here can explain the solvent dependence of the band shape as well. We recorded the Raman spectrum of S_1 tSB in normal alcohol solutions from n=4 (butanol) to n=10 (decanol). The time delay was set to 50 ps so that the observed band shapes are not affected by the time dependent changes. In Figure 4, the observed peak positions are plotted against the known trans-cis isomerization rates, k_{iso} [8]. Again, we obtained a linear relation between the two quantities, the peak position of a Raman band and the isomerization rate, which apparently have no connection with each other. If we use equations (1), (2) and the observed $\Delta\Omega / \Delta\Gamma$ value of 1.1 (figure not shown), however, we obtain a linear relation between $\Delta\Omega$ and W_1:

$$\Delta\Omega / \text{cm}^{-1} = 2.0 \times 10^{-12} W_1 / \text{s}^{-1} \quad (4)$$

From equation (4) and the observed $\Delta\Omega / k_{iso}$ value of -0.79×10^{-10} in Figure 4,

$$k_{iso} = 0.025 W_1. \quad (5)$$

Equation (5) indicates that certain amount of tSB molecules after the hopping to the S_2 state proceed to the isomerization reaction while others are deactivated back to the S_1 state. This situation will be realized if the S_2 state has a "dissociative" character at trans configuration (or the S_2 potential surface has a local maximum at trans). The branching ratio for the reaction, 0.025, can be expressed with an exponential function, $\exp(-\Delta E/kT)$, where ΔE is 2.2 kcal/mol if T=300 K. This ΔE value agrees with the observed activation energy for this isomerization reaction in alcohol solutions, 1.6 to 3.6 kcal/mol [8]. We have reported a similar agreement between the ΔE value deduced from the solvent dependent change of the Raman bands in alkane solutions and the independently observed activation energy of the isomerization reaction [5]. These results support our present approach to interpret the isomerization reaction of S_1 tSB as well as the time and solvent dependent changes of the S_1 Raman band shapes.

References

1.R. M. Hochstrasser, Pure Appl. Chem. **52**, 2683 (1981).
2.H.Hamaguchi,Chem.Phys.Lett. **126**,185(1986); J. Chem. Phys. **89**,2587 (1988).
3.W.L.Weaver,L.A.Huston,K.Iwata,andT.L.Gustafson,J.Phys.Chem.**96**,8956(1992).
4.K. Iwata and H. Hamaguchi, Chem. Phys. Lett. **196**, 462 (1992).
5.H. Hamaguchi and K. Iwata, Chem. Phys. Lett. **208**, 465 (1993).
6 K. Iwata, S. Yamaguchi, and H. Hamaguchi, Rev. Sci. Instrum., in press.
7.R. M. Shelby, C. B. Harris, and P. A. Cornelius, J. Chem. Phys. **70**, 34 (1979).
8.S.K.Kim, S.H.Courtney, and G.R.Fleming, Chem. Phys. Lett. **159**, 543 (1989).

Mode-Specific Effects in the Picosecond Resonance-Raman Spectra of Trans-Stilbene Solutions

J.N. Moore[1], *P. Matousek*[2], *A.W. Parker*[2], *W.T. Toner*[2], *M. Towrie*[2], *and R.E. Hester*[1]

[1]Chemistry Department, University of York, York Y01 5DD, UK
[2]SERC Rutherford Appleton Laboratory, Chilton, Didcot, Oxon OX11 0QX, UK

Abstract. Picosecond time-resolved resonance Raman spectra of singlet excited trans-stilbene have been obtained using several photolysis (pump) and Raman excitation (probe) wavelengths. Vibrational mode-selective changes in peak wavenumbers and bandwidths are reported which are dependent on the sample temperature and the pump wavelength.

1. Introduction

Picosecond time-resolved resonance Raman spectroscopy (ps-TR^{3}S) has been applied successfully to the study of S_1 tS and recent work [1-3] has revealed the existence of mode- specific solvent-dependent changes occurring on a *c.* 10 ps timescale following photoexcitation. Other studies have shown that the bandwidth of the S_1 tS visible absorption spectrum narrows on a similar timescale on photoexcitation at wavelengths significantly shorter than the origin of the S_0- S_1 transition [4]. In this study, ps-TR^{3}S has been used to probe S_1 tS using several pump and probe wavelengths and a range of temperatures in order to reveal further details of this mechanism.

2. Experimental

The ps-TR^{3}S apparatus has been described [5]. Its characteristics are summarised in Fig. 1.

3. Results

Picosecond TR^3 spectra of excellent quality have been obtained from S_1 tS over the range 200-1700 cm^{-1}, with pump wavelengths from 275 - 310 nm and solution temperatures from - 50 to + 50 ^{o}C. Lorentzian profiles gave the best fit to the observed Raman bands. The peak positions, bandwidths, and relative intensities have been measured as functions of the delay time between the pump and probe pulses, of the pump and probe wavelengths, and of the temperature.

Fig. 2(a) shows a part of the ps-TR^3 spectrum of S_1 tS obtained from an n-hexane solution using a 10 ps delay between pump (285 nm) and probe (570

Springer Proceedings in Physics, Vol. 74 Time-Resolved Vibrational Spectroscopy VI
Editors: A. Lau, F. Siebert, W. Werncke

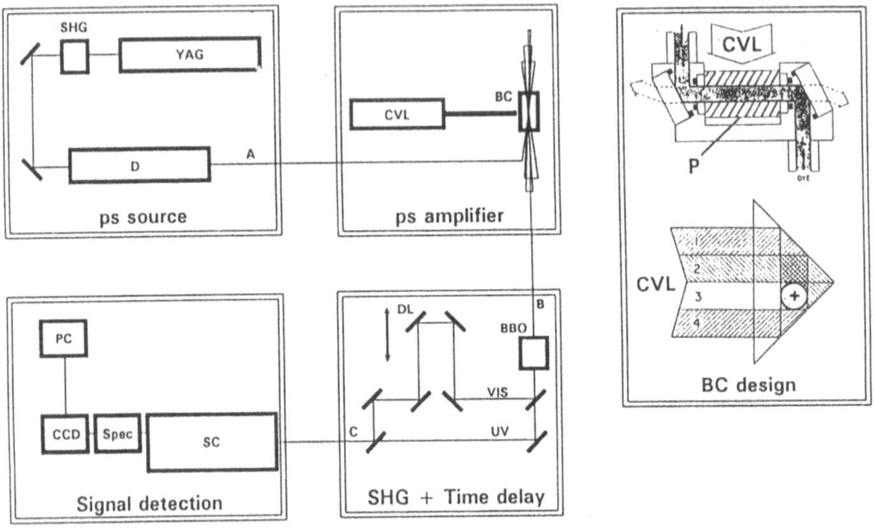

Fig. 1. Layout of ps-TR^{3}S apparatus. **Key:** YAG, mode-locked Nd:YAG laser; SHG, second harmonic generator; D, cavity dumped dye laser; CVL, copper vapour laser; BC, Bethune cell; P, prism with offset dye channel; BBO, b-barium borate crystal; DL, delay line; SC, sample compartment; Spec, spectrometer; A - 610 nm, 4.46 kHz, 20 nJ, 6 ps; B - 610 nm, 4.46 kHz, 55 μJ, 6 ps; C - 610 nm and 305 nm, 10 and 2.5 μJ, respectively.

nm) pulses. Solvent peaks have been subtracted and the bands fitted by Lorentzians. Fig. 2(b) shows the changes resulting from an increase in the delay time from 10 to 80 ps and the difference between the two spectra in 2(b) is shown in 2(c). The largest changes are seen for the *c.* 1570 cm^{-1} band, but changes also are evident in the *c.* 1180 cm^{-1} band and possibly in the *c.* 1240 cm^{-1} band. Both peak wavenumber shifts and bandwidth changes contribute to these effects. Fig. 3 shows that changes in peak position and bandwidth for the *c.* 1570 cm^{-1} band display similar kinetics with both short (*c.* 4 ps) and long (*c.* 10 ps) components. These changes show similar kinetics but different magnitudes with shorter wavelength pulses. Pumping at 285 nm and probing at 570 nm or 605 nm reveals that the effects result primarily from changes in the pump wavelength.

By changing the sample temperature we found that, with fixed pump and probe wavelengths (305 and 610 nm, respectively, chosen so as to minimize the changes to the *c.* 1570 cm^{-1} band arising from excitation above the electronic origin), mode-specific changes were obtained for a fixed pump/probe delay time of 20 ps, i.e. temperature-induced mode-specific changes. By far the largest effects in the 1100 - 1600 cm^{-1} region again were apparent in the *c.* 1570 cm^{-1} band. With increasing sample temperature, the peak shifted to lower wavenumbers and broadened, as shown in Fig. 4. Kinetic measurements made at -33, 20, and 40 ^{o}C, with 285/570 nm pump/probe wavelengths, showed the peak wavenumber changes with delay to be temperature independent.

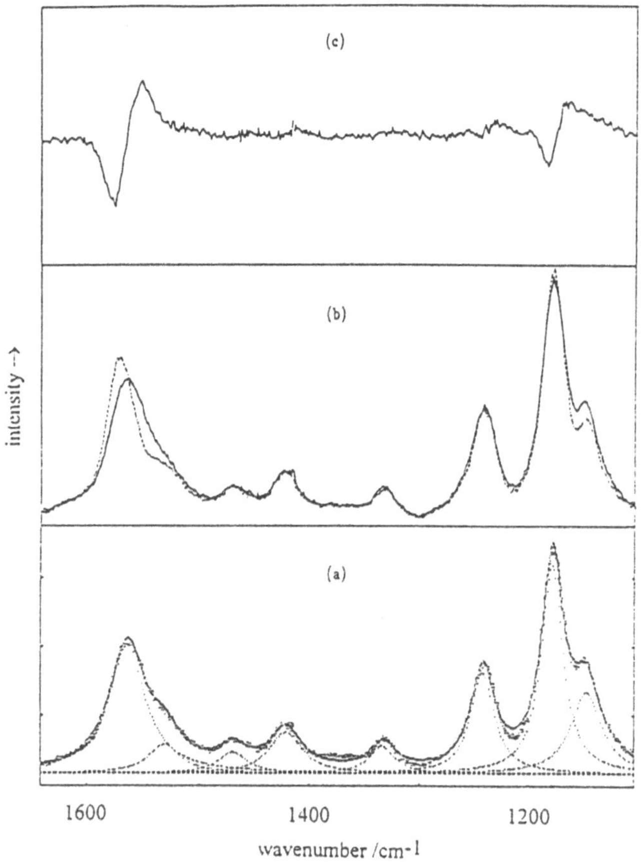

Fig. 2. S_1 tS resonance Raman spectrum, obtained from tS in n-hexane at 18 °C using 285 /570 nm pump/probe pulses with delays of (a) 10 ps and (b) 10 and 80 ps. The difference between the spectra shown in (b) is shown in (c). Solvent bands have been subtracted and the band shapes fitted by Lorentzians in (a).

4. Discussion

Our vibrational mode selectivity data are supportive of the pioneering work of Gustafson et al. [2] and Hamaguchi et al. [3] but go considerably beyond these studies and enable us to gain further insights into the early time behaviour of photoexcited tS in solution. In particular, the mode-specific temperature dependence of S_1 tS is reported here for the first time.

The S_1 tS Raman band at c. 1570 cm^{-1} is associated with a vibrational mode which is of mainly olefinic C_o=C_o bond stretching character in S_0 tS. This shows by far the greatest sensitivity to mode-selective effects associated with changes in solvent, pump-probe delay time, pump wavelength, and temperature. A few other modes show smaller effects, as is illustrated in Fig. 2 for the *c*. 1180 cm^{-1} band which corresponds to a mode involving mainly C_o–C_p (olefinic carbon to phenyl

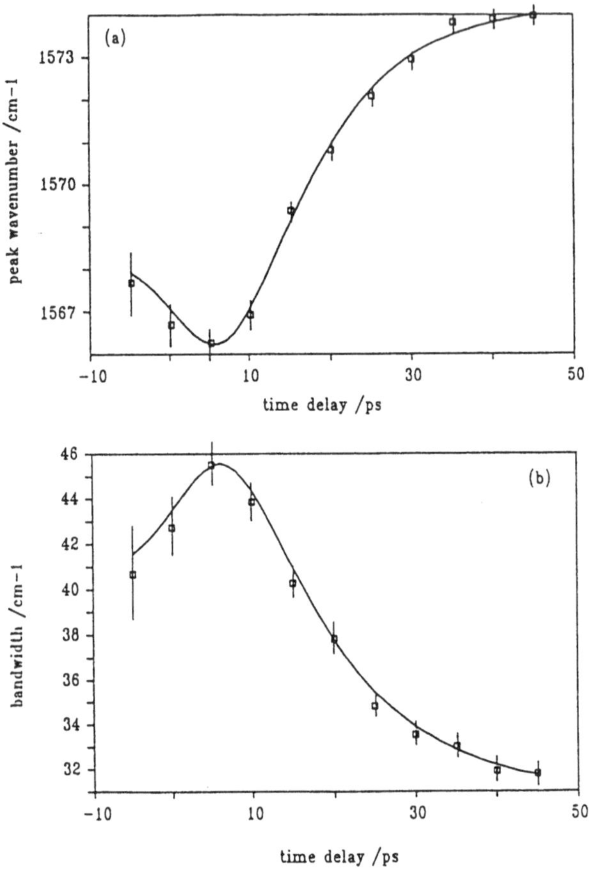

Fig. 3. Peak wavenumbers (a) and bandwidths (b) of the band at *c.* 1570 cm^{-1} at various time delays between pump/probe pulses at 275/550 nm. Data from S_1 tS in n-hexane at 18 °C. The fitted curves are based on the convolution of the instrument response with a double exponential (growth and decay).

carbon bond) stretching in S_0 tS. The mode-selective effects were found to be proportional to the excess photoexcitation energy of *c.* 1000-5000 cm^{-1} (310-275 nm) given to the S_1 tS.

Our temperature-induced mode-specific effects are strikingly similar in form and in magnitude to those changes associated with excess energy effects. It seems reasonable, then, to attribute these to thermal activation of S_1 tS in its local solvent cage. The lack of high wavenumber anti-Stokes Raman band intensity enhancement in our spectra indicates that intramolecular vibrational relaxation processes eliminate any "hot" olefin modes on a timescale which is faster than our measurements. But anharmonic coupling of the olefinic "doorway" modes to low wavenumber "exchange" modes [6], through which process energy is transferred out to the solvent environment, certainly could result in the generation of low wavenumber "hot" modes. Such low energy "hot" modes may alternatively be generated by direct heating of the sample and it seems that the *c.* 10 ps process observed through the

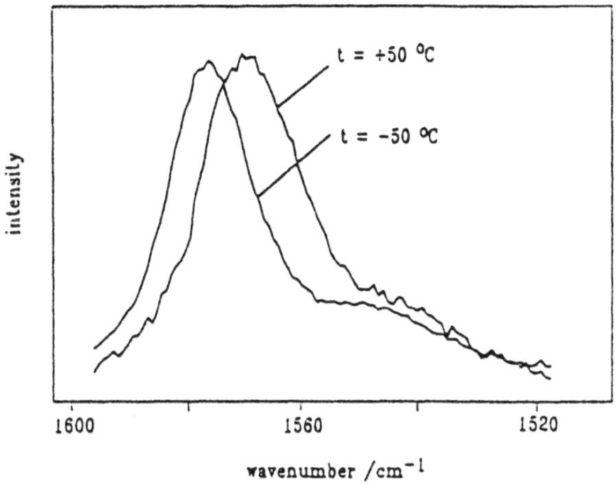

Fig. 4. Effects of changing temperature on the *c.* 1570 cm^{-1} band from S_1 tS in n-hexane at 20 ps delay time between pump (305 nm) and probe (610 nm) pulses.

high wavenumber bands may correspond to relaxation of the "hot" low-wavenumber modes.

Taking all the available evidence together, we may visualise a possible mechanism for the observed mode-selective dynamics as follows. Photoexcitation of S_0 tS creates S_1 tS in an initially "hot" state, with vibrational excitation of a wide range of modes, dependent on the excess energy provided, i.e. on the photoexcitation wavelength. Ultrafast intramolecular vibrational energy redistribution occurs, resulting in an essentially impulsive creation of vibrationally "hot", highly anharmonic, low frequency (deformation) modes which distort the molecular geometry [7]. The observed early-time (0 - 5 ps) exponential decrease in the *c.* 1570 cm^{-1} peak wavenumber and increase in bandwidth both are consistent with such distortions, these being damped by coupling to the local solvent cage. Superimposed on these initial IVR effects is a longer-timescale (0 - 40 ps) "cooling" process, involving energy transfer from these low frequency modes to the reorientating solvent environment which also provides progressively increasing microdielectric stabilization of the S_1 tS [8]. The proposed rôle of vibrationally "hot" low frequency modes is consistent with the fact that raising the temperature produces similar results to those observed at 10 ps on increasing the excess energy in the photoexcitation pulse. However, our measurements of the excess energy and temperature dependence of the *c.* 1570 cm^{-1} peak wavenumber and bandwidth show the increase in width with excess energy to be different from the corresponding increase with temperature. This indicates that at delay times less than *c.* 40 ps S_1 tS is not in true equilibrium either internally or with the solvent.

5. Acknowledgements

We are grateful to S. Umapathy for his assistance in the early stages of this work, to V Kamalov, I Lednev and D Phillips for helpful discussions, to M Russell and A P Tilley for experimental assistance, and to T L Gustafson for preprints of his work.

6. References

1. R. E. Hester, P. Matousek, J. N. Moore, A. W. Parker, W. T. Toner, and M. Towrie, Chem. Phys. Lett., in press (1993).
2. W. L. Weaver, L. A. Huston, K. Iwata and T. L. Gustafson, J. Phys. Chem. 96, 8956 (1992).
3. K. Iwata and H. Hamaguchi, Chem. Phys. Lett. 196, 462 (1992).
4. B. I. Greene, R. M. Hochstrasser and R. B. Weisman, Chem. Phys. Lett. 62, 427 (1979) ; F. E. Doany, B. I. Greene and R. M. Hochstrasser, Chem. Phys. Lett. 75, 206 (1980).
5. P. Matousek, R. E. Hester, J. N. Moore, A. W. Parker, D. Phillips, W. T. Toner, M. Towrie, I. C. E. Turcu and S. Umapathy, Meas. Sci. Tech., in press (1993).
6. R. M. Shelby, C. B. Harris and P. A Cornelius, J. Chem. Phys. 70, 34 (1979).
7. S. A. Asher and J. Murtaugh, J. Am. Chem. Soc. 105, 7244 (1983).
8. K. Yoshihara, A. Nmiki, M. Sumitani and N. Nakashima, J. Chem. Phys. 71, 2892 (1983).

The Cis→Trans Stilbene Photoreaction Probed Through Two-Color UV Picosecond Pump-Probe Raman Spectroscopy

D.L. Phillips, J.-M. Rodier, and A.B. Myers

Department of Chemistry, University of Rochester,
Rochester, NY 14627-0216, USA

Abstract. The formation of hot *trans*-stilbene by photoisomerization of *cis*-stilbene, and its subsequent vibrational cooling, have been monitored by picosecond two-color uv pump-probe anti-Stokes Raman spectroscopy. Hot ground-state *trans*-stilbene appears promptly within our 10 ps experimental time resolution, and the anti-Stokes intensity decays with a time constant of 39 ± 10 ps in cyclohexane and 17 ± 5 ps in methanol. The decay of the anti-Stokes intensity is accompanied by a shift of the C=C stretching vibrations by about 20 cm^{-1} to higher frequencies, due probably to cooling of other vibrational modes anharmonically coupled to the C=C stretches.

1. Introduction

The photoisomerization of *cis*-stilbene is a very fast (≈1 ps) excited state reaction that creates a highly vibrationally excited ground-state product. Despite numerous studies of the dynamics of *cis*-stilbene's ultrafast photochemistry,[1-5] there remain many unanswered questions about the pathway of the excited-state interconversions, the partitioning between isomerization to *trans*-stilbene and ring closure to dihydrophenanthrene, the role of environmental friction, and the rate of randomization of the excess vibrational energy. Time-resolved vibrational spectroscopy can address many of these questions. The transient Raman spectrum of the longer-lived excited state of *trans*-stilbene has already been studied in some detail.[6-8] Here we obtain picosecond anti-Stokes resonance Raman spectra of the hot ground-state *trans*-stilbene product formed following excitation of *cis*-stilbene.[9] Our goal is to determine the conformation of the nascent product, how the excess vibrational energy is partitioned, and how that energy relaxes with time.

2. Experimental

The experiments employ an amplified synchronously-pumped dye laser system operating at 500 Hz.[9] A fraction of the 1053 nm output from a mode-locked Nd:YLF laser seeds a cw-pumped Nd:YLF regenerative amplifier, while the remainder is frequency doubled and pumps a three-mirror dye laser whose output is amplified in a three-stage dye chain quasi-longitudinally pumped by the frequency doubled output of the regenerative amplifier. Pump pulses at 295 nm (3-4 μJ/pulse) are obtained by frequency doubling the amplified dye laser, while probe pulses at 278 nm (0.8-1.2 μJ) are generated by mixing the amplified dye with 10% of the 527 nm output from the regenerative amplifier. The pump and variably delayed probe beams are focused through a 5 cm lens and overlapped within the liquid sample flowing in an open jet. Raman scattering is collected in a backscattering geometry with a fused silica condenser lens and detected with a 0.75-meter single spectrograph and an intensified diode array. The instrument response function has a FWHM of 9.5 ps. Each 100 mL sample of 1.4 mM concentration is replaced every 30 min to minimize accumulation of photoproducts. All experiments are carried out at about 23°C. polarity is the most obvious difference between cyclohexane and methanol, it may not be the only important one. The solvent's mass, its internal vibrational frequencies, and the local solvation structure may also influence the vibrational relaxation times.[12] Finally, since we can monitor only a few of stilbene's 72 vibrations, we do not know whether the vibrational cooling we observe corresponds to an intra- or an intermolecular process. "Conventional wisdom" holds that the decay times we observe are too slow to be due to IVR in a molecule of this size, but this point remains controversial.[5]

3. Results and discussion

Figure 1 shows representative anti-Stokes resonance Raman spectra as a function of pump-probe time delay. The spectra at large negative delay times (-25 ps) are essentially identical to the probe-only spectra, and are dominated by solvent bands and by the 403 and 560 cm^{-1} vibrations of *cis*-stilbene. Since the intensities of the *cis*-stilbene lines relative to solvent are nearly independent of delay, they must be due to thermal population in the unpumped ground state *cis*-stilbene molecules. The bands in the 1590-1640 cm^{-1}, 1150-1200 cm^{-1}, and 950-1000 cm^{-1} regions, which reach their maximum intensity near time zero, arise from vibrationally hot *trans*-stilbene molecules formed subsequent to absorption of a pump photon. The ring-closure product, dihydrophenanthrene, does not absorb strongly at the probe wavelength and does not appear in the transient spectra.[10]

Since the excited state lifetime of *cis*-stilbene is shorter than our pulse durations, and the *trans*-stilbene product absorbs strongly at the pump wavelength, the vibrationally ex-

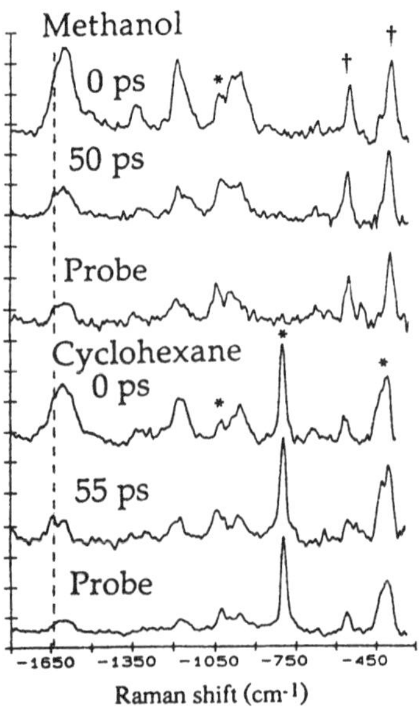

Figure 1. Anti-Stokes resonance Raman spectra of *cis*-stilbene photoproducts in cyclohexane and methanol at the indicated pump-probe delay times. Asterisks mark solvent lines; daggers label vibrations of the parent *cis*-stilbene. "Probe" is the probe-only spectrum. The dashed vertical line indicates the position of the strongest C=C stretch in the room temperature Stokes spectrum.

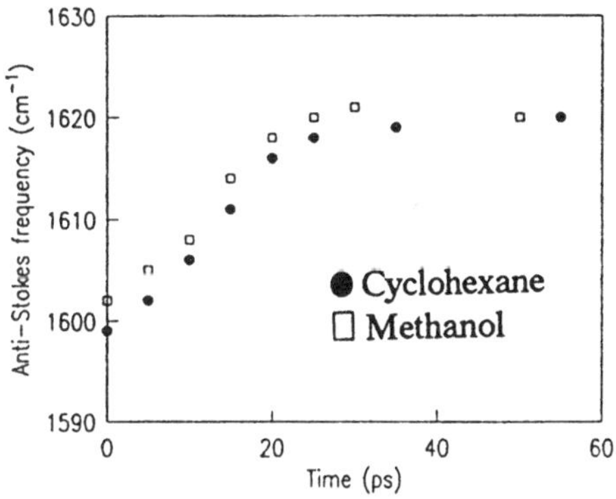

Figure 2. Anti-Stokes frequencies of the C=C stretches as a function of pump-probe time delay. The spectra from 0 to 20 ps were fit to a single band, while the spectra at longer delays were fit to two lines whose mean is plotted.

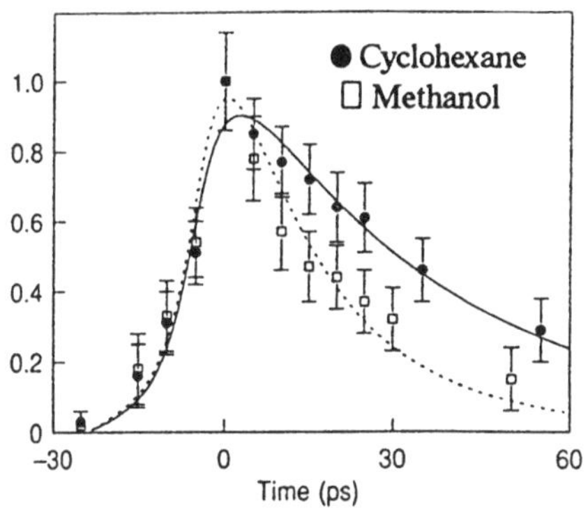

Figure 3. C=C stretching anti-Stokes intensity as a function of pump-probe time delay. Band areas measured relative to solvent are scaled to 1.0 at t=0. The probe-only intensity has been subtracted. Curves are theoretical fits (see text).

cited product molecules could absorb a second pump photon. We have quantitated the contribution of double-pumped molecules to the signal by examining the dependence of the $t = 0$ anti-Stokes intensity on pump energy. Increasing the pump energy from 1.0 to 4.0 μJ increases the C=C anti-Stokes intensity by a factor of 3.8. This shows that depletion of the *cis*-stilbene ground state is not severe and that contributions to the spectra from molecules that have absorbed more than one pump photon are minimal.

The strongest lines in the Stokes spectrum of room temperature *trans*-stilbene are the C=C stretches at 1638 and 1599 cm^{-1}. At the longest time delays the anti-Stokes spectra in both solvents exhibit two resolved peaks near these frequencies, but at shorter times the bands are broadened and shifted to lower frequencies. The average C=C stretching frequency increases by nearly 20 cm^{-1} between 0 and 25 ps in both solvents (Figure 2). The simplest explanation is that at short delay times there are contributions to the spectrum from molecules excited to more than one quantum; that is, not only $1 \rightarrow 0$ transitions, which would appear at the same frequency as the Stokes $0 \rightarrow 1$, but also $2 \rightarrow 1$ and/or $(1,n) \rightarrow (0,n)$ transitions, where the second index refers to another mode that is anharmonically coupled with the C=C stretch(es). Based on the anharmonicities determined from the Stokes overtone spectra, the observed anti-Stokes frequency shift of about 20 cm^{-1} at short times would require excitation of 3-4 quanta in the C=C stretches if the shift were due to excitation of these stretches alone. Coupling of the highest frequency, ethylenic-localized C=C stretch with lower-frequency modes, particularly torsional modes, is also expected to be significant since the ethylenic stretching frequency should depend strongly on bond order and be sensitive to skeletal distortions.

Figure 3 shows that the anti-Stokes intensity decays significantly faster in methanol than in cyclohexane. In these experiments the pump and probe pulses have the same polarization, so the anti-Stokes signal decay has contributions from both rotational motion and vibrational cooling. The time dependence of the anti-Stokes signal was fit to the function $I(t)=I(0)\exp(-t/\tau_{pop})[2r(0)\exp(-t/\tau_{rot})+1]$ convolved with the instrument function, where r(0) is the initial product anisotropy of 0.20[5] and τ_{rot} is the rotational time. The best-fit "population decay time" τ_{pop} is 17 ps in methanol and 39 ps in cyclohexane. This parameter cannot be directly equated to the population of excited vibrational levels because the Raman cross sections depend on the populations in all of the vibrational modes, which are changing with time.[9] However, since both the room-temperature absorption spectrum and the Stokes resonance Raman spectra are only slightly solvent dependent,[11] comparison of the kinetics in different solvents should still be valid.

Our anti-Stokes decay time in methanol is in good agreement with the cooling time of 17.5 ps inferred by Hochstrasser *et al.* from red-edge absorption measurements,[5] but those workers observed little difference between the red-edge decays in methanol and in alkanes. Our decay times in cyclohexane are considerably longer than those observed for smaller trienes in the same solvent.[13] It should be stressed that the anti-Stokes intensities do not correspond directly to vibrational populations. Also, the decays seem to be nonexponential, consisting of a fast initial phase followed by a somewhat slower decay. Biphasic decays have been attributed to initial heating of the first solvent shell followed by slower conduction to the bulk solvent,[12] or to fast IVR followed by slower cooling to solvent.[14]

The anti-Stokes intensity in the C=C stretches reflects the total population in states that are excited to at least one quantum in a C=C stretching mode, while the bandshape and frequencies reflect the population in levels higher than n=1 and/or the population in other modes that are anharmonically coupled to the C=C stretches. The data of Figures 2 and 3 suggest that the difference between the vibrational relaxation process in methanol and in cyclohexane may lie mostly in the loss of the last vibrational quantum in a C=C stretch rather than in the initial cooling of a molecule that is excited to many quanta in several modes.

Solvent polarity effects on the excited state dynamics of both *cis-* and *trans-*stilbene are well established[2-4,15,16] and are often attributed to the dipolar or zwitterionic character of the twisted intermediate state. It is not obvious, however, why solvent polarity should affect the rate of vibrational relaxation on the ground state surface. Measurements in a wider range of solvents are needed to determine which physical parameters of the solvent dictate the vibrational relaxation rates in S_0 *trans-*stilbene.

4. References

1. A.B. Myers and R.A. Mathies, *J. Chem. Phys.* **81**, 1552 (1984).
2. J.K. Rice and A.P. Baronavski, *J. Phys. Chem.* **96**, 3359 (1992).
3. L. Nikowa, D. Schwarzer, J. Troe and J. Schroeder, *J. Chem. Phys.* **97**, 4827 (1992).
4. D.C. Todd and G.R. Fleming, *J. Chem. Phys.* **98**, 269 (1993).
5. R.J. Sension, S.T. Repinec, A.Z. Szarka and R.M. Hochstrasser, *J. Chem. Phys.* **98**, 6291 (1993).
6. W.L. Weaver, L.A. Huston, K. Iwata and T.L. Gustafson, *J. Phys. Chem.* **96**, 8956 (1992).

7. H. Hamaguchi and K. Iwata, *Chem. Phys. Lett.*, in press.
8. R.E. Hester, P. Matousek, J.N. Moore, A.W. Parker, W.T. Toner and M. Towrie, *Chem. Phys. Lett.* 208, 471 (1993).
9. D.L. Phillips, J.-M. Rodier and A.B. Myers, *Chem. Phys.*, in press.
10. J.-M. Rodier, X. Ci and A.B. Myers, *Chem. Phys. Lett.* 183, 55 (1991).
11. X. Ci and A.B. Myers, *Chem. Phys. Lett.* 158, 263 (1989).
12. T. Elsaesser and W. Kaiser, *Ann. Rev. Phys. Chem.* 42, 83 (1991).
13. P.J. Reid, S.J. Doig, S.D. Wickham and R.A. Mathies, *J. Am. Chem. Soc.* 115, 4754 (1993).
14. H.J. Bakker, P.C.M. Planken and A. Lagendijk, *J. Chem. Phys.* 94, 6007 (1991).
15. D.H. Waldeck, *Chem. Rev.* 91, 415 (1991).
16. J.-M. Rodier and A.B. Myers, *J. Am. Chem. Soc.*, submitted.

Photoisomerization Studies of Cyanine Dyes by Time-Resolved Resonance CARS

W. Werncke, M. Pfeiffer, T. Johr, A. Lau, and W. Freyer

Max-Born Institute for Nonlinear Optics and Short Pulse Spectroscopy, Rudower Chaussee 6, D-12489 Berlin-Adlershof, Germany

Abstract. Time resolved Resonance CARS spectra of a series of bisdimethylaminomethine dyes and of their isomers recorded after photoexcitation have been obtained. For all dyes a very intense characteristic "photoisomer CARS line" at 1070cm^{-1} is observed which can be assigned to a mode localized at one end of the π-chain of the 1,2 mono-cis photoisomer. The observed frequency and intensity differences between the ^{14}N and ^{15}N- substituted heptamethine and their photoisomers are reproduced by our calculations fairly well.

1. Introduction

Photoisomerization is one of the fundamental molecular processes occurring after light irradiation.

Since the early paper of Scheibe it has been known that bisdimethylaminomethine dyes show a trans-cis photoisomerization with an bathochromic shift of appr. 30nm of the absorption band with respect to the parent molecules [1]. It has been shown that in the case of the trimethine the 1,2 mono-cis is generated [1], but for the dyes with the longer methine chains numerous E,Z photoisomers have to be expected and could occur even simultaneously. Following quantum chemical calculations in lengthening the methine chain generation of 2,3-mono-cis isomers should become more probable [2]. Contrary we suggested that even in the case of the heptamethine and of the nonamethine the 1,2 mono-cis photoisomer is generated [3,4].

Here we report on recent results obtained by time resolved Resonance CARS of the series of bisdimethylaminomethine dyes $[(CH_3)_2N(CH)_nN(CH_3)_2]^+X^-$, n=3,5,7,9 $X^-=ClO_4^-$ exhibiting different lengths of the π-chain and on a comparison with normalcoordinate calculations of Raman frequencies as well as of the Raman intensities.

2. Experimental

Within in the series the dyes with n=3,5,7,9 (dissolved in ethanol) exhibit absorption bands at 315nm, 414nm, 514nm, 615nm respectively. The wavelengths used for photoexcitation λ_{EXC} and for CARS probing λ_P were optimized to get the best spectra from the different species. CARS spectra were recorded using a ns-excite and probe CARS spectrometer described elsewhere [3].

Springer Proceedings in Physics, Vol. 74 Time-Resolved Vibrational Spectroscopy VI
Editors: A. Lau, F. Siebert, W. Werncke

Table 1 Observed and calculated CARS frequencies and intensities of the ^{14}N heptamethine and of the photoisomer. Measured intensities are classified by: s-strong, m-medium, w-weak, normalized to the strong unshifted peak at $1133cm^{-1}$ either of the parent molecules or of the photoisomers. Calculated intensities indicated by parentheses are given in arbitrary units. Calculated frequencies of the parent molecules are fitted to reproduce the observed frequencies.

photoisomer λ_{exc} = 510nm λ_P = 580nm				parent molecule λ_P =580nm			assignment
frequencies $[cm^{-1}]$, (intensities) [a.u.]							
observed		calculated		observed		calculated	
1426	w	1428	(35)				
1414	m	1402	(37)	1405	s	(50)	s CH_3 def + CN str
1320*	m	1356	(45)				CC str + CH_3 def
		1315	(67)				
1300	w			1303	s	(175)	CC str
1205	m	1220	(67)	1200	m	(215)	CH_3 -N str
1161	w	1199	(68)	1178	m	(61)	
1133	s	1181	(149)	1133	s	(206)	CH_3 rock, CCstr
1072	s	1094	(212)	1070	w	(5)	C=C=N band + CH def + NC_2 str

(*) at least two overlapping not resolved lines

The most striking feature in the spectrum of the photoisomers is the appearance of the strong new CARS line at appr. $1070cm^{-1}$ being nearly unaffected by the chain lengths [4]. All the photoisomer lines of the heptamethine decay at room temperature within the same time of appr. 5ms and are therefore assumed to originate from the same species. As an example the CARS frequencies and intensities of the ^{14}N- substituted heptamethine (parent molecules and photoisomers) as well as corresponding calculated values in the most instructive frequency range of $1430cm^{-1}$ - $1000cm^{-1}$ are summarized in table 1.

3. Discussion

The nearly equal appearance of the characteristic photoisomer mode at about $1070cm^{-1}$ for all photoisomers in the investigated series suggests that always as in the case of the trimethine the 1,2-mono-cis photoisomer is generated.

This assumption is proven by normal coordinate calculations including a force constant adjustment to reproduce the observed infrared and Raman spectra of the series of oligomers and frequency shifts due to ^{15}N substitution of the all trans

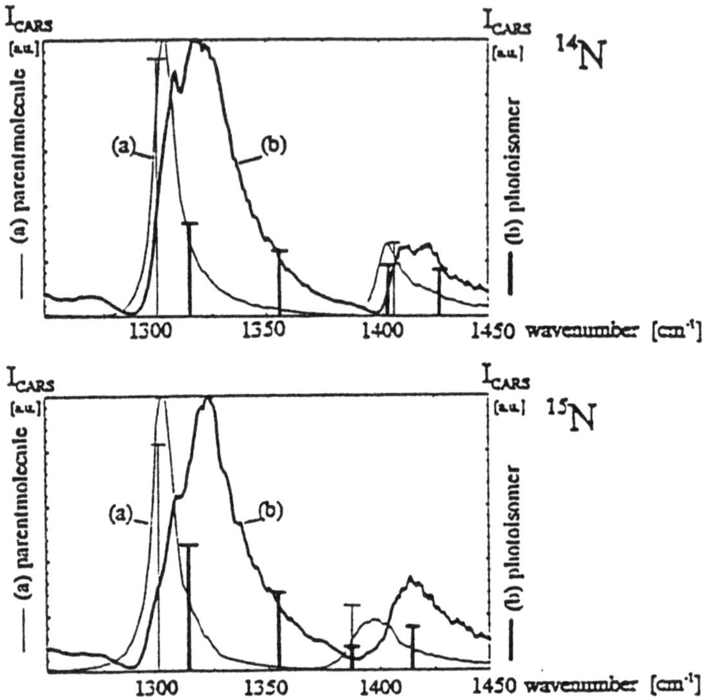

Fig. 1 Experimental and calculated CARS spectra of the parent molecules and of the photoisomers of ^{14}N- and ^{15}N- heptamethine. Bars indicate calculated frequencies and relative intensities.

parent molecules. The Raman intensities were calculated taking into account only those contributions to the molecular polarizability which are caused by a distortion of the π-bonds during the vibrations [4]. The polarizability increments $(\partial\alpha_{\pi}/\partial S_j)$ were calculated within the Hückel frame which is known to work well to explain the electronic absorption of the methine dyes.

Most of the calculated Raman intensities of the irradiated sample (see table 1) show qualitatively the observed intensity relations. For the intensity of the 1200cm^{-1} mode a too large value is calculated which may be due to small deviations of the calculated from the true normal coordinates.

As these calculations reproduce the spectra satisfactorily and as main vibrational frequencies remain unchanged by photoisomerization we transferred both force constants and bond polarizabilities to the photoisomers taking into account their changed geometry's only.

From the different possible planar mono-cis configurations only the 1,2 mono-cis photoisomer results into spectra corresponding to the experimental findings of a characteristic photoisomer mode near to 1070cm^{-1}. According to our normal coordinate calculation this mode is localized in the bent CH=CH=N=$(CH_3)_2$ end group of the 1,2 monocis photoisomer. Its vibrational energy results from a NC_2-stretching mode combined with a deformation of the C=C=N-angle plus a

CH-bending. This vibration does not essentially couple to further stretching movements of the π-chain, thus being nearly independent of its length. The calculated drops of intensities in the photoisomer spectrum of the lines at 1300cm^{-1} and 1200cm^{-1} are also observed (see table 1). For the ^{14}N- and the ^{15}N-substituted heptamethine the recorded band modifications in the 1300cm^{-1} and 1400cm^{-1} range can be explained by mode splitting due to the lowering of symmetry of the photoisomers.

Details off the calculations show that the isotopically unshifted 1303cm^{-1}-line of the parent molecules splits into two lines with approximately no isotopic shifts. The ^{14}N-heptamethine Raman vibration at 1405cm^{-1} splits into two lines (1402/1428cm^{-1}) with nearly equal intensities. Contrary, for the isotopic downshifted line at 1396cm^{-1} of the ^{15}N-heptamethine a more intense line at 1415cm^{-1} and a second but less intense line at 1387cm^{-1} is calculated. This behavior corresponds to the experimental spectrum shown in Fig. 1.

In conclusion our new and more detailed results of experimental and of calculated CARS spectra further support that in the series of bisdimethylaminomethine dyes the 1,2 mono-cis photoisomer is generated exclusively. They also demonstrate that Raman intensity calculations based on normal coordinate calculations connected with Hückel theory are useful to investigate those rather complex molecules.

References

[1] G. Scheibe, J. Heiss, K. Feldmann; Ber. Bunsenges. Phys. Chem. **70**, 52 (1966)

[2] H. Hartmann, P. Wähner; Abstracts "Fourth Symposium Optical Spectroscopy" Oct. 1986; Reinhardsbrunn, GDR

[3] A. Lau, W. Werncke, M. Pfeiffer; Spectrochim. Acta Rev. **13**, 191 (1990)

[4] W. Werncke, M. Pfeiffer, A. Lau, L. Holz, T. Hasche; Inst. Phys. Conf. Ser. No 126, 557 (1992)

[4] M. Pfeiffer, W. Werncke, A. Lau, W. Freyer; "Electronic properties of polymers", Proceedings of the IWPP, Kirchberg 1991; Springer Series in Sold-State Sciences 107, Eds. H. Kuzmany, M. Mehring, S. Roth

Vibrationally Hot (dd) Excited States of Ni^{II}-Porphyrins Probed by Picosecond Time-Resolved Resonance-Raman Spectroscopy

S. Sato[1] *and T. Kitagawa*[2]

[1]Institute for Molecular Science, Myodaiji, Okazaki 444, Japan
[2]Okazaki National Research Institutes, Myodaiji, Okazaki 444, Japan

1. Introduction

Ni^{II}-porphyrins have served as an exemplary system for elucidating dynamical features involved in photoexcitation and relaxation of metalloporphyrins [1-5]. The recent subpicosecond time-resolved absorption study [4] of Ni^{II}-porphyrins in non-coordinating solvents has demonstrated that the $(d_{z^2},d_{x^2-y^2})$ states are generated within 350fs of the $(\pi\pi^*)$ excitation of the macrocycle and decays with biphasic kinetics as reported (τ_1 = ~10ps and τ_2 = 250-300ps) [1,3]. While the τ_2 relaxation is established to be an electronic decay process of the (dd) excited state, the τ_1 relaxation is a concern of current photodynamic investigations. Earlier studies proposed this to be an electronic relaxation from $(\pi\pi^*)$ to (dd) [1], or from ^{1}(dd) to ^{3}(dd) [3], but recent studies assigned it to non-electronic relaxations within the (dd) manifolds [4-6], since the value of τ_1 is varied with monitoring wavelengths of the transient absorption. The value of τ_1 (~5ps) at the red-edge of the transient absorption is almost the half of the value (~10ps) measured at the blue-edge[4]. Rodriguez et al. proposed it to be a vibrational cooling process [4,5] of the vibrationally hot (dd)-excited states generated when the potential of $(\pi\pi^*)$ crosses that of (dd), while Courtney et al. proposed it to be an conformational relaxation on the basis of time-resolved resonance Raman (TR^3) spectra measured in the τ_2 region [6]. Ni^{II}-porphyrins are known to have planar- and ruffled-conformers in solution, and several Raman lines of the in-plane skeletal vibrations of porphyrin macrocycle (ν_{10}, ν_2 and ν_3) are apparently broadened (Γ = 16-25cm^{-1}) in the ground electronic state [7], although Courtney et al. observed that these modes become sharp (10-14cm^{-1}) in the TR^3 spectra of the τ_2 region [6]. This fact suggests that Ni^{II}-porphyrins will have only a single conformer in the (dd) state in the τ_2 region, and so as to make it possible, there should be some conformational relaxation in the earlier time range. Courtney et al. attributed the τ_1 process to it [6]. However, the problem seems to be still under debates [5]. We report here the measurement of TR^3 spectrum in the τ_1 region which is indicative of the vibrational cooling for the τ_1 process, although the possibility that the conformational relaxation coexists cannot be ruled out completely.

2. Experimental

The ps pump/probe TR^3 measurement was carried out by using an active-passively mode-locked Nd:YAG laser with 10Hz repetition and 40ps pulse duration. Ni^{II}-octaethylporphyrin (Ni^{II}OEP) in benzene or THF was excited by a pumping pulse at 532nm. Resonance Raman spectra of excited species were observed by probe pulses at 416nm or 436nm which were generated through stimulated Raman scattering

Springer Proceedings in Physics, Vol. 74 Time-Resolved Vibrational Spectroscopy VI
Editors: A. Lau, F. Siebert, W. Werncke

of H_2 gas excited by the 2nd or 3rd harmonic of the Nd:YAG laser. Powers of the pump (~75μJ/pulse) and probe (~18μJ/pulse) pulses were made as low as possible. Spacial and temporal overlaps of the pump and probe pulses were confirmed by monitoring the bleaching signal of the S_0-species of Fe^{III}OEPCl for which the lifetime of the excited state was reported to be less than 13ps [8]. Raman scattered light was dispersed by a triplepolychromator and detected by an intensified photodiode array. The pump/probe and probe-only spectra were accumulated quasi-simultaneously [9].

3. Results and Discussion

Figure 1 displays TR^3 difference spectra of Ni^{II}OEP in benzene probed at 416nm between Δt = -40 and 860ps. The contribution from the S_0 species was subtracted. The TR^3 spectra after Δt = 60ps were in agreement with the reported one [6] and decayed with the reported decay constant (~280ps) without any changes of the spectral pattern. However, the RR spectrum at Δt = 0ps is distinct from others with regards to slight downshifts of frequencies and broadening of bands, although the spectral pattern is very similar to those of the later time delays. This fact clearly ruled out the possibility of $(\pi\pi^*) \rightarrow (dd)$ transition for the τ_1 process since the

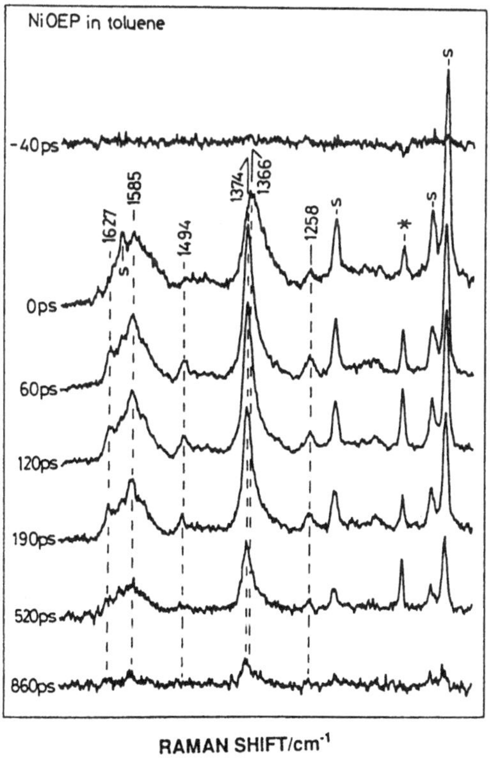

Figure 1. Pump/probe minus probe-only TR^3 difference spectra of Ni^{II}OEP in toluene at room temperature probed at 416nm.

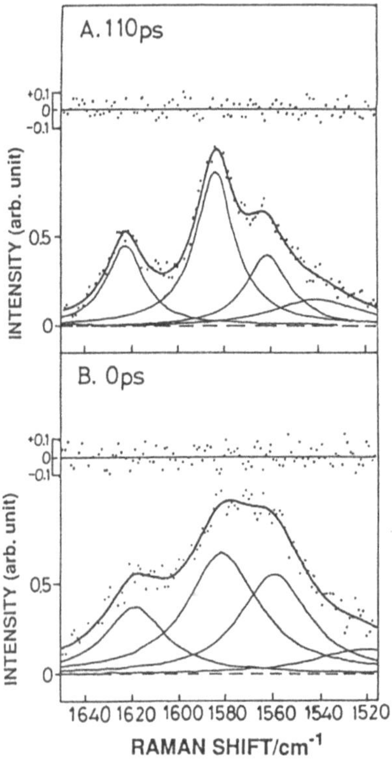

Figure 2. Expanded presentation of the pump/probe minus probe-only TR^3 difference spectra in the 1650-1520cm^{-1} region of $Ni^{II}OEP$ in THF and calculated bands for Δt = 110ps (A) and Δt = 0ps (B) fitted with Lorentzian band shape .

RR spectra of H_2OEP and $Zn^{II}OEP$ in the $(\pi\pi^*)$ excited states exhibited the intensity pattern distinct from that of the (dd)-excited state. It is to be noted for the (dd)-excited state that the excitation is localized at the central metal leaving the π-electron sytem of the macrocycle in the ground state and that resonance enhancement pattern is very similar between (dd) and S_0 states.

Since Raman bands of sample in the the 1600-1550cm^{-1} region were overlapped with those of solvent in Fig.1, TR^3 measuremet was carried out with THF. Figure 2 shows the expanded spectra in the 1650-1500cm^{-1} region for the THF solution and the results of band fitting calculations for the spectra at Δt = 0 and 110ps by assuming Lorenzian band shape. Table 1 summarizes the result of band fitting, from which it is clear that all the Raman bands (ν_{10}, ν_2, ν_{11}, ν_3 and ν_4) in the high frequency region at Δt=0ps are rather uniformly downshifted and broadened. This behavior can not be explained by the idea of the conformational relaxation for the τ_1 process because, if the conformers excited from the S_0 state were observed in the spectrum at Δt = 0ps, the broadening should be especially large for the ν_{10}, while the ν_4 exhibit no broadening. The observed phenomenon can be explained well by the vibrational cooling process if we assume the intra-molecular vibrational

TABLE 1 Frequencies(ν) and line-widths(Γ) of Raman bands for NiOEP

	Ground state (A_{1g})	Excited state ($^{*}B_{1g}$)		Hot excited state	
ν_i	$\nu(\Gamma^a)/cm^{-1}$	$\nu(\Gamma^a)/cm^{-1}$	$\Delta\nu^b/cm^{-1}$	$\nu(\Gamma^a)/cm^{-1}$	$\Delta\nu^b/cm^{-1}$
ν_{10}	1652(14)	1622(9)	-30	1619(18)	-33
ν_2	1601(11)	1584(10)	-17	1580(18)	-21
ν_{11}	1575(11)	1562(11)	-13	1558(17)	-17
ν_3	1519(8)	1490(14[c])	-29	1479[c](14[c])	-40
ν_4	1382(8)	1372(8)	-10	1366(16)	-16

[a]Full width at half maximum obtained by assuming Lorentzian band shape. [b]Frequency shifts accompanied with the electronic transition from the ground state. [c]These values are less accurate due to lower S/N ratio.

redistribution (IVR) in the subpicosecond time scale [10] followed by vibrational energy relaxation.

To confirm the vibrational cooling for the τ_1 process, we measured the TR^3 spectra in the anti-Stokes region and compared the band intensities between the Stokes and anti-Stokes scatterings. The anti-Stokes scattering was greatly enhanced at $\Delta t = 0ps$. Vibrational temperature was calculated to be >1000K under the assumption of Boltzman distribution.

However, it should be noted that the phenomenon in the τ_1 region may not be a simple vibrational cooling and the occurrence of the conformational relaxation in the τ_1 region can not be eliminated, because there remains a feature that can not be simply explained by the vibrational cooling for the τ_1 process; the maximum of transient absorption in Soret region is red-shifted in the τ_1 region compared with the τ_2 [4-6], although it is blue-shifted upon raising temperature for Ni^{II}-porphyrins in the S_0 state [11]. This fact is rather easily explained by conformational relaxation since the ruffled conformer has red-shifted Soret absorption in comparison with the planar one [7], and the planar conformer is considered to be electronically more stable [3] in the (dd) state. In order to investigate this point, we tuned the probing wavelength to 436nm. Although the very similar phenomena (i.e. broadening, downshifts of Raman bands and the enhancement of anti-Stokes scattering at $\Delta t = 0ps$) was observed in the τ_1 region, their appearance seems to be less extent compared with those probed at 416nm. Especially it is to be noted that the frequency of the ν_4 was observed at $1370cm^{-1}$ with Stokes scattering at 436nm contrary to $1366cm^{-1}$ with Stokes scattering at 416nm, although they were at $1365cm^{-1}$ with anti-Stokes scattering. This observation would be explained by the better resonance with the higher vibrational energy levels when λ_{probe} = 416nm than λ_{probe} = 436nm, if we assume that the seperation of vibrational energy levels is larger in the resonant intermediate state than in the (dd)-state. However, this assumption cannot account for the red-shift of transient Soret absorption maximum in the τ_1 region. To explain the red-shift, the coexsistence of another faster prcess, presumably conformational relaxation, with the vibrational cooling will be necessary.

References

[1] T.Kobayashi, K.D.Straub and P.M.Rentzepis, Photochem. Photobiol. **29**, 925 (1979).
[2] V.S.Chrivonyi, B.M.Dzhagarov, Yu V.Timinskii and G.P.Gurinvich, Chem. Phys. Lett. **70**, 79 (1983).
[3] D.H.Kim, C.Kirmaier and D.Holten, Chem. Phys. **75**, 305 (1983).
[4] J.Rodriguez and D.Holten, J. Chem. Phys. **91**, 3525 (1989).
[5] J.Rodriguez, C.Kirmaier and D.Holten, J. Chem. Phys. **94**, 6020 (1991).
[6] S.H.Courtney, T.M.Jedju, J.M.Friedman, R.G.Alden and M.R.Ondrias, Chem. Phys. Lett. **164**, 39 (1989).
[7] R.G.Alden, B.A.Crawford, R.Doolen, M.R.Ondrias and M.R.Shelnutt, J. Am. Chem. Soc. **111**, 2070 (1989).
[8] I.Fujita, T.Netzel, C.K.Chang and C.-B.Wang Proc. Natl. Acad. Sci. U.S.A. **79**, 413 (1982).
[9] S.Sato, K.Kamogawa, K.Aoyagi and T.Kitagawa, J. Phys. Chem. **96**, 10676 (1992).
[10] R. Lingle,Jr., X.Xu, H.Zhu, S.-C.Yu and J.B.Hopkins, J. Am. Chem. Soc. **113**, 3992 (1991).
[11] S.A.Asher and J.Murtaugh, J. Am. Chem. Soc. **105**, 7244 (1983).

Time-Resolved IR Spectroscopy of Excited States

J.J. Turner and M.W. George

Department of Chemistry, University of Nottingham,
Nottingham NG7 2RD, UK

Abstract. Fast time-resolved infrared spectroscopy (TRIR) can be used to probe the excited states of transition metal species. In particular the behaviour of fairly long-lived MLCT states can be monitored as they either decay back to the ground state or undergo further chemistry.

1. Introduction

A knowledge of the properties of excited states is a prerequisite to an understanding of photochemical processes. The most common techniques have employed absorption or emission spectroscopy in the UV/visible region, or resonance Raman spectroscopy, either on the ground state or, in time-resolved mode, on the excited state. More recently it has become possible to apply time-resolved infrared spectroscopy (TRIR) to excited states, and this greatly extends the available information [1].

1

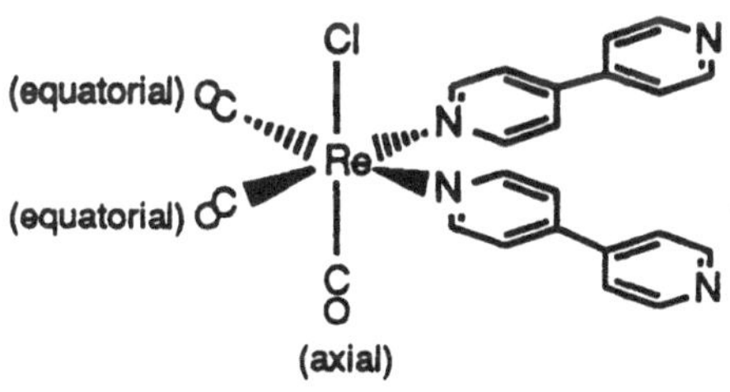

2

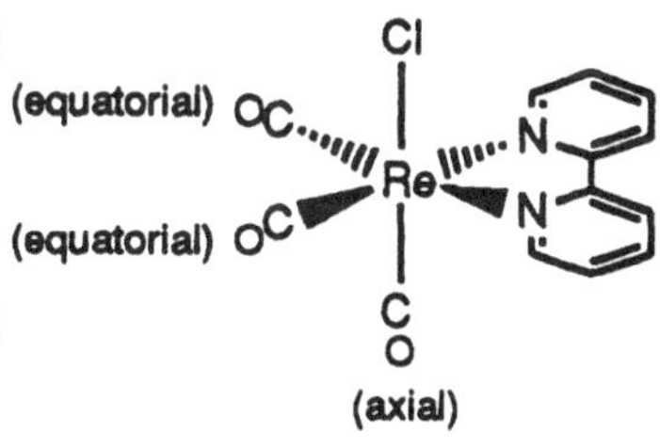

Work in Nottingham started some years ago [2a] on the complex $ClRe(CO)_3(4,4'$-bipyridyl$)_2$ (**1**). This compound was selected for several reason: it was known from UV spectroscopy [3] to have a long-lived (~1 μs) MLCT state with charge transfer localised in a bipyridyl ligand; it is thermally and photochemically stable; it has strong IR bands in the ν(CO) region. Preliminary experiments [2a] with **1** demonstrated that ν(CO) bands could be detected in the excited state; because of the limitations in the frequency range of the CO laser used as the IR spectral source, only two bands were observed instead of the three to be expected. However the data were sufficient to demonstrate that in the MLCT excited state there is an increase in frequency in the ν(CO) bands; this is as anticipated since in the MLCT state there is effective oxidation of the metal. More recently, following instrumental developments, and in collaboration with colleagues at Mülheim [2b], the

Table. *Summary of data for 1 in ground and excited states*

ν(CO) frequencies (cm^{-1})				EFFF force constants (Nm^{-1})				Δr_{CO} (Å)
Ground	Excited	Δ	Mean		Ground	Excited	Δ	
2027	2055	+28		k_{ax}	1492	1566	~+74	-0.008
1926	1992	+66	53	k_{eq}	1555	1644	~+89	-0.009
1891	1957	+66						

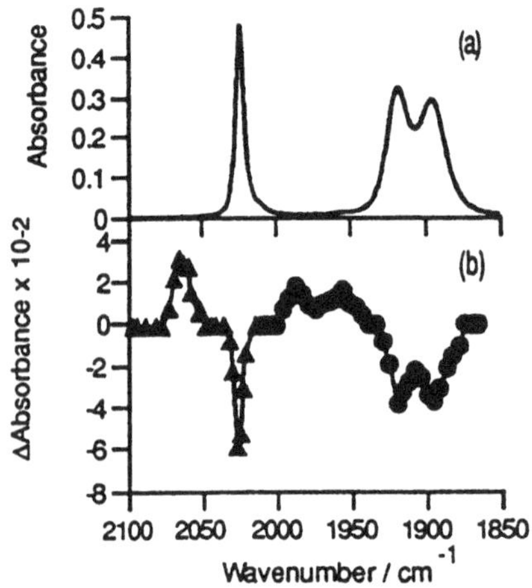

Figure: (a) FTIR spectrum in ν(CO) region of **2** in CH_2Cl_2 (~5 x 10^{-4} M); (b) TRIR spectrum at ~50 ns after flash from excimer laser flash; data points below zero represent loss of parent, while those above represent generation of the excited state.

spectroscopic data has been completed, and the complete energy factored force field (EFFF) has been solved for **1** in the ground state, and an approximate solution obtained for the excited state. The results are summarised in the Table.

The calculation of the changes in the C-O bond lengths uses a recently developed relationship [4]:

$$r_{CO} = 1.621 - 0.174\ln(k_{CO}) \quad \text{(with } r_{CO} \text{ in Å and } k_{CO} \text{ in mdynes/Å)}$$

This relationship, which is similar to one proposed by Woodruff [5], is probably accurate to ~0.001 Å, when calculating **differences** in bond length on excitation.

With improvement in the time resolution of our apparatus, we have recently [6] examined the TRIR spectrum of $ClRe(CO)_3$(2,2'-bipyridyl) (**2**) in the lowest MLCT excited state, which has a lifetime of ~50 ns: see figure.

As expected, the ν(CO) bands shift up in frequency on excitation; the mean shift is very close to that observed for **1**, (55 cm^{-1} compared with 53 cm^{-1}), although without a detailed force field analysis it is not possible to be more specific. It would be useful if the values of the ν(CO) shifts could be directly correlated with the degree of charge transfer; it will be recalled that the TR^3 spectra of bpy-type ligands can be directly compared with the vibrational data of the appropriate isolated anions. Unfortunately there is no direct calibrant for the ν(CO) shift and there is not yet sufficient data to provide an accurate assessment.

2. Reactions of excited states

The excited states being probed in the above experiments are fairly long-lived, "thexi" (thermally equilibrated excited) states in Adamson's nomenclature [7]. Usually they will not be the states directly excited; for instance, where spin labelling is appropriate, if the ground state is a singlet, MLCT excitation will involve a singlet excited state, whereas the thexi state will be a triplet. This has the important consequence that care is needed in comparing thexi state studies with structural techniques directly related to the excitation process such as resonance Raman spectroscopy.

These thexi states can simply decay to the ground state or, more interestingly, lead to chemical reaction, either directly or via quenching by some other species in solution. An example of "direct" chemistry is provided by compounds of type $W(CO)_5L$. If L is pyridine, the lowest excited state is ligand field (LF) in character, it is very short lived, and has a high quantum yield for dissociation to $W(CO)_5$ + L; however if L is, for example 4-CNpyridine or 4-acetylpyridine, the lowest excited state is MLCT, it is longer lived, and the dissociation has a low quantum yield since this occurs via an equilibrium with the higher-lying LF state [8]. Because UV/visible absorption bands are usually fairly broad with little fine structure, it is difficult by conventional flash photolysis to follow the whole process, ie ground state to MLCT, dissociation to $W(CO)_5$ and L, followed by reaction of $W(CO)_5$. On the other hand the ν(CO) IR bands are fairly narrow and very characteristic, and it has proved possible to monitor the whole of this process for $W(CO)_5$(4-CNpyr) [9a] and $W(CO)_5$(4-acetylpyr) [9b]. Such TRIR experiments rely heavily on matrix isolation spectroscopy to provide definitive identification of the appropriate intermediates.

The complex $[Ru(bpy)_3]^{2+}$ is the classic "indirect" example. In its lowest MLCT excited state, $[Ru(bpy)_3]^{2+*}$ is a more powerful oxidant and reductant than in its ground state, and this property has been exploited in a wide range of experiments, many related to solar energy conversion [10]. As yet there have been no reports of the IR spectra of this excited state, presumably because of experimental difficuties. On the other hand $ClRe(CO)_3$(bpy) (2), mentioned above, has been shown to be an effective photocatalyst in the reduction of CO_2 [11]. The first stage in the reduction cycle involves the excitation of 2 into the MLCT state. In this state it reacts with an electron transfer agent such as NEt_3 and is reduced to $[ClRe(CO)_3(bpy)]^{-\cdot}$; it is this species that is almost certainly substituted with CO_2, although the exact mechanism is not known. These first two stages can be monitored by TRIR [6]; future work in this laboratory will involve TRIR combined with electrochemistry to probe the subsequent chemistry. A more recently discovered CO_2 reductant is $[Re(CO)_2(bpy)\{P(OEt)_3\}_2]^+$ [12], and preliminary experiments with this species are described elsewhere in this volume [13].

3. Acknowledgments

Work in Nottingham has been supported by the SERC, the EEC, the Paul Instrument Fund of the Royal Society, Mütek GmbH and NATO.

4. References

[1] For a recent review see J. J. Turner, M. W. George, F. P. A. Johnson and J. R. Westwell, Coord. Chem. Rev. **125**, 101 (1993).

[2] (a) P. Glyn, M. W. George, P. M. Hodges and J. J. Turner, J. Chem. Soc. Chem. Commun. 1655 (1989); (b) D. Gamelin, M. W. George, P. Glyn, F-W. Grevels, F. P. A. Johnson, W. Klotzbücher, S. L. Morrison, G.Russell, K.Schaffner and J. J. Turner, submitted for publication.

[3] P. J. Giordano and M. S. Wrighton, J. Am. Chem. Soc. **101**, 2888 (1979).

[4] S. L. Morrison and J. J. Turner, J. Mol. Struct. in press.

[5] V. M. Miskowski, R. F. Dallinger, G. G. Christoph, D. E. Morris, G. H. Spies and W. H. Woodruff, Inorg. Chem. **26**, 2127 (1987)

[6] M. W. George, F. P. A. Johnson, J. R. Westwell, P. M. Hodges and J. J. Turner, submitted for publication.

[7] A. W. Adamson, in *Concepts of Inorganic Photochemistry,* eds. A. W. Adamson and P. D. Fleischauer, Wiley (1975).

[8] M. S. Wrighton, H. B. Abrahamson and D. L. Morse, J. Am. Chem. Soc. **98**, 4105 (1976).

[9] (a) P. Glyn, F. P. A. Johnson, M. W. George, A. J. Lees and J. J. Turner, Inorg. Chem. **30**, 3543 (1991); (b) F. P. A. Johnson, M. W. George and J. J. Turner, Inorg. Chem. in press.

[10] See eg A. Juris, V. Balzani, F. Barigelletti, P. Belser and A. von Zelewski, Coord. Chem. Rev. **84**, 85 (1988).

[11] J. Hawecker, J-M. Lehn and R. Ziessel, J. Chem. Soc., Chem. Commun. 536 (1983); J. Hawecker, J-M. Lehn and R. Ziessel, Helv. Chim. Acta. **69**, 1990 (1986).

[12] C. Pac, S. Kaseda, K. Ishii, S. Yanagida and O. Ishitani in *Photchemical Processes in Organized Molecular Systems*, ed. K. Honda, Elsevier (1991); unpublished results.

[13] J. R. Westwell, M. W. George, O. Ishitani, F. P. A. Johnson and J. J. Turner *Time-resolved infrared spectroscopy of a CO_2 reductant*, this volume.

Excitation Profiles of the Triplet-State Resonance-Raman Spectrum of p-Terphenyl in the Contour of the 400–480 nm $T_n \leftarrow T_1$ Absorptions

G. Buntinx[1], *N. Locoge*[1], *O. Poizat*[2], *N. Legue*[2], *and C. Sourisseau*[3]

[1]Laboratoire de Spectrochimie Infrarouge et Raman, CNRS, USTLFA, Bât. C5, F-59655 Villeneuve-d'Ascq, France
[2]Laboratoire de Spectrochimie Infrarouge et Raman, CNRS, 2, rue Henri-Dunant, F-94320 Thiais, France
[3]Laboratoire de Spectroscopie Moléculaire et Cristalline, CNRS, Université Bordeaux I, F-33405 Talence, France

The $T_n \leftarrow T_1$ absorption of p-Terphenyl presents a strong band at 445nm and a shoulder around 425nm. A constant decay is observed at all wavelengths. We thus assume that this spectrum characterizes a single T_1 species, the 445 and 425nm components being due either to a single electronic transition showing a vibronic structure, or to distinct transitions.

The T1 Resonance Raman Spectra of TP and of its derivatives deuteriated on the central ring ($-d_4$), on the external rings ($-d_{10}$), or fully deuteriated ($-d_{14}$), have been recorded for a series of probe wavelengths in the 400-480nm range. Although important vibrational couplings between the three phenyl rings make the assignments difficult in the absence of force field calculation, two groups of two bands appear approximately ascribable to modes specific of the central (1590 and 1225 cm^{-1}) and external (1541 and 1167 cm^{-1}) rings. In each one of these groups the high and low frequency components can be correlated to the totally symmetric CC stretch 8a and CCH bend 9a of benzene, respectively, which are quinoidal-type vibration. comparable resonance effects characterize the 8a and 9a components of each group but the behaviour of the 8a/9a pair related to the external rings appears rather different from the behaviour of the 8a/9a pair ascribed to the central ring.

Components 9a at 1167 and 1225 cm^{-1} are clearly defined and have been found appropriate for measuring excitation profiles. Figure 1 shows these profiles as compared to the $T_n \leftarrow T_1$ absorption. They have been realized carefully from repetitive measurements in cyclohexane and ether solutions (10^{-3} M and 10^{-4} M in the regions of weak and strong resonance, respectively). However the fact that four different laser dyes were necessary to provide probe wavelengths covering the 400-480nm region is certainly a source of experimental error due to the difficulty in reproducing identical pump/probe optical coincidence in the sample. Depopulation of the T1 state by the probe beam in the region of high resonance may also be a source of error in the measurements of relative Raman intensities in this region. Accordingly, realizing reliable REPs for transient species produced and monitored using a pump/probe arrangement is extremely difficult and the profiles in Figure 1 are only tentative. They show nevertheless that the component 9a related to the external rings is much more enhanced, i.e. much more involved in the resonant absorption, than that related to the central ring. The same conclusion could be

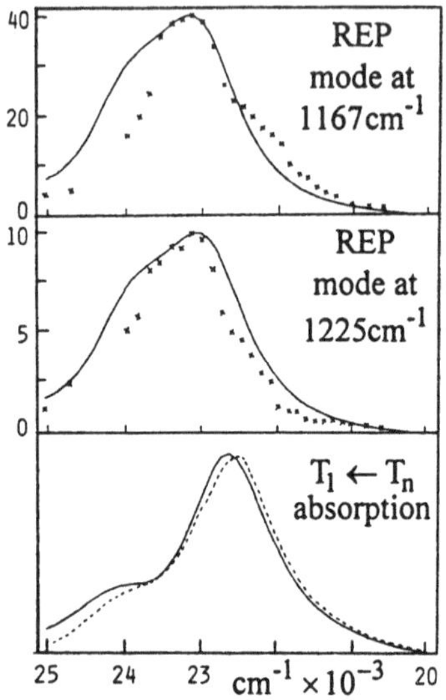

Figure 1 - Experimental (crosses or dashed line) and calculated (full line) Raman excitation profiles (REP) and absorption spectrum for the T_1 state of p-terphenyl.

stated for modes 8a. Therefore the visible $T_n \leftarrow T_1$ absorption leads mainly to a distortion of quinoidal type of the triplet TP molecule, which is more localized on the external than on the central phenyl rings.

An attempt to reproduce the REPs using a Franck-Condon model including homogenous and inhomogenous broadening effects [6] (Figure 1) suggests that these profiles result from interferences between at least two resonance effects due to two distinct $\pi\pi*$ transitions (0←0 energies at 22750 and 24050 cm^{-1}, relative transition moments 1.0 and 0.65, displacement parameters + 0.38 and -0.20 for the 1167 cm^{-1} mode and + 0.16 and -0.10 for the 1225 cm^{-1} mode, respectively. Homogeneous width 250 cm^{-1} and inhomogeneous width 450 cm^{-1}). A third resonance contribution, which assumes the presence of another weaker transition around 460 nm, is probably responsible for the shoulder observed on the low energy side of the 1167 cm^{-1} REP and for the rapid drop of this profile on its high energy side as compared to the calculated profile.

[1] I.B. Berlman "Handbook of fluorescence Spectra of Aromatic Molecules" Academic Press, New York, 1971.
[2] H.W. Furumoto and H.L. Ceccon, IEEE J. Quantum Electron. 6, 262 (1970).
[3] R.Wilbrandt, N.H. Jensen, P. Pagsberg, A.H. Silleson and K.B. Hansen, Nature 276, 167 (1978).
[4] S. Matsunuma, S. Yamaguchi, C. Hirose and S. Maeda, J. Phys. Chem. 92, 1777 (1988).
[5] W. Heinzelmann, and H. Labhart, Chem. Phys. Lett. 3, 399 (1969).
[6] R.S. Cataliotti, S.M. Murgia, G. Paliani, A. Poletti, and M.Z. Zgierski, J. Raman Spectrosc. 16, 251 (1985); 16, 258 (1985).

Mode-Specific Temperature Dependence of the Resonance-Raman Spectrum of Trans-Stilbene in the Ground State

W.T. Toner[1], *R.E. Hester*[2], *P. Matousek*[1], *J.N. Moore*[2], *A.W. Parker*[1], *and M. Towrie*[1]

[1]SERC Rutherford Appleton Laboratory, Chilton, Didcot, Oxon OX11 0QX, UK
[2]Chemistry Department, University of York, York YO1 5DD, UK

Abstract. New data on the steady state (cw) resonance Raman spectrum of S_0 trans-stilbene are presented which show temperature dependent shifts of lines whose analogues in S_1 trans-stilbene show similar but larger effects and also exhibit transient shifts and width changes [1,2].

Resonance Raman spectrum of S_0 trans-stilbene

The near equivalence of the shift in the *c.* 1570 cm^{-1} line in the picosecond transient resonance Raman spectrum of S_1 trans-stilbene produced by increases in thermal energy to that produced by excess energy of excitation [1,2] led us to investigate the temperature dependence of the cw resonance Raman spectrum of S_0 trans-stilbene (tS), whose relaxation timescale [3] is similar to that of S_1 tS. An ion laser at 363 nm was used to excite the spectrum and discharge lamp marker lines were recorded throughout the experiment to eliminate errors due to small spectrometer drifts. A Triplemate spectrometer with a 700 element diode array detector was used to record the spectrum from 1000 cm^{-1} to 1700 cm^{-1} with a resolution of 6 cm^{-1}. A small portion of a typical spectrum is shown in Figure 1 and the temperature dependences of the *c.* 1643 cm^{-1} and *c.* 1195 cm^{-1} lines (analogues of *c.* 1570 cm^{-1} and *c.* 1180 cm^{-1}

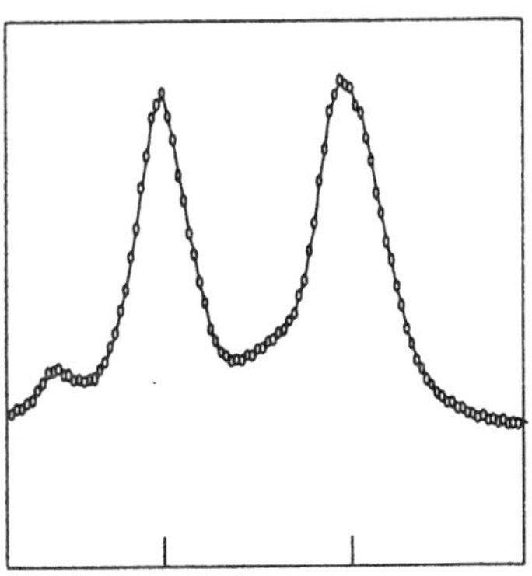

Fig. 1. S_0 tS resonance Raman spectrum in n-hexane in the region 1600 cm^{-1}.

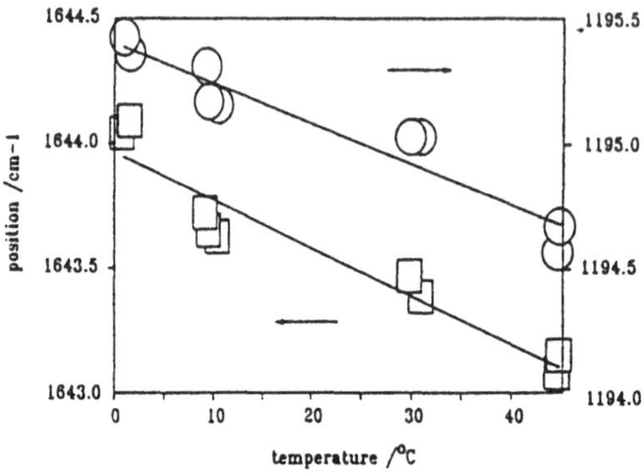

Fig. 2. Temperature dependence of S_0 tS bands at *c.* 1643 and *c.* 1195 cm^{-1} in n- hexane.

in S_1) are shown in Figure 2, for tS in n-hexane The *c.* 1600 cm^{-1} line shows a temperature dependence which is approximately half that of the *c.* 1643 cm^{-1} line. Straight lines fitted to the positions of the *c.* 1643 cm^{-1} and *c.* 1195 cm^{-1} S_0 bands give temperature dependences of (0.019 ± 0.002) and (0.016 ± 0.002) cm^{-1}/K, respectively. The corresponding values for tS in acetonitrile are (0.0187 ± 0.0007) and (0.0125 ± 0.0008) cm^{-1}/K showing a possible small solvent dependence for the *c.* 1195 cm^{-1} line but none for the *c.* 1643 cm^{-1}. Small increases in linewidth with temperature were observed, particularly in the case of the *c.* 1643 cm^{-1} line. The temperature coefficients of the line shift are factors of 5.3 and 1.7 smaller than the equivalent values for the *c.* 1570 cm^{-1} and *c.* 1180 cm^{-1} bands of S_1 tS in n-hexane.

Since this work was presented at the TRVS meeting we have received a preprint from A.B.Myers [4] which shows a change in the position of the *c.* 1640 cm^{-1} anti-Stokes line of S_0 tS by nearly 20 cm^{-1} between 0 and 25 ps following isomerisation from photoexcited cis-stilbene, in both cyclohexane and methanol solvents. To the extent that solvent dependences for this line in S_0 tS seem to be weak we may compare this change of ~ 0.6 cm^{-1} per 1000 cm^{-1} excess energy with the 0.38 cm^{-1} per 1000 cm^{-1} increase in thermal energy given by the temperature coefficients above. The approximate equivalence of the thermal and dynamical shifts in the position of the *c.* 1570 cm^{-1} line in S_1 tS [1,2] seems therefore to be reproduced in the shift of the analogous ground state line. We note, however, that this equivalence did not extend to the change in width of the *c.* 1570 cm^{-1} S_1 line, by factor of 3.

Acknowledgements

We are grateful to A.B.Myers for a preprint of her work and to H.Hamaguchi, R.M.Hochstrasser and J.Turner for useful discussions on dynamical effects..

References

1. J.N.Moore, P.Matousek, A.W.Parker, W.T.Toner, and R.E.Hester, these proceedings.

2. R.E.Hester, P.Matousek, J.N.Moore, A.W.Parker, W.T.Toner and M.Towrie, Chem.Phys.Lett. 208 471 (1993).

3. D.L.Phillips, J.M.Rodier and A.B.Myers, Technical Digest of the Eighth International Conference on Ultrafast Phenomena, (Antibes-Juan-Les-Pins, France 8-12 June 1992) p. 106.

4. A.B.Myers, J.M.Rodier, D.L.Phillips, to appear in *Reaction Dynamics in Clusters and Condensed Phases,* B.Pullman, J.Jortner, & R.O.Levine, eds. (Kluwer Academic Publishers, The Netherlands), 1993.

Singlet and Triplet (π, π^*) Excited States of Zinc(II) Octaalkylporphyrins: A Time-Resolved Resonance-Raman Study

R. Kumble, G.R. Loppnow, S. Hu, S.E. Vitols, and T.G. Spiro

Department of Chemistry, Princeton University, Princeton, NJ 08544, USA

The excited states of porphyrins and hydroporphyrins act as primary intermediates in processes ranging from biological energy transfer and electron transfer to photocatalysis and photodynamic therapy. We report resonance Raman (RR) spectra of the singlet S_1 and triplet T_1 excited states of zinc(II) octaethylporphyrin (ZnOEP) and zinc(II) etioporphyrin-I (ZnEtio), using picosecond and nanosecond RR techniques. Picosecond time-resolved resonance Raman experiments were performed using 532 nm photoexcitation pulses and 436 nm probe pulses (both 100 ps, 50 Hz); the experimental apparatus consisted of a mode-locked Nd:YAG laser (Coherent) in combination with a Nd:YAG regenerative amplifier (Continuum) and an H_2 cell for stimulated Raman shifting. The apparatus used for nanosecond time-resolved RR experiments has been described previously [1]. Porphyrin samples were prepared in a nitrogen atmosphere at a concentration of 0.5 mM (tetrahydrofuran solvent).

Figure 1 compares the RR spectra of the S_0 (ground) and S_1 (singlet excited) states of ZnOEP. Enhancement of totally symmetric porphyrin skeletal vibrations is seen in the S_0 state: the modes ν_2 (C_β-C_β), ν_4 (C_α-N) and ν_5 (C_α-C_β/C_β-C_1) are seen at 1583, 1371 and 1136 cm^{-1} respectively. RR spectra of the S_1 state (Δt = 100 ps) reveal features at 1567, 1354 and 1130 cm^{-1}; we tentatively assign these to ν_2, ν_4 and ν_5 respectively in the S_1 excited state. An ethyl -CH_2 twisting mode is observed at 1258 cm^{-1} in both the S_0 and S_1 states. The observed frequency shifts imply that (π,π^*) excitation to the S_1 state results predominantly in expansion of the C_β-C_β and C_α-N bonds, while leaving the substituent bonds unaffected.

RR spectra of the T_1 states (Δt = 100 ns) of ZnOEP, ZnOEP *meso*-d_4 (d_4-ZnOEP), ZnEtio and ZnEtio-2,7,12,17-β-^{13}C (β-^{13}C ZnEtio) are shown in figure 2. No effect is observed upon meso-deuteration, indicating that none of the modes contain significant contributions from C_m-H bending or stretching motions. The feature at 1564 cm^{-1} for ZnEtio and 1559 cm^{-1} for ZnOEP is assigned to C_β-C_β stretching modes (ν_2/ν_{11}) based on the 22 cm-1 downshift seen for this feature in β-^{13}C ZnEtio.

The electronic configuration of the S_1 state results from configuration interaction (CI) between two singly excited (π,π^*) configurations, $^1(a_{1u}e_g)$ and $^1(a_{2u}e_g)$ [2]. In contrast, the T_1 state is described by a pure excited configuration, $^3(a_{1u}e_g)$ in the case of octaalkylporphyrins. Both states possess E_u electronic

Springer Proceedings in Physics, Vol. 74 Time-Resolved Vibrational Spectroscopy VI
Editors: A. Lau, F. Siebert, W. Werncke

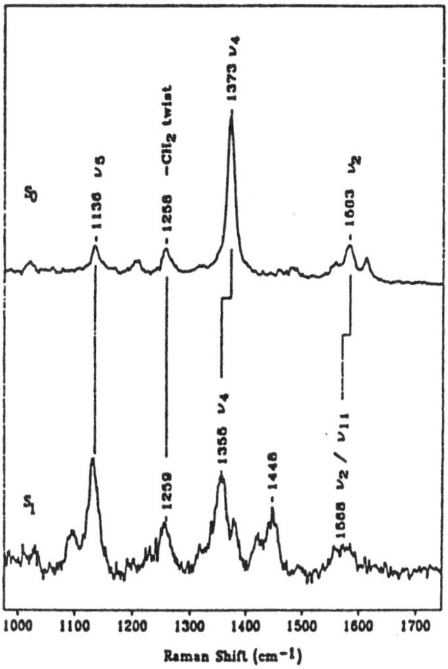

Figure 1: RR spectra of the S_0 and S_1 states of ZnOEP

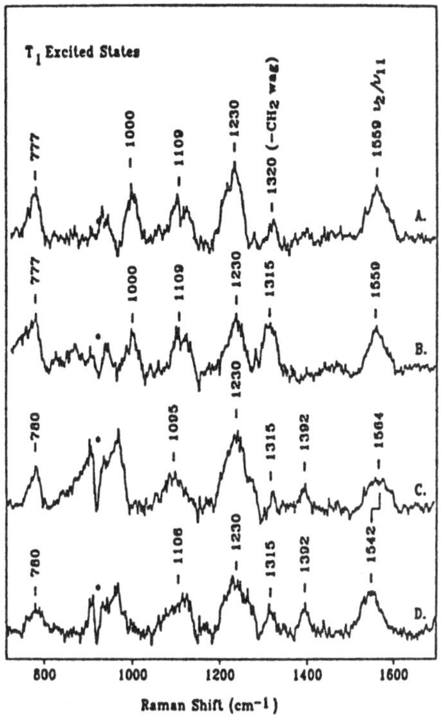

Figure 2: RR spectra of the T_1 states of A. ZnOEP B. d_4-ZnOEP C. ZnEtio D. β-^{13}C ZnEtio

symmetry and are expected to be susceptible to a symmetry-lowering Jahn-Teller (JT) distortion; however, this effect is predicted to be of smaller magnitude in the S_1 state as a result of configuration interaction [2]. Recent calculations predict different vibrational frequencies for the S_1 and T_1 states, resulting from both factors described above [3]. Our studies reveal minor differences in the C_β-C_β stretching frequencies for the S_1 and T_1 states, possibly reflecting these factors. The results also indicate stronger C_β-C_β bonds in the T_1 state of ZnOEP and ZnEtio than in the T_1 state of zinc(II) tetraphenylporphyrin [1], consistent with the predicted differences in electronic configuration [3].

References

[1]. Reed, R. A.; Purrello, Roberto; Prendergast, Kristine; Spiro, Thomas G. *J. Phys. Chem.* **1991**, *95*, 9720. [2]. Gouterman, M. in *The Porphyrins*; Dolphin, D.; Academic Press: New York **1978**; Vol. III, Chapter 1 and references therein. [3]. Prendergast, K; Spiro, T. G. *J. Phys. Chem.* **1991**, *95*, 9728.

Resonance-Raman Study of Intermolecular Interaction Between Electronically Excited Copper Porphyrins and Molecular Environment in Solutions

P.A. Apanasevich, V.V. Ermolenkov, S.G. Kruglik, V.V. Kvach, and V.A. Orlovich

Institute of Physics, Belarus Academy of Sciences,
70 F. Skaryna Ave., 220072 Minsk, Belarus

1. Introduction

The study of interaction between metalloporphyrins (MeP) and environment molecules permits to make an insight into diverse dynamic processes occurring in biological species at molecular level. Of the particular interest are copper porphyrins (CuP) having as a central atom copper ion with unfilled $3d^9$ electronic shell. CuP are on the margin between nonluminescent porphyrins with $3d^4 \ldots 3d^8$ configuration of inner d-shell of the central ion and luminescent ones with the filled shell. The half-filled $d_{x^2-y^2}$-orbital determines the high CuP's sensitivity to the axial binding of the solvent molecules which can donate the electronic density. Such types of complexes were extensively studied, for example, the interactions between CuP and nitrogen(N)-containing molecules were considered in [1-3]. The studies revealed that CuP in N-containing solvents can form intermolecular complex both in the ground and excited electronic states. Simultaneously the lifetimes of the electronic excitations are strongly decreased comparing to the noncomplexed porphyrin, dropping to the value of 100 ps. The effect is driven by the perturbation of the porphyrin π-electronic system resulting from the axial binding of N-containing molecule by CuP. It puts the charge transfer (CT) state near the lowest triplet one thus facilitating the dramatic decrease of the $^{2,4}T_1$ excited triplet states lifetime.

2. Experimental results and discussions

Previously we carried out the study of transients in Resonance Raman (RR) spectra of CuP in solvents having as a component molecules, including oxygen heteroatoms like tetrahydrofuran (THF), 1,4-dioxane, cyclohexanon, DMSO, DMFA [4,5]. The samples excited with the Q and/or Soret absorption band give rise to new Raman peaks in RR spectra. The peak frequencies have negative shifts up to 46 cm^{-1} in comparison with their counterparts in the ground electronic state. It was shown that the transient bands can be assigned to the intermolecular complex of CuP and oxygen(O)-containing solvent molecule. The number of the porphyrin molecules in complex is assumed to be dependent on both the geometry of the solvent molecule and the number of oxygen atoms in binded molecule.

A few mechanisms can be proposed to explain new spectral features in RR spectra, and the two of them are most probable.

Springer Proceedings in Physics, Vol. 74 Time-Resolved Vibrational Spectroscopy VI
Editors: A. Lau, F. Siebert, W. Werncke

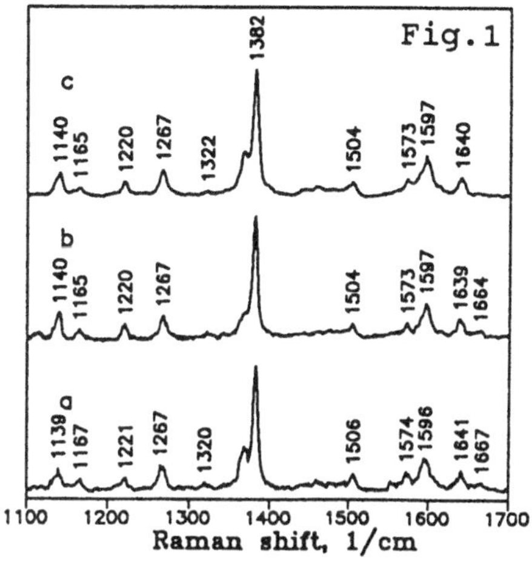

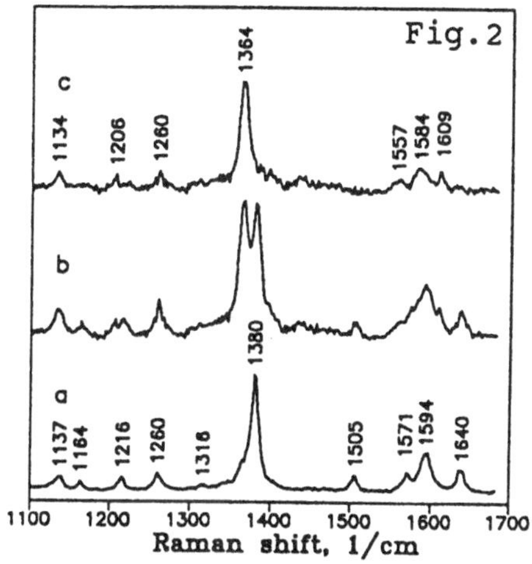

Fig.1 Raman spectra of CuOEP in THF/diethyl ether at 77 K. Excitation wavelength: (a) 400 nm; (b) 402 nm; (c) 405 nm.

Fig.2 Raman spectra of CuOEP in THF recorded at 293 K (λ_{ex}=400 nm) using low (a) and high (b) excitation power. Spectrum (c) represents excited state scattering (c=b-a).

The first one suggests that the new Raman bands correspond to the porphyrin-ligand complex in the ground electronic state of CuP. Another mechanism is related to the formation of electronically excited intermolecular complex (exciplex) with the porphyrin molecule, being excited to a long-lived excited electronic state. To clarify the problem there were performed the measurements of absorption and RR spectra at liquid nitrogen temperature (77 K). One should expect that the decrease of the solution temperature had to be followed by the shift of intermolecular dynamics equilibrium to the direction of the complex formation with CuP molecule in the ground state because of the diminishing of thermal energy excess. And this is supported by the absorption spectra measurements: for CuP in THF/diethyl ether mixture at 77 K Soret band is splitted on two adjacent bands of near equal intensity at 397 and 403 nm [6]. The observed features in absorption spectra show that two different molecular forms of CuP exist in solution: complexed and noncomplexed.

Figure 1 depicts Raman spectra of CuOEP (T= 77 K) recorded at low excitation intensities to prevent the substantial population of the excited electronic states. Three excitation wavelengths (λ_{ex}= 400, 402, 405 nm) were used to cover the Soret band absorption region. The small (less then 7 cm^{-1}) positive shifts of Raman bands are observed in comparison with the ground state spectra at the room temperature (Fig.2, spectrum a). The observed shifts are similar to both the noncomplexed (λ_{ex}= 400 nm) and complexed form (λ_{ex}= 402, 405 nm) of CuOEP. Small shifts of Raman bands prove the weak effect of the molecular environment on vibrational modes of porphyrin skeleton in the ground electronic state, although the π-conjugated system displays prominently the effect of intermolecular binding reflected in Soret band splitting. The phenomenon differs strongly from that at intense photoexcitation (Fig. 2, spectrum b), followed by large negative shifts of vibrational modes. Thus the above described photoinduced complex can not be related to the ground electronic state of complexed CuP.

A few excited states of CuP can be proposed as a candidate to explain the above phenomena:

1. (π,π^*) state of the excited π-conjugated system of the porphyrin macrocycle .

2. (π,d) charge transfer (CT) excited state arising from the transferring of electronic density from the highest occupied molecular orbital a_{2u} to the half-filled metal orbital $d_{x^2-y^2}$.

3. (d,d) excited state related to the promotion of the electron from d_{z^2}-orbital into the lowest half-filled $d_{x^2-y^2}$ orbital of the central metal under nonexcited π-electronic system of the porphyrin macrocycle.

The variety of experimental data obtained supports the idea that the most probably the exciplex exist with CuP in the (d,d) excited state [5,6].

The described phenomenon is relative to that observed in biological systems, studied by P.-Y.Turpin et al. [7]. They observed similar regularities in RR spectra of water soluble Cu(TMPy-P4) complexed with different nucleic acids: transient bands arising with the increase of excitation power were shifted to the low-frequency region about 15-17 cm^{-1}. To understand the nature of photochemical process underlying the effect observed we have undertaken the study of copper tetraphenylporphyrin (CuTPP) in O-containing solvents being the model system for the solutes studied in [7,8]. The transients were observed in Raman spectra of CuTPP only under intense excitation, and the frequency shifts for vibrational modes detected were close to that observed for CuP-DNA exciplex [7].

We have also carried out experiments on measuring Raman excitation profiles (REP) in the Soret absorption band region for a certain prominent Raman bands of CuTPP. The second harmonic output of pulsed nanosecond Nd:YAG laser at 532 nm was used as intense laser pulse to produce sufficient percentage of electronically excited CuTPP molecules. The low intensity output from narrowband tunable dye laser (C440 in ethanol pumped by the YAG third harmonic) was used as a probe beam to measure RR spectra of the excited region in solute. The most prominent transient band $\nu_4^* = 1346\ cm^{-1}$ is explicitly seen for all the probe wavelengths used (Fig. 3). This band intensity related to that of solvent band changes significantly when probe wavelength is scanned. The highest ratios achieved at $\lambda_{probe} = 428...430$ nm (Fig. 3) thus the REP maximum of transient peak is slightly red-shifted to the maximum of Soret absorption band at 416 nm. This fact is consistent with the measurements of REP for the above discussed biological systems [8]. The variety of common regularities taking place for both CuTPP in THF and aqueous solution of Cu(TMPy-P4) with DNA, e.g. the similar frequency shifts, the excitation power-dependent behavior of transient peaks and the same regularities of scattering efficiency against excitation wavelengths (REP) prove that the excited electronic (d,d) state is the origin of the transient peaks not only in model CuP solution, but in biological system too.

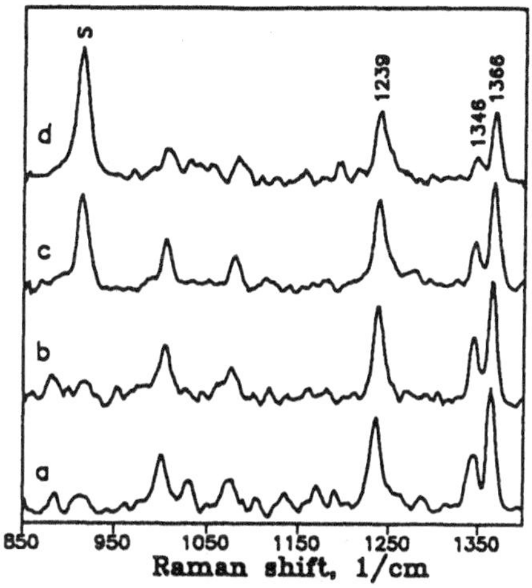

Fig.3. Raman spectra of CuTPP in THF, recorded with excitation 532 nm and probing at (a) 424 nm; (b) 428 nm; (c) 440 nm; (d) 448 nm. "S" indicates solvent band. All spectra have been recorded with the same excitation power.

The authors would like to thank Prof. P.-Y.Turpin for giving his experimental results before publication.

References

1. V.S.Chirvony, B.M.Dzagarov, B.M.Timinskii and G.P.Gurinovich, Chem. Phys. Lett. 70 (1980) 79
2. D.Kim, D.Holten and M.Gouterman, J. Am. Chem. Soc. 106 (1984) 2793.
3. J.A.Shellnutt, K.D.Straub, P.M.Rentzepis, M.Gouterman and R.E.Davidson, Biochemistry 23 (1984) 3946.
4. P.A.Apanasevich, R.Gadonas, V.V.Kvach, V.Krasauskas, V.A.Orlovich and V.Chirvony, Sov. Chem. Phys. 7 (1988) 21.
5. P.A.Apanasevich, V.S.Chirvony, S.G.Kruglik, V.V.Kvach and V.A.Orlovich, *Laser Applications in Life Sciences*, S.A.Akhmanov, M.Y.Poroshina, Editors, Proc. SPIE Vol.1403, p.195-211 (1991)
6. P.A.Apanasevich et al., to be published.
7. P.-Y.Turpin, L.Chinsky, A.Laigle, M.Tsuboi, J.R.Kincaid and K.Nakamoto, Photochem. Photobiol. 51 (1990) 519.
8. L.Chinsky, P.-Y.Turpin, A.Al-Obaidi, S.E.J.Bell and R.E.Hester, J. Phys. Chem. 95 (1991) 5754.

Photochemical Reaction of Furocoumarins: Time-Resolved Absorption and Resonance-Raman Studies

Y. Uesugi and H. Takahashi

Department of Chemistry, School of Science and Engineering, Waseda University, Tokyo 169, Japan

Time-resolved absorption spectra of psoralen, 5-methoxypsoralen and 8-methoxypsoralen revealed that the conversion of the T_1 state into the radical anion occurs much more efficiently in 8-methoxypsoralen than in the other two. Time-resolved resonance Raman spectra of these furocoumarins showed that the psoralen skeleton is considerably weakened in the T_1 state and radical anion.

1. Introduction

Furocoumarins are important drugs used in the photochemotherapy of skin diseases such as psoriasis and vitiligo. The biological activity of furocoumarins has been correlated with their photoreactivity towards the pyrimidine bases of DNA to form a crosslink between the two separate strands and inhibit the overactive DNA synthesis. The photoreactive sites in furocoumarins are considered to be the C=C double bonds at the 3,4 position of the pyron moiety and the 4',5' position of the furan part [1], and possible implication of the T_1 state has been suggested in the crosslink formation [2].

In this study the structures and dynamics of the T_1 state and the radical cation and anion of psoralen (PS), 5-methoxypsoralen (5-MOP) and 8-methoxypsoralen (8-MOP) have been investigated by laser flash photolysis and time-resolved resonance Raman spectroscopy.

2. Experimental

Our experimental system for time-resolved Raman spectroscopy has been reported previously [3] and that for laser flash photolysis will be described elsewhere. PS, 5-MOP and 8-MOP were purchased from Sigma Chemical Co. and were used without further purification.

Springer Proceedings in Physics, Vol. 74 Time-Resolved Vibrational Spectroscopy VI
Editors: A. Lau, F. Siebert, W. Werncke

3. Results and Discussion

Time-resolved absorption spectra of PS, 5-MOP and 8-MOP in deoxygenated acetonitrile are shown in Fig. 1. The bands at around 450 nm are quenched by the presence of oxygen indicating that these bands are attributable to the T_1 state. These bands are also drastically quenched by addition of electron donating agents such as DABCO with concomitant intensity increase of the bands at around 390 nm, indicating that the bands at around 390 nm are assignable to the radical anion produced from the T_1 state by electron abstraction.

It is seen that the spectra of the transients of 8-MOP are markedly different from those of the other two: the peak at around 390 nm is more intense than that at around 450 nm for 8-MOP while opposite is the case for other

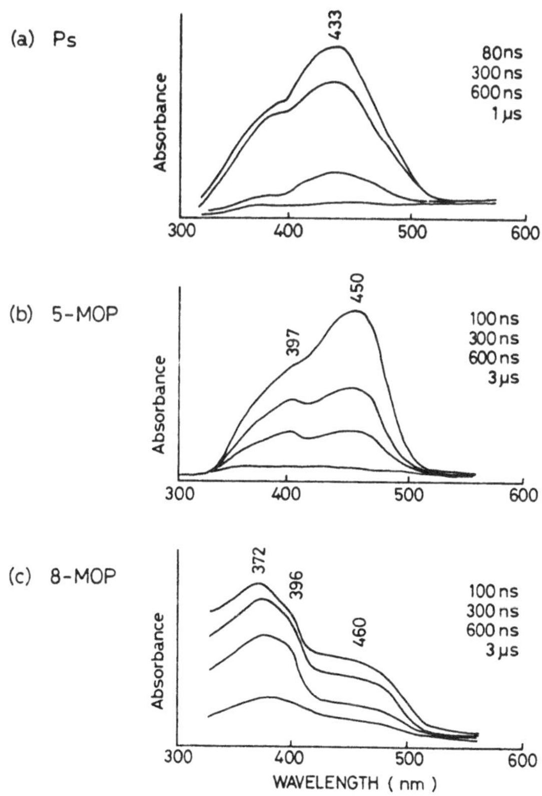

Fig. 1 Time-resolved absorption spectra of furocoumarins in deoxygenated acetonitrile. Concentration, 5.0×10^{-4} mol dm^{-3}. Pump wavelength, 308 nm.

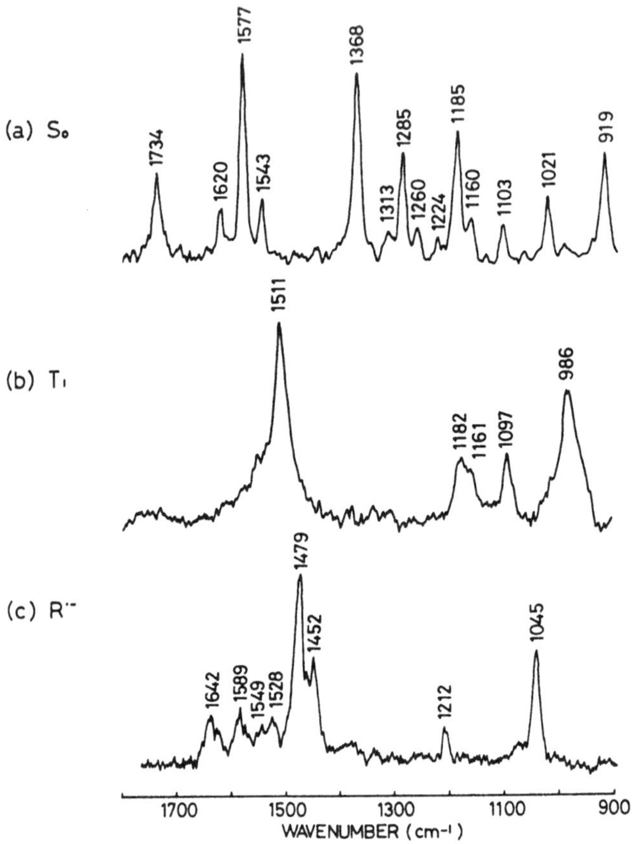

Fig. 2 Raman spectra of the S_0 state (probe wavelength, 530 nm; concentration, nearly saturated) and the resonance Raman spectra of the T_1 state (concentration, 2.0×10^{-3} mol dm^{-3}; probe wavelength, 460 nm) and the radical anion (concentration, 3.0×10^{-3} mol dm^{-3}; probe wavelength, 400 nm) of psoralen.

furocoumarins. This indicates that the quantum yield of the radical anion is larger in 8-MOP than in the other two. It is also seen that a intense peak is observed at around 370 nm in the spectrum of 8-MOP which is not distinctly seen for other furocoumarins. This band is probably attributable to the radical cation. These photochemical singularities of 8-MOP may be related to its wide use as a drug for the skin diseases.

Raman spectrum of the ground state S_0 and the transient resonance Raman spectra of the T_1 state and the radical anion of PS in deoxygenated acetonitrile shown in

Fig. 2 reveal that the C=O stretch at 1734 cm^{-1} and the $^{3}C=^{4}C$ stretch at 1620 cm^{-1} of the S_0 state are both downshifted respectively to 1642 cm^{-1} and 1589 cm^{-1} in the radical anion and are probably hidden under the strong band at 1511 cm^{-1} in the T_1 state. The low-frequency shifts of these vibrations indicate that the skeleton of psoralens is considerably weakened in the radical anion and T_1 state. This is consistent with the attribution of the $^{3}C=^{4}C$ bond to one of the photoreactive sites. Similar results are also obtained for 5-MOP.

References

1. L. Musajo, F. Bordin, G. Caporale, S. Marciani and G. Rigatti, Photochem. Photobiol. **6**, 711 (1967); L. Musajo, F. Bordin and R. Bevilacqua, Photochem. Photobiol. **6**, 927 (1967).
2. P.S. Song, M.L. Harter, T.A. Moore and W.C. Herndon, Photochem. Photobiol. **14**, 521 (1971).
3. S. Hashimoto, A. Shimojima, T. Yuzawa, H. Hiura, J. Abe and H. Takahashi, J. Mol. Struct. **242**, 1 (1991).

Part IV

Chemical, Photochemical, and Photophysical Reactions

Solvent/Solute Interactions Probed by Picosecond Transient-Raman Spectroscopy

T.L. Gustafson[1], *D.L. Morris, Jr.*[1], *L.A. Huston*[1], *R.M. Butler*[2], *and I. Noda*[3]

[1]Department of Chemistry, The Ohio State University, 120 West 18th Avenue, Columbus, OH 43210-1173, USA
[2]Eli Lilly & Company, Clinton Laboratories, P.O. Box 99, Clinton, IN 47842-0099, USA
[3]Corporate Research Division, The Procter & Gamble Company, Miami Valley Laboratories, P.O. Box 398707, Cincinnati, OH 45239-8707, USA

Abstract. We present the picosecond time-resolved resonance Raman spectra of *trans*-4,4'-diphenylstilbene. Based on the solvent dependent transient Raman spectra, we suggest that dielectric stabilization of the S_1 state controls the dynamics of the solute conformational relaxation, but solvent viscosity controls the final equilibrium geometry. We also present the first picosecond transient Raman two-dimensional correlation spectrum. The asynchronous correlation spectrum accentuates the subtle changes we observe in the time-resolved spectra.

1. Introduction

While much is understood on a macroscopic scale about the interactions that occur between solute and solvent in solution, the microscopic details, involving the interaction potentials, transport processes, and energy transfer, are not well understood. The last decade has seen considerable effort, both experimental and theoretical, aimed at elucidating the role of solvent dynamics in chemical reaction rates. Much of this work has been reviewed.[1][2][3] The primary experimental methods used to probe solvation dynamics have been based on electronic spectroscopies. Typically, the rates of a reaction are compared to bulk solvent properties, since the solvent is viewed from the perspective of statistical mechanics as "non-interacting." Recent experimental studies suggest that the concept of a "non-interacting" solvent is not completely rigorous, depending on the specific molecular system and the solvent involved.[4][5]

While the rates of the reactions have provided significant insight into solvent/solute interactions, it is clear that direct structural information about the specific interactions between a solute and its solvent environment would be of interest. We are using ps transient Raman spectroscopy to probe directly the structure of a reacting molecule in different solvent environments. We have obtained results for several systems where we observe *mode specific, solvent dependent* interactions between a reacting solute molecule and its solvent

Editors: A. Lau, F. Siebert, W. Werncke

environment.[6][7][8] The mode specificity occurs because we are monitoring the redistribution of energy from the initially excited Franck-Condon modes of the molecule to the rest of the molecule and to the solvent environment. Other workers have recently confirmed these effects.[9][10][11][12] In general, we observe two types of variations in the ps transient Raman spectra. The first is the change in the nominal peak positions, bandwidths, and intensities of the various bands in different solvents ("static" effects). The second is the change in the peak positions, bandwidths, and intensities of certain bands as the delay is changed ("dynamic" effects). Each of these types of effects provides insight into the way that the solvent accommodates the photoexcited molecule.

In this work we present the ps transient Raman spectra of *trans*-4,4'-diphenylstilbene (DPS) in methylene chloride. We also present the first ps transient Raman two-dimensional (2D) correlation spectra. The 2D correlation spectra provide a convenient way to observe the mode specific changes in the solvent dependent ps transient Raman spectra.

2. Experimental

The MHz amplified laser system used in this work has been described previously.[6][7] Briefly, the second harmonic (532 nm) of a cw mode locked Nd:YAG laser pumps two synchronously pumped dye lasers. One dye laser is amplified at MHz repetition rates using a six pass amplifier pumped by a cavity dumped argon ion laser.[13] Using 3 ps pulses at a 1 MHz repetition rate, we obtain 200 to 400 nJ/pulse. The amplified output is frequency doubled to generate the actinic pulse at ~300 nm (20 to 40 nJ/pulse). The second dye laser is used to probe the excited state population. The probe traverses an optical delay and is combined colinearly with the UV pump pulse at a dichroic beam splitter. The jitter between the two dye lasers is measured by cross-correlation to be ~5 ps. It should be noted that both dye lasers are independently tunable, allowing the pump and probe wavelengths to be adjusted to suit the molecule of interest. The polarization of the pump and probe pulses can also be independently controlled, and, in this work, the polarization was set to the magic angle (54.7°) in order to remove contributions from rotational reorientation to the observed transient Raman signals.[14] The pump and probe beams are focused onto a 1 cm quartz sample cell with a 5 cm lens at a ~45° angle. The sample cell is spun at ~60 Hz to remove the photoexcited volume before the next pulse pair arrives. Transient Raman spectra are obtained by collecting the backscattering from the sample cell and dispersing it through a single spectrograph. The Raman signal was detected using a CCD optical multichannel analyzer.

3. Results and Discussion

In previous studies, the spectral changes observed in *trans*-stilbene (tS) have been attributed to vibrational relaxation and to conformational changes in the S_1

state.[6][9][10][11][12] In order to address the issue of the relative contribution of conformational changes and vibrational relaxation to the solvation process, we have obtained the S_1 Raman spectra of DPS, a structural analogue of tS, in methylene chloride and dioxane.[7] DPS has a longer fluorescence lifetime and higher fluorescence quantum yield than tS. The higher fluorescence quantum yield reflects a change in the relative contribution of conformation (*i.e.*, photoisomerization quantum yield) to the lifetime and structure of S_1 DPS. In Figure 1 we show the Raman spectra of S_1 DPS in methylene chloride (CH_2Cl_2) over the range from 1100 to 1700 cm^{-1}. As we observed in tS, we observe mode specific, solvent dependent changes in the S_1 Raman spectra of DPS. However, unlike tS, the primary spectral changes involve changes in the relative intensities of the S_1 bands; there are slight shifts in the peak positions and no discernible changes in the bandwidths of the S_1 bands. The "static" solvation effects in the spectra involve solvent dependent changes in the intensities of several vibrational modes associated with the biphenyl moiety of DPS. We attribute the changes in intensity to a variation in the Franck-Condon overlap between S_1 and S_n (the state from which the resonance enhancement occurs) caused by differences in the planarity of the biphenyl portion of DPS in the two solvents. There is precedent in the work of Proutiere *et al.* for correlating the intensity of S_0 Raman bands, specifically the ν_8 band at 1607/1590 cm^{-1}, to the planarity of BP.[15] They observed that more planar BP derivatives had a greater intensity in the ν_8 band. In comparing the results between CH_2Cl_2 and dioxane, S_1 DPS appears to assume a more planar structure in CH_2Cl_2, the lower viscosity solvent. Relative intensity changes also dominate the "dynamic" solvation effects. Based on our interpretation that the bands become more intense as DPS becomes more planar, the dynamics suggest that it takes longer for S_1 DPS to reach its equilibrium conformation in CH_2Cl_2 than it does in dioxane. And the final structure in CH_2Cl_2 is more planar than it is in dioxane. In comparing the "static" and "dynamic" changes in the ps transient Raman spectra between dioxane and CH_2Cl_2, we suggest that the solvent viscosity controls the final equilibrium geometry of DPS, but dielectric stabilization of S_1 DPS by the solvent controls the conformational dynamics.[7]

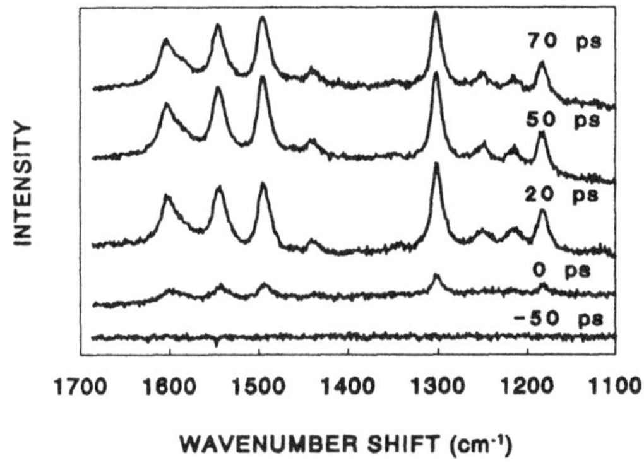

Figure 1. The ps transient Raman spectra of S_1 DPS in methylene chloride: Pump, 303 nm; Probe, 650 nm.

The typical approach for analyzing ps transient Raman spectra, such as shown in Figure 1, is to fit the data to a sum of Lorentzian lineshapes.[6][7][9][10][11][12] The peak positions, bandwidths, and intensities are then plotted as a function of delay and/or solvent. 2D correlation techniques provide a complementary approach for visualizing changes in the transient spectra. Noda and co-workers have used 2D correlation methods extensively as an aid in interpreting perturbation-induced time-dependent changes in infrared spectra[16][17] and, more recently, in ns transient Raman spectra.[18] In the ps transient Raman 2D correlation spectra, the synchronous spectrum has both autopeaks and cross-peaks, since the bands all grow in and decay away following the actinic pulse. Because all the peaks are present, the synchronous correlation spectrum is not particularly informative for this specific ps transient Raman experiment. However, the asynchronous spectrum, corresponding to the unsynchronized changes in the spectra, is quite informative. In Figure 2 we show a portion of the asynchronous 2D correlation spectrum for the data shown in Figure 1. We know that there are changes in the peak position and the relative intensities, but little change in the bandwidths of the transient spectra of DPS.[7]

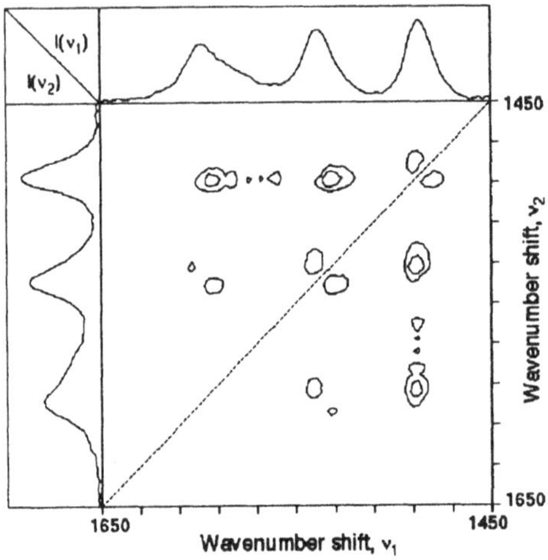

Figure 2. The asynchronous two-dimensional correlation spectrum of *trans*-4,4'-diphenylstilbene in methylene chloride.

There are no autopeaks in the asynchronous spectrum, but there are small features near the diagonal. The bands at 1496 and 1547 cm^{-1} have small shifts to higher wavenumber shift.[7] This is evidenced by the cross-peaks near the diagonal, as seen in Figure 2; one side of the peak is unsynchronized with the other side. It is also of interest to note that the low wavenumber side of the 1547 cm^{-1} band and the peak of the band at 1603 cm^{-1} are asynchronous with the high wavenumber side of the 1496 cm^{-1} band. This is shown by the cross-peaks in Figure 2. We suggest that the cross-peaks arise from the changes in relative intensity with delay. We are currently using model data to evaluate the relative contributions of peak position shift and relative intensity change to the 2D correlation spectra. While there is no intrinsic new information in the 2D correlation spectra, it is clear that the method accentuates the subtle effects that we observe in the spectra, making it easier to visualize the changes that occur. The 2D correlation spectra of DPS between two solvents will also provide a unique way to monitor solvent dependent changes in the ps transient Raman spectra. This work is in progress.

4. References

1. M. Maroncelli, J. MacInnis, and G. R. Fleming, *Science* **243**, 1674 (1989).
2. J. D. Simon, *Acc. Chem. Res.* **21**, 128 (1988).
3. P. F. Barbara and W. Jarzeba, *Adv. in Photochem*. **15**, 1 (1990).

4. (a) D. M. Zeglinski and D. H. Waldeck, *J. Phys. Chem.* **92**, 692 (1988); (b) N. S. Park and D. H. Waldeck, *J. Phys. Chem.* **94**, 662 (1990).
5. (a) S. P. Velsko, D. H. Waldeck, and G. R. Fleming, *J. Chem. Phys.* **78**, 249 (1983); (b) S. K. Kim and G. R. Fleming, *J. Phys. Chem.* **92**, 2168 (1988).
6. W. L. Weaver, L. A. Huston, K. Iwata, and T. L. Gustafson, *J. Phys. Chem.* **96**, 8956 (1992).
7. R. M. Butler, M. A. Lynn, and T. L. Gustafson, *J. Phys. Chem.* **97**, 2609 (1993).
8. D. L. Morris, Jr. and T. L. Gustafson, *J. Phys. Chem.*, submitted for publication.
9. K. Iwata and H. Hamaguchi, *Chem. Phys. Lett.* **196**, 462 (1992).
10. H. Hamaguchi and K. Iwata, *Chem. Phys. Lett.*, submitted for publication.
11. R. E. Hester, P. Matousek, J. N. Moore, A. W. Parker, W. T. Toner, and M. Towrie, *Chem. Phys. Lett.*, accepted for publication.
12. J. Qian, S. L. Schultz, G. R. Bradburn, and J. Jean, *J. Phys. Chem.*, submitted for publication.
13. W. L. Weaver, K. Iwata, and T. L. Gustafson, *J. Opt. Soc. Am. B* **10**, 852 (1993).
14. K. Iwata, W. L. Weaver, and T. L. Gustafson, *Chem. Phys. Lett.*, in press.
15. A. Proutiere, D. Legoff, M. Chabanel, and E. Megnassan, *J. Molec. Struct.* **178**, 49 (1988).
16. (a) I. Noda, *J. Am. Chem. Soc.* **111**, 8116 (1989); (b) *Appl. Spectrosc.* **44**, 550 (1990); (c) *Appl. Spectrosc.*, in press.
17. (a) I. Noda, A. E. Dowrey, and C. Marcott, in *Time-Resolved Vibrational Spectroscopy*, H. Takahashi, Ed., (Springer-Verlag, New York, 1992), pp. 331-334; (b) *Polym. Prepr.* **31**, 576 (1990).
18. K. Ebihara, H. Takahashi, and I. Noda, *Appl. Spectrosc.*, in press.

Dual-Track Picosecond Infrared Spectroscopy of Metal-Carbonyl Photochemistry

T.P. Dougherty and E.J. Heilweil

Molecular Physics Division, National Institute of Standards and Technology, Building 221, Room B268, Gaithersburg, MD 20899, USA

Abstract. A novel spectroscopic method incorporating broadband infrared down and upconversion with 0.4 ps time resolution is described. Broadband (ca. 120 cm^{-1}) IR pulses near 2000 cm^{-1} (5 μm) are difference frequency generated in $LiIO_3$ between 20 Hz Nd^{+3}:YAG regeneratively amplified, synchronously pumped ps and compressed fs dye lasers. Beamsplit IR pulses are upconverted in $LiIO_3$ producing two broadband visible pulses that are dispersed on a rapid-scanned CCD camera as two spectral tracks (4 cm^{-1} resolution). Dividing probe by reference tracks on each shot allows detection of transient difference absorptions as small as 0.001 ΔOD (8000 shots) without laser or crystal tuning. Photolysis of $Rh(CO)_2$(acetylacetonate) and $Co(CO)_2$(cyclopentadiene) in room temperature n-hexane, 1-hexene, CCl_4 and mixed solvent solutions yield spectra suggesting vibrational cooling or complex formation with solvent occurs in these systems.

1. Introduction

Time-resolved infrared (IR) methodology has made tremendous advances over the last few years. With new femtosecond lasers and detector technologies now available, it is possible to generate ultrashort pulses throughout most of the mid-IR wavelength region (3-11 μm) to obtain IR spectra of transient chemical species and determine vibrational energy dynamics in condensed-phase systems. Although high spectral and temporal resolution is attainable through fs upconversion gating of infrared CW gas and diode laser sources [1], transient spectra can only be obtained by scanning the probe laser frequency at a particular time delay between the excitation and probe pulse. Even with high repetition rate apparatus, obtaining a complete sampling of time-dependent spectra over an appropriate time and frequency range with sufficient signal-to-noise sensitivity becomes an arduous and time-consuming task.

Given the ultimate goal of measuring detailed spectral changes with femtosecond time resolution to understand transient species kinetics, energy exchange and related phenomena, we have set out to develop an optical system capable of collecting wide-bandwidth IR spectra with high temporal resolution. Such a system, described and demonstrated here, takes advantage of multichannel detection of broadband IR probe pulses which effectively pushes low repetition-rate experiments into the kHz acquisition time regime. By splitting a difference-frequency generated broadband IR probe pulse into two beams (probe and reference), signal-to-noise baseline sensitivity has been increased by about a factor of seven over the original single beam design [2]. Multiple IR difference spectra with low baseline noise are thus obtained without probe laser frequency tuning or physical alignment of the beams throughout an experiment. These conditions allow for rapid time-domain spectral data acquisition under computer control.

The optical approach described here was implemented to monitor transient vibrational absorption spectra in the CO-stretch region (2000 cm^{-1}) of metal-dicarbonyl species in liquid solutions at 298 K. Such prototypical reactions are of particular interest because short-time

Springer Proceedings in Physics, Vol. 74 Time-Resolved Vibrational Spectroscopy VI
Editors: A. Lau, F. Siebert, W. Werncke

metal-solvent bonding dynamics and photoproduct equilibration rates are ill determined in homogeneous catalytic systems. We present new detailed time-dependent spectra showing that monocarbonyl formation and thermal equilibration occurs within 10's of picoseconds and evidence for solvent substitution reactions also take place on this timescale [3].

2. Dual Beam Broadband IR Probing Method

The general optical scheme employed in this work is an extension of that used in a single IR probe beam approach [2]. Briefly, broadband (ca. 150 cm^{-1} FWHM) sub-picosecond IR pulses near 2000 cm^{-1} (5 μm) are generated by focused angular difference frequency mixing in a 2 mm thick $LiIO_3$ crystal between two 20 Hz Nd^{+3}:YAG regeneratively amplified, synchronously pumped dye lasers. In this work, a narrowband 4 ps, 658 nm dye laser (DCM, 350 μJ) is mixed with a pulse compressed, broadband 150 fs duration R6G laser at 578 nm (550 μJ). The resulting IR pulse (ca. 100 nJ) is split and both halves are focused (beams ca. 0.3 mm diameter, displaced 2 mm at the focus) through a flowing sample cell where an ultraviolet pump beam (R6G SHG λ=289 nm, 20 μJ/pulse) is spatially overlapped with one of the IR probe beams (see Fig. 1). The IR beams are frequency upconverted with residual DCM pulses in a 14x14 mm aperture, 1 cm long $LiIO_3$ crystal to yield two broadband visible (589 nm) pulses. At this point, the IR spectral changes of interest are imposed on the visible probe pulse. The visible pulses are spectrally dispersed by a double holographic grating (2400 groove/mm) spectrograph onto a cooled, rapid-scanned 384x576 pixel CCD array as two spectral tracks with 4 cm^{-1} FWHM spectral resolution. Dispersed spectral maxima approach the CCD's full 16 bit A/D dynamic range.

Reference and probe beam spectra are highly correlated but have large shot-to-shot spectral shape and intensity variations. Dividing the probe signal track by the reference track on each laser shot (ignoring shots below a threshold intensity level) and chopping the pump beam on alternate shots, produces laser drift-free detection of small transient difference absorption changes (baseline uncertainty=0.001 ΔOD or less) over the upconverted IR bandwidth in <10 minutes (20 Hz, 8000 total shots). This routinely achievable baseline noise is 5-10 times better than the single beam approach where a baseline ΔOD of ca. 0.009 was typically obtained. A direct comparison of the single versus dual track approach was made by appropriate treatment of the same spectral data. The dual track baseline was flatter (lacking structure) with a standard deviation about seven times smaller than for the single track approach (containing large, misleading features).

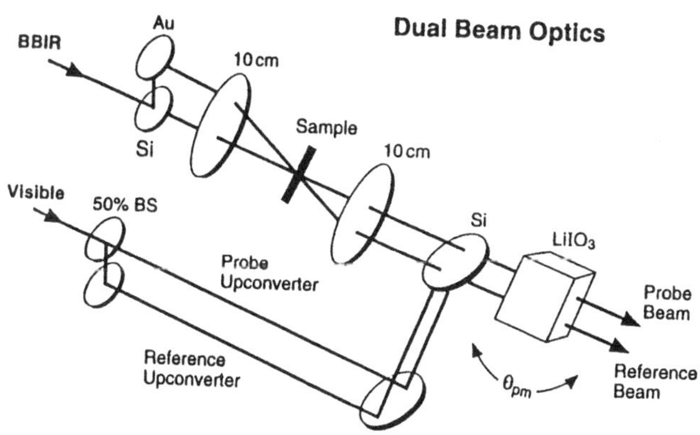

Fig. 1: Dual track optical layout for femtosecond broadband IR transient spectroscopy

3. Experimental

The time resolution of the apparatus was determined by replacing the sample with a polished silicon window. Monitoring the average spectral change in IR transmission of these materials (ca. 5-10%) as a function of pump-probe delay yields a single-sided function with baseline to maximum change of 0.9 ± 0.1 ps. From this response, we extracted an effective experimental time resolution of 0.39 ± 0.04 ps. Assuming the UV pump and IR probe have the same gaussian time profile, the pulse durations are approximately 280 fs FWHM.

Metal dicarbonyls $Rh(CO)_2acac$ (acac ≡ acetylacetonate) and $Co(CO)_2cp$ (cp ≡ cyclopentadiene) were obtained from commercially available sources and used without further purification. Spectroscopic grade n-hexane, 1-hexene and CCl_4 solvents were used after having been dried overnight over molecular sieves. Concentrations were in the range 1-$5x10^{-4}$ M so that an optical density of ca. 1.0 for each sample at its infrared absorption peaks was achieved in the 0.8 mm pathlength flow cell. The cell windows were 3 mm thick CaF_2 and the cell assembly was translated by a stepping motor driven stage to avoid product build-up on the entrance window. Tests were conducted using $Cr(CO)_6$:n-hexane which gave ca. 30:1 signal-to-noise, 10 ps delayed appearance of new absorptions and ca. 50 ps equilibration time of $Cr(CO)_5$-hexane photoproduct.

4. Results and Discussion

Experiments were conducted to evaluate whether: (1) different metal-carbonyls show varying dynamics in the same solvent, (2) the same molecule exhibits differing behavior in solvents known to quench excess vibrational energy with varying rates, and (3) a molecule in a mixed solvent system may show evidence for intermediate bonding to the weaker solvent adduct.

$Rh(CO)_2acac$ and $Co(CO)_2cp$ in room temperature n-hexane solvent were first examined. Representative time-dependent spectra (Fig. 2) exhibit pulsewidth-limited removal of $Co(CO)_2cp$ parent (1975 and 2035 cm^{-1} bleach features) and the delayed growth and sharpening of a new absorption (1990 cm^{-1}) corresponding to monocarbonyl photoproduct [4]. Similar behavior is observed for $Rh(CO)_2acac$ except the equilibrated monocarbonyl absorption is found at 2000 cm^{-1} (bleaches at 2014, 2084 cm^{-1}). The time-dependence of the monocarbonyl absorption bandwidths for both molecules crudely fits an exponential time constant of 9.0 ± 0.6 ps. These results imply that despite different internal degrees of freedom, the two systems react with hexane and equilibrate (undergoing conformational changes or vibrational cooling) with the surroundings on the 10 ps timescale.

To investigate possible solvent-dependent vibrational energy cooling rates, $Rh(CO)_2acac$ was also studied in CCl_4. Specific vibrational modes of many polyatomic solutes exhibit

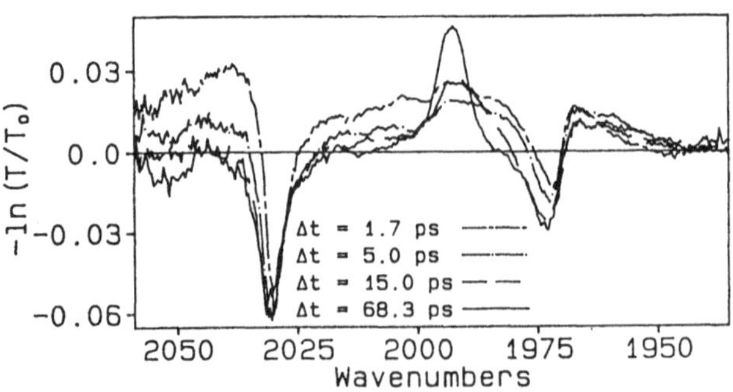

Fig. 2: Composite time-dependent spectra for $Co(CO)_2cp$ in room temperature n-hexane

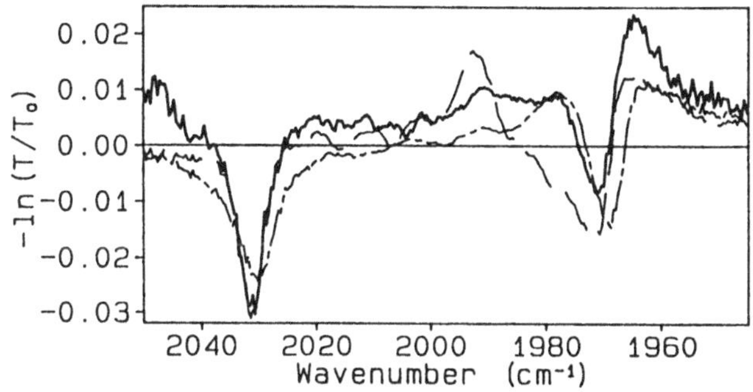

Figure 3: $Co(CO)_2cp$ in (----) n-hexane for $t_d > 35$ ps, (-- -) in 1-hexene for $t_d > 35$ ps, and (—) in 1:1 Mixture of n-hexane and 1-hexene (same concentrations) at $t_d = 22$ ps

relatively long vibrational relaxation times in this solvent [5]. Despite minor broadening and spectral overlap of the monocarbonyl absorption feature with the 2013 cm^{-1} parent bleach, we found that the photo-product equilibrates with a time constant significantly longer (ca. 33 ps) than for the same molecule in n-hexane. From these results, we deduce that after CO photoejection, the oscillator strength of the newly-formed monocarbonyl is spread over a large spectral range and solvent re-arrangement and/or vibrational energy transfer to the solvent bath dominates the equilibration processes that occur on the 10's of ps timescale.

Next, competition between weakly binding saturated hydrocarbons (n-hexane, Fig. 3) and an electron donating alkene (1-hexene) was examined. For $Co(CO)_2cp$ in room temperature 1-hexene, a series of transient difference spectra (Fig. 3) were obtained which show immediate parent depletion at 2030 and 1969 cm^{-1}, and appearance of solvent-complexed monocarbonyl absorption at 1975 cm^{-1} (unchanged for delay times $t_d > 35$ ps). This band overlaps the low frequency parent bleach and is apparently the stable photoproduct absorption since mercury lamp irradiation (5 min, 100 W) produces a similar FTIR difference spectrum. During early times ($0 < t_d < 35$ ps), complex blue-shifting absorption is observed above and below the 1969 cm^{-1} parent bleach feature, similar to the shifting and narrowing observed for the two dicarbonyl species in the other solvents.

$Co(CO)_2cp$ in n-hexane:1-hexene (1:1 by volume) was studied to look for transient absorptions arising from both solvent adducts. Under acquisition conditions identical to those above, transient spectra taken around $t_d = 22$ ps delay exhibit two weak absorption bands centered at approximately 1975 and 1990 cm^{-1} (Fig. 3). These band frequencies closely match the monocarbonyl band positions observed for the neat solvents, and for long delay times, the spectra resemble those obtained in pure 1-hexene solvent. Our interpretation of these observations is that there are transitory solvent-ligated structures where the metal interacts with saturated parts of the two solvent species at early times, but during the 20-30 ps time-frame, the more stable 1-hexene product eventually forms.

5. Conclusions

We have described a versatile broadband multichannel transient IR spectrometer with sub-picosecond time resolution and spectral resolution adequate for many condensed-phase dynamics studies. The optical system operated near 2000 cm^{-1} using $LiIO_3$ frequency converters which perform at 1750 cm^{-1} and higher. Using tunable fs visible pulses in this solid-state frequency conversion scheme should allow studies at lower IR frequencies (using other nonlinear optical materials) and continue to take advantage of the high quantum efficiency of the CCD detector. The overall advantage of this system is reasonably high signal sensitivity and rapid spectral acquisition rates.

Novel spectra obtained during photochemical decomposition of two metal-dicarbonyls in room temperature solvents support the appearance of monocarbonyl-solvent product with subsequent vibrational cooling or structural equilibration. Competitive solvent adduct formation was also observed in the IR, perhaps for the first time, for $Co(CO)_2cp$ in a n-hexane:1-hexene mixture. Related studies are planned to investigate the dynamics of metal-carbonyl insertion reactions in hydrocarbon solvents.

References

[1] J. N. Moore, P. A. Hansen, R. M. Hochstrasser, Chem. Phys. Lett., 138, 110 (1987).
[2] E. J. Heilweil, Optics Lett., 14, 551 (1989).
[3] J. D. Simon, X. Xie, J. Phys. Chem., 91, 5538 (1987).
[4] E. P. Wasserman, R. G. Bergman, C. B. Moore, J. Am. Chem. Soc., 110, 6076 (1988).
[5] H. J. Bakker, J. Chem. Phys., 98, 8496 (1993), and references therein.

Femtosecond Intramolecular Proton Transfer and Vibrational Relaxation in the Liquid Phase

*T. Elsaesser**

Physik Department E 11, Technische Universität München, James-Franck-Str., D-85748 Garching, Germany

*Present address: Max-Born-Institut für Nichtlineare Optik und Kurzzeitspektroskopie, Rudower Chaussee 6, D-12489 Berlin, Germany

Abstract. Photoinduced intramolecular proton transfer in benzotriazoles is directly monitored by femtosecond spectroscopy. Proton transfer in the excited state occurs on a 100 fs time scale and is followed by radiationless deactivation of the reaction product with time constants of about 150 fs. The reaction cycle is completed by the subpicosecond proton back-transfer in the electronic ground state. The back-reaction creates vibrationally hot molecules which cool down within 40 ps by collisional interaction with the surrounding solvent.

Intramolecular proton transfer represents a chemical reaction that has mainly been studied in electronically excited molecules. For a number of compounds forming intramolecular hydrogen bonds, recent femtosecond experiments gave evidence of excited state proton and deuterium transfer on a time scale of 100 to 200 fs [1-4]. In 2-(2'-hydroxy-5'-methylphenyl)-benzotriazole (trade name: TINUVIN P, TIN), the ultrafast enol→keto reaction (see structures (I) and (II)) forms keto-type TIN in the first excited state which shows a very short lifetime of 150 fs and decays by radiationless relaxation to the ground state [2]. This species eventually undergoes a proton back-transfer to the ground state of the original enol structure. Until now, there was little information on the ground state reaction of TIN and of other compounds undergoing a closed reaction cycle. In addition, vibrational relaxation processes randomizing the large amount of excess energy from the initial photoexcitation ($\simeq$ 4 eV per molecule) have not been observed. In this paper, we present a study of those processes in the femtosecond regime, directly revealing the subpicosecond proton transfer in the ground state of TIN and the strong vibrational excitation of the reaction product.

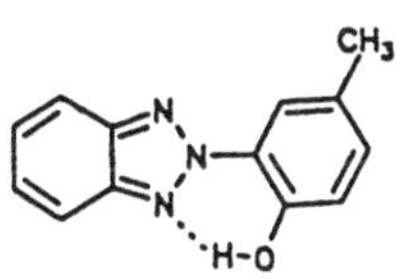

(I) enol

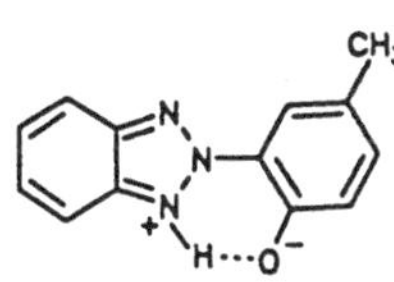

(II) keto-type

In our experiments, we studied TIN dissolved in cyclohexane (concentration c=2 × 10^{-3} M). Femtosecond pump and probe pulses are derived from the amplified output of a colliding-pulse-modelocked dye laser working at 620 nm. The second harmonic at 310 nm serves for excitation of the molecules. Probe pulses tunable between 325 nm and 400 nm were generated by frequency doubling the recompressed output of a femtosecond traveling wave dye laser. Probe light in the wavelength range from 400 to 800 nm was derived from a femtosecond white light continuum. The time resolution was 100 fs.

The kinetics of ground state proton transfer is studied via transient changes of (i) the absorption originating from the ground state S_0' of the keto-type reaction product and

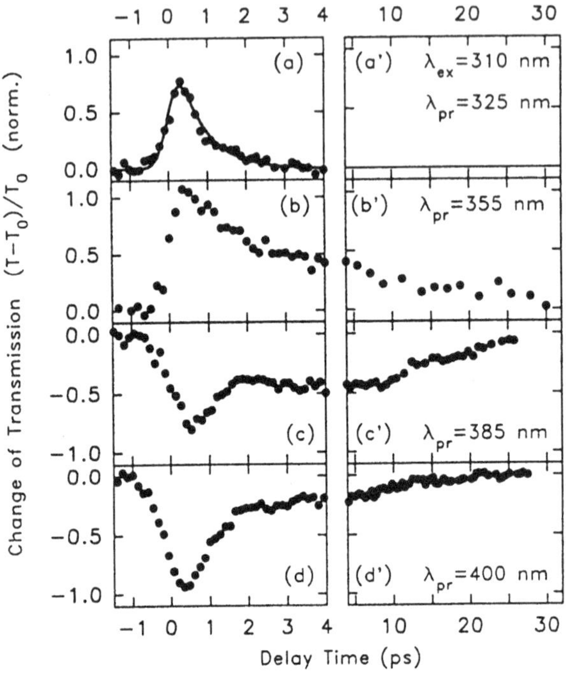

Figure 1 Time resolved absorption changes in the range of the $S_0 \rightarrow S_1$ absorption band of TIN after femtosecond excitation at λ_{ex} = 310 nm for probe wavelengths of (a,a') λ_{pr} = 325 nm, (b,b') 355 nm, (c,c') 385 nm, and (d,d') 400 nm. The normalized change of transmission $(T\text{-}T_0)/T_0$ is plotted as a function of delay time for the first 4 ps after excitation (left hand side) and from 4 to 30 ps (right hand side, points); T_0 and T represent the transmission before and after excitation. The solid line in Fig. 1 (a) is calculated from a numerical simulation.

(ii) of the original enol $S_0 \rightarrow S_1$ absorption band. The absorption band of the S_0' state is located between 500 and 600 nm with a broad maximum around 520 nm. This band decays by depopulation of the S_0' level with a time constant of 700 fs as was discussed in Ref. [2]. Here we concentrate on measurements in the range of the enol ground state absorption between 300 and 400 nm [5]. In Fig. 1, the change of transmission $(T\text{-}T_0)/T_0$ at four different wavelengths λ_{pr} is plotted for early (left-hand-side) and late delay times (right hand side, T,T_0: transmission with and without excitation). At 325 nm and 355 nm, we observe an increase of transmission due to the depletion of the enol ground state. At 325 nm (Fig. 1 a), this bleaching rises with the excitation pulse and decays by repopulation of the ground state with a time constant of 700$\pm$100 fs (solid line), i.e. identical to the decay time of the keto ground state absorption. Thus keto-TIN directly transforms to the enol geometry by the back-transfer of the proton on a subpicosecond time scale. At a probe wavelength of 355 nm (Fig. 1 b), the fast bleaching is followed by a slower change of transmission that decays within 40 ps. At longer wavelengths λ_{pr} between 375 and 400 nm (Fig. 1 c,d), we observe an induced absorption exhibiting both the fast and the slow kinetics. To visualize the overall behavior of TIN, we derived transient absorption spectra from the time resolved data. The results presented in Fig. 2 demonstrate a well-pronounced reshaping of the enol $S_0 \rightarrow S_1$ absorption band with a

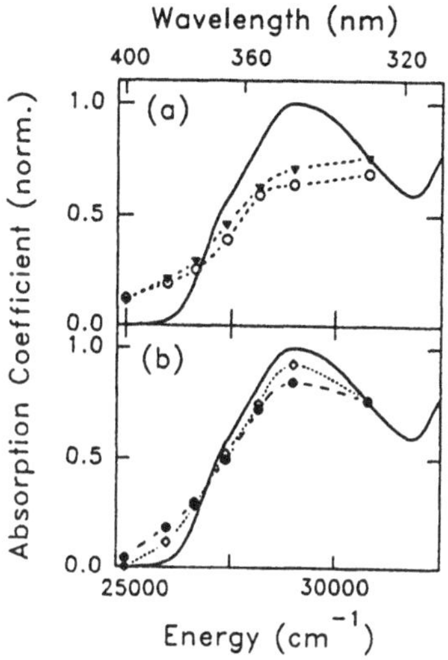

Figure 2 Transient absorption spectra of enol-TIN after femtosecond excitation (λ_{ex}=310 nm). (a) The normalized absorption coefficient is plotted vs photon energy for delay times of 2 ps (open circles), and 5 ps (triangles). The solid line gives the steady state $S_0 \rightarrow S_1$ band. (b) Transient spectra for delay times of 10 ps (points) and 15 ps (diamonds).

strong enhancement of the low-energy absorption tail and a reduced extinction around the maximum.

The fast proton motion points to a ground state potential with a low or even negligible barrier along the reaction coordinate. Recent theoretical calculations for molecular structures comprising the key elements of the TIN geometry give a potential energy surface in the electronic ground state that shows a single broad minimum for the enol species [6]. The subpicosecond time scale of proton transfer corresponds to vibrational frequencies in the range of 50 to 200 cm^{-1}, i.e. low-frequency vibrations of large amplitude are relevant for the reaction step whereas the OH stretching motion is of minor importance.

The fast reaction cycle of TIN transfers the energy supplied by the exciting ultraviolet photon to the vibrational manifold of the molecules, creating enol tautomers with a highly excited vibrational system. Intra- as well as inter-molecular processes are important for the subsequent vibrational relaxation. The transient absorption spectra of enol-TIN (c.f. Fig. 2) exhibit no distinct vibrational structure, even at very early delay times. This fact suggests that non-thermal vibrational populations of TIN created by the back-reaction equilibrate on a subpicosecond time scale similar to that of proton transfer. The hot vibrational distribution in the enol ground state gives rise to the transient broadening of the $S_0 \rightarrow S_1$ absorption band found in our measurements. The increase of absorption at low frequencies is caused by vibrational excess populations of high lying vibrational states which couple to the electronic transition. The decrease of absorption at higher frequencies is due to a reduction of population density at the bottom of the vibrational distribution. The hot molecules subsequently cool down by inelastic collisions with the surrounding solvent. An amount of energy of several hundred wavenumbers per collision is transferred to the first solvent shell around the hot solute and from there to the remainder solvent. At early times after excitation, the rate of energy flow from the TIN molecules to the first solvent shell is considerably higher than that out of the first shell, leading to a strong heating of the solvent molecules neighbouring the solute. A quantitative theoretical simulation of the different energy transfer processes which was presented in Ref. [7], gives cooling times of several tens of picoseconds. Cooling of the hot TIN molecules reduces the broadening of the transient absorption bands that evolve towards the steady state spectrum (Fig. 2 b).

In conclusion, the subpicosecond intramolecular proton transfer in the electronic ground state of TINUVIN P was directly monitored by femtosecond spectroscopy. The ultrafast reaction cycle and the rapid vibrational cooling of the molecules account for the high efficiency of TIN as a photostabilizer of polymers.

Acknowledgements

I am greatly indebted to W. Kaiser, F. Laermer, W. Frey, M. Wiechmann, and H. Port for their important contributions to this work. Valuable discussions with W. Lüttke, University of Göttingen, are gratefully acknowledged.

References

[1] F. Laermer, T. Elsaesser, and W. Kaiser, Chem. Phys. Lett. **148**, 119 (1988)
[2] M. Wiechmann, H. Port, W. Frey, F. Laermer, and T. Elsaesser, J. Phys. **95**, Chem. **95**, 1918 (1991)
[3] B.J. Schwartz, L.A. Peteanu, and C.B. Harris, J. Phys. Chem. **96**, 3591 (1992)
[4] T. Arthen-Engeland, T. Bultmann, N.P. Ernsting, M.A. Rodriguez, and W. Thiel, Chem. Phys. **163**, 43 (1992)
[5] W. Frey, and T. Elsaesser, Chem. Phys. Lett. **189**, 565 (1992)
[6] X. Duan, and S. Scheiner, Chem. Phys. Lett. **204**, 36 (1993)
[7] U. Sukowski, A. Seilmeier, T. Elsaesser, and S.F. Fischer, J. Chem. Phys. **93**, 4094 (1990)

Femtosecond Raman-Induced Kerr-Effect Study of Polar Solvent Dynamics: Amides

Y.J. Chang and E.W. Castner, Jr.

Chemistry Department, Brookhaven National Laboratory,
Building 555A, Upton, NY 11973-5000, USA

Abstract. We have measured the ultrafast pure solvent dynamics of highly polar liquids, formamide (FA), N-methylformamide (NMF), N-methylacetamide (NMA), N-methylpropionamide (NMP) and N,N'-dimethylformamide (DMF) using femtosecond optical-heterodyne-detected Raman-induced Kerr effect spectroscopy (OHD-RIKES). The effects of deuteration and temperature-dependence were studied to characterize in detail both the inertial (or non-diffusive) and diffusive intermolecular motions in these liquids.

1. Introduction

Detailed characterization of neat solvent dynamics is an important step in elucidating the role that the solvent plays in determining condensed phase chemical reaction dynamics. Time-resolved degenerate four-wave mixing techniques such as optical Kerr effect [1-5] and impulsive stimulated Raman scattering [6] spectroscopies have been used recently to observe directly the ultrafast intermolecular motions in various solvent systems.

We have studied the pure solvent dynamics of the amides FA, NMF, NMA, NMP and DMF using the electronically non-resonant femtosecond OHD-RIKES technique. These amides are highly polar molecules, and all except DMF form highly structured liquids that have an extensive intermolecular H-bonding network. DMF liquid is less structured because of the absence of intermolecular H-bonds. We present the results of temperature-dependent and deuterium-substitution OHD-RIKES experiments on the formamide liquid.

The time-resolved OHD-RIKES is a Fourier-transform nonlinear optical polarization spectroscopy that measures the decay of a transient birefringence induced in a transparent liquid sample by an intense ultrashort optical pulse. Through the Fourier transform relationship, the heterodyne detection provides a linear relationship between the measured signal decay and the third order nonlinear susceptibility, $\chi^{(3)}$, and spontaneous Raman cross section, σ_r, of the sample [7]. We have taken advantage of the nearly transform-limited 50 fs optical pulses to analyze our data both in the time and frequency domains. The details of our experimental set-up have been discussed previously [5].

2. Results and Discussion

The nonlinear optical response of a sample detected in the OHD-RIKES experiment is separable into electronic and nuclear contributions. The former originates from

Springer Proceedings in Physics, Vol. 74 Time-Resolved Vibrational Spectroscopy VI
Editors: A. Lau, F. Siebert, W. Werncke

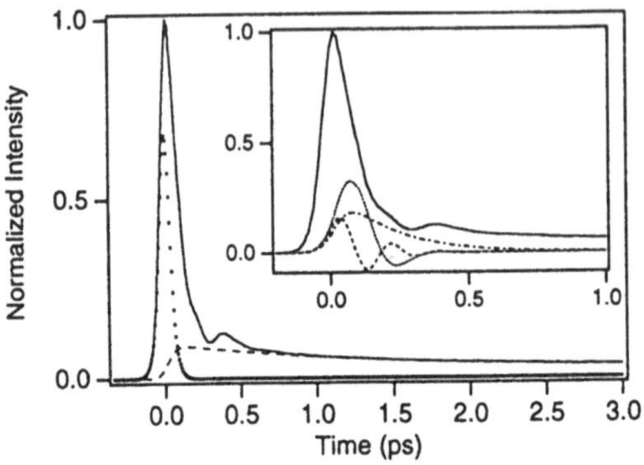

Figure 1. OHD-RIKES data for FA at 299K shown with various sub-component responses: —— total response, • • • electronic, - - diffusive rotational. Inset: ······ 100 cm^{-1} librational, ---- 190 cm^{-1} librational, -·- collisional interaction-induced responses.

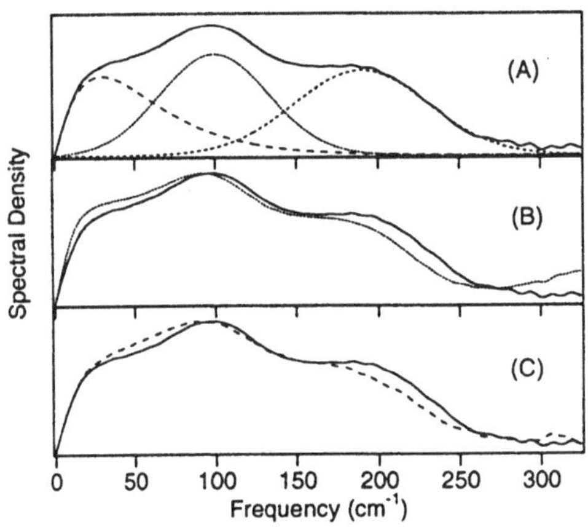

Figure 2. Frequency domain data for: (A) FA at 299K fit with two anti-symmetrized Gaussian lineshapes at ~100 and 190 cm^{-1} and an Ohmic lineshape peaked at ~30 cm^{-1}, (B) —— FA and ······ deuterated FA, both at 299K, (C) FA at —— 299K and at - - 350K.

the electronic hyperpolarizability of the sample and is effectively instantaneous. The time delayed response arises from nuclear dynamics that include both the inertial and diffusive types of molecular motions. The electronic and nuclear responses for FA are illustrated in Figure 1. The surprising finding is that in addition to the long diffusive nuclear response caused by molecular reorientation, we observe inertial responses that have both oscillatory and exponentially-decaying components on the sub-ps timescale. We have further analyzed the data in the frequency domain using the Fourier transform/deconvolution procedure described previously by McMorrow and Lotshaw [1]. The low frequency spectrum obtained for FA is shown in Figure 2(A). We were able to adequately fit the frequency data using two anti-symmetrized

Gaussian lineshapes and an Ohmic lineshape of the form

$$I(\omega)=\omega\cdot\exp(-\omega/\omega_0). \quad (1)$$

The inhomogeneously broadened Gaussian bands centered at about 100 and 190 cm^{-1} have been observed previously in the low-frequency Raman study by Faurskov-Nielsen et al. [8], and are assigned to out-of-plane intermolecular H-bond bending motions in the FA liquid. The oscillatory responses we observe in the time domain are correlated exactly to these Raman band frequencies, and we assign these responses to the underdamped intermolecular H-bond librational dynamics. The sub-ps exponentially-decaying component has been observed previously in other liquids and is assigned to collisional interaction-induced effects such as multipole-induced multipole interactions. These collision-induced dynamics correspond to the lowest frequency band peaking at about 30 cm^{-1} in the spectrum. The rotational diffusion response is found to decay biexponentially (τ_1=1.45, τ_2=11.30ps @ 299K), and we attribute this to the rotations about the dissimilar inertial axes of the FA molecule [5].

We have carried out deuteration effect and temperature dependence studies to further characterize the inertial and diffusive nuclear motions occurring in FA liquid. In Fig. 2(B), we observe that the low-frequency bands of $HCOND_2$ red-shift relative to those of $HCONH_2$. We would expect a factor of ~1.4 red-shift with deuteration for intramolecular stretching vibrations. However, the bands are observed to red-shift by a factor of ~1.1. We attribute this to the fact that the nuclear motions that give rise to these bands are intermolecular in nature and therefore involve complex motions including librations and translations.

The temperature dependence of the longer rotational diffusion time constant, τ_2, shows an Arrhenius behavior, and we obtain an enthalpy of activation, ΔH_a, of 4.3±0.2 kcal/mol. This is in excellent agreement with the ΔH_a of 4.2 kcal/mol obtained from the temperature dependent study of depolarized Rayleigh linewidths carried out by Whittenburg, et al. [9]. In view of the fact that this ΔH_a is correlated with the ΔH_a obtained from the temperature dependence of viscosity and to the enthalpy of H-bond formation, ΔH_f=-5.2 kcal/mol [10], we conclude that the diffusive rotational motions in FA liquid are strongly correlated with the making and breaking of the intermolecular H-bonds.

All three low-frequency intermolecular Raman bands in Fig. 2(A) are also observed to red-shift upon increasing the temperature, and this is shown in Fig. 2(C). The local minima in the intermolecular interaction potential become less pronounced with increasing temperature, and this is reflected in the lowering of the frequencies of these bands. We are currently analyzing the results we have obtained for the amides. Future work will involve pressure-dependent OHD-RIKES studies on these highly polar liquids.

This research was carried out at Brookhaven National Laboratory under contract DE-AC02-76CH00016 with the U. S. Department of Energy and supported by its Division of Chemical Sciences, Office of Basic Energy Sciences.

3. References

[1] D. McMorrow and W. T. Lotshaw, J. Phys. Chem. **95**, 10395 (1991).
[2] T. Hattori, A. Terasaki, T. Kobayashi, T. Wada, A. Yamada and H. Sasabe, J. Chem. Phys. **95**, 937 (1991).
[3] K. Wynne, C. Galli and R. M. Hochstrasser, Chem. Phys. Lett. **193**, 17 (1992).
[4] M. Cho, M. Du, N. F. Scherer, G. R. Fleming and S. Mukamel, J. Chem. Phys., submitted, 1993.
[5] Y. J. Chang and E. W. Castner, Jr., J. Chem. Phys. **99**, 113 (1993); ibid., submitted May, 1993.
[6] S. Ruhman, A. G. Joly and K. A. Nelson, IEEE J. Quant. Elect. **24**, 46 (1988).
[7] G. L. Eesley, M. D. Levenson and W. M. Tolles, IEEE J. Quant. Elect. **14**, 45 (1978).
[8] O. Faurskov-Nielsen, P.-A. Lund and E. Paestgaard, J. Chem. Phys. **77**, 3878 (1982).
[9] S. L. Whittenberg, V. K. Jain and S. J. McKinnon, J. Phys. Chem. **90**,1004 (1986).
[10] J. N. Spencer, S. K. Berger, C. Powell, B. D. Henning, G. S. Furman, W. M. Loffredo, E. M. Rydberg, R. A. Neubert, C. E. Shoop and D. N. Blauch, J. Phys. Chem. **85**, 1236 (1981).

Spectroscopy of Charge-Transfer Dynamics: Application to Electron and Proton Self-Exchange in Solutions

D.E. Khoshtariya[1,2], *R. Billing*[1], *T.A. Marsagishvili*[2], *and J. Ulstrup*[3]

[1]Universität Leipzig, Fachbereich Chemie/Kristallographie, Talstraße 35, D-04103 Leipzig, Germany

[2]Georgian Academy of Sciences, Institute of Inorganic Chemistry and Electrochemistry, Jikia 7, 380086 Tbilisi, Georgia

[3]The Technical University of Denmark, Chemistry Dept. A, Building 207, DK-2800 Lyngby, Denmark

Abstract. Elementary physical mechanisms for rapid optically-induced and related radiationless charge (electron, proton) self-exchange steps in solutions were analyzed on the basis of fundamental Marcus-Hush-Dogonadze theoretical approach and experimental spectroscopic band-shape data. Contributions of Franck-Condon (reorgational) and electronic exchange factors to both optical molar extinction coefficient and thermal rate constant were estimated with a special attention to fine effects connected with configurational distributions of "reactive associates".

1. Introduction

Phenomenological rationalization and theoretical generalization of elementary physical mechanisms of fast and ultrafast charge transfer processes in solutions requires collecting the complete set of kinetic data including study of precursor and successor complex formation, drawing free energy relationships etc. One well-tried way for obtaining such the data lies via direct following rapid kinetics using highly time-resolved pulse spectroscopy.

Alternative spectroscopic approach for gethering desirable kinetic information is in turn, based on theoretical background founded by Marcus [1], Hush [2] and developed by others [3,4]. According to the fundamental result of modern cherge transfer rate theory adopted here, new optical absorption bands directly connected with elementary charge phototransfer (or photoexchange) would arise. Such the bands carry explicit information regarding both optical and related thermal elementary processes due to close physical and formal similarities between them, and could be observed either in ultraviolet, visible or infrared spectral regions depending on structural and energetic parameters of reactant system. Charge-transfer absorption bands have been found and well documented earlier for binuclear mixed-valence complexes and related ion pairs where electrpon trapping centers are fixed by covalent bridges or strong electrostatic interactions, see e.g. [5,6].

Recently attention was paid also to charge phototransfer phenomena with the participation of relatively free and mobil reactants, viz. the $[Fe(CN)_6]^{4-}$ and

$[Fe(CN)_6]^{3-}$ ions. Particularly, the explicit band-shape analysis for a new near-infrared absorption band (800 nm) was performed [7,8] based on a "bimolecular" theoretical treatment developed by Dogonadze et al. [9,10]. Meanwhile the approach was extended to ultrafast optical and thermal proton self-exchange processes occuring in hydrogen-bonded groupings formed by lyate ions [11]. In the present paper further analysis of abovementioned mechanisms is presented with a special attention to configurational effects of reactant systems.

2. Electron self-exchange between free hexacyanoferrate (II) and (III) ions

Transition in this system was regarded as optical outer-sphear electron transfer from $[Fe(CN)_6]^{4-}$ to $[Fe(CN)_6]^{4-}$ through the "reactive associate" of type

$$K^{+}\cdots NC\text{-}Fe^{II}\text{-}CN\cdots K^{+}\cdots NC\text{-}Fe^{III}\text{-}CN\cdots K \qquad (1)$$

In the framework of most general theoretical treatment [9,10], using Condon approximation the following two equations for the "bimolecular" molar absorption coefficient and the similar thermal rate constant were derived associated with optical and related thermal charge transfer steps, respectively

$$\varepsilon^{(2)}_{ct} = (4\pi\nu\, M_{fi}^2/3h)[f(R_c)2\pi\, R_c^2\, \delta R_c/\Psi''(\theta^*)]\exp\{\beta\theta^*[h\nu-\Delta G_o-\Psi(\theta^*)]\} \qquad (2)$$

$$k^{(2)}_{ct} = (L_{fi}^2 / h)\,[f(R_c)\, 2\pi\, R_c^2\, \delta R_c \,/\, \Psi''(\theta^*)]\exp\{\beta\theta^*[-\Delta G_o - \Psi(\theta^*)]\} \qquad (3)$$

Here $\beta=1/k_bT$, M_{fi} is the transition dipole for the optical, and L_{fi} is the electronic transition matrix element for the thermal processes; R_c is the charge transfer distance, δR_c is the characteristic length of overlap integral decay, ΔG_o is the free energy gap, and θ^* is a transfer coefficient analogous to Bronsted parameter; $f(R_c)$ is, in general, configurational distribution function of reactants in the initial state of transition and $\Psi(\theta^*)$ is a band-shape function, both to be specified depending on features of reactant system under study. The band-shape function was considered to have the form (here $\Delta G_o=0$ due to symmetrical character of processes to be described)

$$\Psi(\theta^*) = (1-\theta^*)\Delta G_s + (1-\theta^*)\Delta G_r\,\gamma\,\{\, 1/[(1-\theta^*)\gamma^2+\theta^*] + 1/[(1-\theta^*)+\theta^*\gamma^2]\,\} \qquad (4)$$

which describes reorganization in both the solvent (ΔG_s) and two low-frequency displaced and distorted nuclear modes (ΔG_r) representing ion pair motion in the $[Fe(CN)_6]^{4-}\cdot K^+$ and $[Fe(CN)_6)]^{3-}\cdot K^+$ units, and $\gamma=\omega_{red}/\omega_{ox}$ is the ratio of these mode frequencies (included as an empirical parameter).

Within the model used the charge-transfer band maximum and half-width are [8]

$$h\nu_{max} = \Delta G_s + \Delta G_r\,(\gamma+\gamma^{-1}) \qquad (5)$$

$$\Delta_{band} = \{4k_bT\ln 2\,[\Delta G_s+\Delta G_r\,(\gamma^2+\gamma^{-4})]\}^{1/2} \qquad (6)$$

as opposed to those of Hush model [2]. Using modified method of Job we have recently estimated effective bimolecular equilibrium constant for the "reactive associate" formation and found $K_{ct} \sim 1$ $l.mol^{-1}$ [12]. Thus, for free energy of formation of the associate $\Delta G_{ct} \approx 0$ and bimolecular "collision complex" approach adopted earlier for this system seems to be consistent. At the same time, translational-rotational distribution function $f(R_c)$ was considered as totally identical for optically induced and thermal pathways under study [7,8]. Under this assumption dyrect comparison of Franck-Condon and electronic exchange parameters estimated from our band-shape analysis with those obtained from earlier thermal data of Shporer et al. [13] display following conclusions:

a) Thermal data, even if treated as totally adiabatic process ($L_{fi} \geq 0.02$ eV) yields activation free energy of 0.47 eV which is still considerable lower than value estimated using eqs. (2) and (3), 0.54 eV.

b) Optical transition dipole M_{fi} estimated from the eqn. (2) (~ 0.4 D), according to well-known equation [2,3]

$$L_{fi} = h\nu_{max} M_{fi} / e R_c \tag{7}$$

yields $L_{fi} \leq 0.02$ eV.

Thow discripancy between compared values is not so pronounced, it could be easily eliminated suggesting that functions $f(R_c)$ for optical and thermal processes somewhat differ. Indeed, optical transfer is dominated by contributions from narrow range of R_c most likely to be the sum of reactant ion radii and bridging cation diameter. On the other hand, effective thermal transfer would be associated with motion along the slope work-term [1] leading to significant decreasing of resulting activation free energy and approaching of L_{fi} to the adiabatic limit due to smaller distance for the electron transfer.

3. Proton self-exchange through lyate ion hydrogen bridges

On the basis of same general theoretical approach a new dynamic elementary act model for thermal and related optical proton self-exchange through "symmetrical" very short hydrogen bonds, formed by lyate ions in aqueous solutions (viz. $H_5O_2^+$, $H_3O_2^-$) was produced [11]. It was demonstrated that "continuum" absorption observed for these systems in the range of 500-3500 cm^{-1} is nothing else but an extremely broad band unusual for infrared spectral region, and could be considered as a charge-transfer band associated with the proton self-exchange. The elementary act model which rests also on most reliable quantum-chemical and structural data includes reorganization of classical part of medium polarization (described by single harmonic mode) followed by the proton shift along proton self-exchange co-ordinate. The proton subsystem is considered as essentially quantum [14] thow no underbarrier motion is required here due to small proton transfer distance detected [11].

The equilibrium free energy of "reactive associate" formation in this case is much higher compared with previous system ($\Delta G_{ct} \sim 1$ eV for $H_5O_2^+$ or $H_3O_2^-$) thus favouring application of unimolecular version of theoretical approach. For the same reason here the function $f(R_c)$ could be considered as totally identical for optical and thermal pathways (most likely $f(R_c)=1$) yielding for molar extinction coefficient and rate constant "unimolecular" equations derived in earlier papers [3,14] (for pure electron transfer versions see [1,2]. Participation of proton in this system is considered accepting $L_{fi}=L_eV$ where L_e is the electron resonance energy and V is the overlap integral of proton vibrational wavefunctions [14] (In our case V=1 by the reason of small distance for the proton shift).

The charge-transfer band maxima and half-widths directly produced the value of $\Delta G_s=0.27$ eV according to Marcus-Hush model used [1-3], for both the groupings $H_5O_2^+$ and $H_3O_2^-$. Putting this value in the corresponding equation for the thermal rate constant, along with the critical value of $L_{fi}=L_e=0.02$ eV (which in turn, is setting validity of nonadiabatic approach) we obtain $k_{ct}=10^{12}\ s^{-1}$ - a reasonable value for the elementary rate constant of ultrafast proton self-exchange.

Acknowledgement. D.E.Kh. kindly acknowledges a fellowship from the Alexander von Humboldt Foundation.

References

1. R.A. Marcus: J. Chem. Phys. **24**, 966 (1956)
2. N.S. Hush: Electrochim. Acta **13**, 1005 (1968)
3. R.R. Dogonadze, E.M. Itskovich, A.M. Kuznetsov, M.A. Vorotyntsev: J. Phys. Chem. **79**, 2827 (1975)
4. J.J. Hopfield: Biophys. J. **18**, 311 (1977)
5. C. Creutz: Progr. Inorg. Chem. **30**, 1 (1983)
6. R. Billing, D. Rehorek, H. Hennig: Topics Curr. Chem. **158**, 151 (1990)
7. T.A. Marsagishvili, D.E. Khoshtariya: Khim. Fiz. **6**, 1511 (1987)
8. D.E. Khoshtariya, A.M. Kjaer, T.A. Marsagishvili, J. Ulstrup: J. Phys. Chem. **95**, 8797 (1991); **96**, 4154 (1992)
9. R.R. Dogonadze, T.A. Marsagishvili: In *The Chemical Physics of Solvation*, Part B, ed. by R.R. Dogonadze, E. Kalman, A.A. Kornyshev, J. Ulstrup (Elsevier, Amsterdam 1986) p.189
10. E. Itskovich, J. Ulstrup, M.A. Vorotyntsev: *Ibid*, p.223
11. D.E. Khoshtariya, N.O. Berdzenishvili: Chem. Phys. Lett. **196**, 607 (1992)
12. D.E. Khoshtariya, R. Billing: (to be published)
13. M. Shporer, G. Ron, A. Lowenstein, G. Navon: Inorg. Chem. **4**, 361 (1965)
14. R.R. Dogonadze, A.M. Kuznetsov, T.A. Marsagishvili: Electrochim. Acta **25**, 1 (1980)

Picosecond-Resonance Coherent Anti-Stokes Raman Scattering in Biophysics: Power Dependence in the Bacteriorhodopsin Photocycle

G.H. Atkinson, A. Popp, and L. Ujj

Department of Chemistry and Optical Science Center,
University of Arizona, Tucson, AZ 85721, USA

Abstract. Picosecond resonance coherent anti-Stokes Raman (PR/CARS) spectra in the C=C stretching region (1400-1640 cm^{-1}) of the BR photocycle are recorded with 8-ps pulses and measured as a function of ω_1 laser power. These PR/CARS data are quantitatively fit to third-order susceptibility ($\chi^{(3)}$) relationships to obtain the band origins, bandwidths, relative amplitudes, and electronic resonant phase factors of the observed features. Variations in the ω_1 power alters the relative populations of the primary BR photocycle intermediates from which PR/CARS is generated. For ω_1 pulse energies of 0.7 nJ or lower, PR/CARS scattering occurs from BR-570 only. An analysis of the BR photocycle species which may contribute to the PR/CARS spectra recorded at higher ω_1 energies is also presented. These results provide insight into how the ω_1 power controls these relative populations, an understanding of importance in the design of picosecond time-resolved CARS spectroscopy from the BR photocycle.

1. Introduction

The changes in the vibrational degrees of freedom of the retinal chromophore during the primary (femto/picosecond) events in the bacteriorhodopsin (BR) photocycle are important elements in elucidating the molecular mechanism underlying energy storage and transduction. The BR photocycle is well studied as a photosynthetic system in which light absorption initiates a complex series of structural changes in both the chromophore and the protein that result in transmembrane proton pumping [1,2]. BR also serves as a molecular model for the related photoreaction of rhodopsin which operates as part of the visual process.

Although spontaneous vibrational resonance Raman scattering has been used extensively to determine the structural changes in the BR photocycle [3-6], coherent Raman scattering has generally not been applied to such biophysical systems. There are several reasons including (i) the complexity of the experimental instrumentation and procedures, (ii) the difficulties in separating resonant and non-resonant coherent Raman signals, and (iii) the requirement that the resultant coherent Raman spectra be quantitatively fit with non-linear susceptibility relationships to characterize the vibrational modes.

Recently, PR/CARS has been measured from the BR photocycle utilizing ω_1 ($\Delta\omega_1 \approx 4$ cm^{-1}) and ω_s ($\Delta\omega_s \approx 300$ cm^{-1}) pulses of picosecond duration

[7-9]. Since these PR/CARS data are recorded with 8-ps pulses (FWHM), the coherent Raman scattering recorded is assignable only to those species present within the duration of the optical excitation. An important experimental aspect of these measurements, therefore, is the relationship between the ω_1 pulse energy and the BR photocycle intermediates present within the pulse duration. For example, the threshold for ω_1 pulse energies below which the PR/CARS signal can be assigned to BR-570 alone is of special interest. The ω_1 pulse energy simultaneously controls the BR species from which PR/CARS is generated through the degree of photochemistry it initiates and the S/N in the resultant PR/CARS scattering. The second issue is addressed in detail elsewhere [9], but data pertinent to the first issue are presented here.

2. Experimental

The instrumentation and analysis procedure utilized for these PR/CARS measurements are described in detail elsewhere [9]. The wavelengths of the ω_1 (596 nm) and ω_s (650 nm - 660 nm) pulses are selected to be in resonance with the absorption spectrum of BR-570 and simultaneously to generate $\approx$300 cm^{-1} of the CARS spectrum which is monitored after spectral dispersion by a multichannel detector. The 500 kHz repetition rates of the two picosecond dye lasers and the flow velocity of the liquid jet sample are selected to ensure that each pair of ω_1 and ω_s pulses encounter a fresh BR sample volume.

3. Results and Discussion

PR/CARS spectra from BR in the 1400-1640 cm^{-1} region are presented in Fig. 1 for three different ω_1 pulse energies (0.7 nJ, 1.5 nJ, and 2.5 nJ). Parameters describing the CARS features in this region, obtained from quantitative $\chi^{(3)}$ fits are presented in Tab. I. The effects of changing the ω_1 power on the 1400-1640 cm^{-1} PR/CARS features are most evident for the C=C stretching bands near 1527 cm^{-1} (Fig. 1). Increases in both asymmetric broadening to higher wavenumber displacements and background scattering are observed as the ω_1 power increases. In addition, the scaling factors (μ) are found to change (Tab. I). These are general phenomena since these same spectral changes are observed for the band origins (Ω) and phase factors (Θ) obtained from the $\chi^{(3)}$ fitting for all the 1400-1640 cm^{-1} features. In the case of the C=C stretching bands, the $\chi^{(3)}$ fitting results (Tab. I) show these spectral effects are attributable to a shift in the phase factor and a slight shift of the band origin to a lower wavenumber displacement.

Since only fresh BR sample is excited by each pair of ω_1 and ω_s pulses, the observed spectral changes can be associated solely with CARS scattering at ω_a from a new set of BR photocycle intermediates present during the 8-ps pulses. In order to correctly interpret these PR/CARS

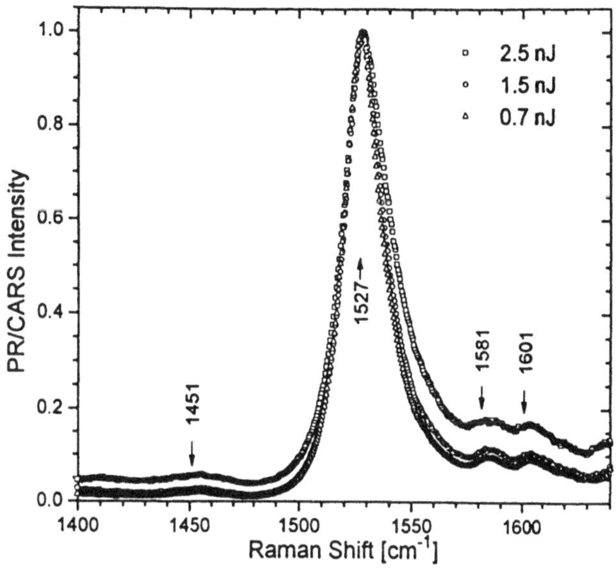

Fig. 1. PR/CARS spectra of the BR in the 1400-1640 cm^{-1} region correspond to different ω_1 pulse energies (2.5 nJ (□), 1.5 nJ (○), 0.7 nJ (△)). The maxima signals at 1527 cm^{-1} are normalized.

Table I: Parameters describing the bands observed in the PR/CARS spectrum of BR (Fig. 1). These results are obtained from fittings of $\chi^{(3)}$ relationships to the PR/CARS data recorded for different powers of ω_1. The amplitudes are normalized to the C=C stretching bands at 1527 cm^{-1}. The fitting procedures are described elsewhere [9].

0.7 nJ			1.5 nJ			2.5 nJ		
Ω	Γ	A	Ω	Γ	A	Ω	Γ	A
1601.6	11.8	0.08	1601.2	11.0	0.07	1601.2	9.4	0.07
1580.7	8.9	0.09	1581.0	9.5	0.09	1581.7	11.1	0.09
1526.7	10.0	1.00	1526.7	10.6	1.00	1526.1	12.3	1.00
1451.6	15.8	0.08	1452.0	17.7	0.09	1450.3	18.1	0.08
$\theta = -0.16$ / $\mu = 0.14$			$\theta = -0.28$ / $\mu = 0.14$			$\theta = -0.50$ / $\mu = 0.2$		

spectra in terms of the vibrational degrees of freedom of a specific BR species, it is important to determine the relationship between the ω_1 pulse energy and the BR photocycle intermediates from which the PR/CARS signal is generated.

Most of the PR/CARS originates with the vibrationally-relaxed BR-570 which has the largest population and an absorption spectrum that is in direct resonance with ω_1 excitation and ω_a scattering. The PR/CARS spectrum measured with an ω_1 energy of 0.7 nJ/pulse is assigned to BR-570 alone (lower ω_1 energy excitation gave identical PR/CARS data [10]).

The spectral broadening, increased background, and changed phase observed for higher ω_1 energies are used (Fig. 1) must be assignable to BR photocycle species, four of which might contribute to the 8-ps PR/CARS signal measured here: two photophysical species (electronically-excited, vibrationally-relaxed BR-570 (BR*-570(I)) and vibrationally-excited BR-570 (BR'-570)) and two photochemical species (J-625 and K-590). The degree to which CARS scattering from each species contributes to the measured PR/CARS signal can be evaluated in terms of two issues: (i) their respective populations within the 8-ps ω_1/ω_s excitation pulsewidths (controlled by the photocycle kinetics and the pulse energies) and (ii) the respective absorption coefficients at ω_1/ω_a which determines the specific degree of resonance enhancement in the PR/CARS signal.

The contribution of BR*-570(I) to the PR/CARS signal is negligible since its ≈450 fs lifetime ensures that its relative population during the 8-ps pulse is small. In addition, absorption from BR*-570(I) appears at <450 nm which is far from ω_1 at 596 nm.

The relative populations of both BR'-570 and J-625 are controlled kinetically by the ≈450-fs lifetime of BR*-570(I), their common precursor, and their respective decay rates. BR'-570 and J-625 decay respectively with time constants of 2-7 ps [6,11] and ≈3.5 ps [12]. The relative yield of J-625 from BR*-570(I) is about double that of BR'-570 [1,2]. The relative population of J-625 within the 8-ps excitation pulses, therefore, is about four times larger than that of BR'-570. Due to the large red-shifted absorption of J-625, however, there is little spectral overlap between ω_1/ω_a and J-625 and the resonance enhancement effect is small. As a consequence, the overall contribution of J-625 to the PR/CARS signal reported here must be small, if any. By contrast, absorption from the vibrationally-excited BR'-570 is shifted only slightly to the red from that of BR-570 and the spectral overlap with ω_1/ω_a is large. At low ω_1 power, the population of BR'-570 is small as is its contribution to the PR/CARS signal, but as ω_1 power increases, both the BR'-570 population and its PR/CARS contribution increase.

The population of K-590 during the 8-ps excitation remains comparable to that of J-625 due to (i) the 3.5-ps time constant observed for the J-625 to K-590 transformation and (ii) the low ω_1 energy used to initiate the BR photocycle. Based on kinetic analysis of time-resolved absorption and fluorescence data from the BR photocycle [13], the K-590 population initially formed is mostly vibrationally-excited (i.e., K'-590). Vibrational relaxation from K'-590 to K-590 is estimated to require ≈8-ps [12,13]. Unlike the J-625 case, the absorption spectrum of K-590 is near the 596-nm ω_1 and therefore, CARS scattering from K-590/K'-590 is resonantly enhanced to a significant degree.

From the kinetic and spectroscopic viewpoints, therefore, BR'-570 and K'-590 are two species other than BR-570 likely to contribute to the PR/CARS spectra recorded with ω_1 energies > 0.7 nJ/pulse. The energy threshold above which the BR'-570 and K'-590 contributions become significant lies between 0.7 nJ/pulse and 1.5 nJ/pulse (Fig. 1). PR/CARS spectra taken with energies ≤ 0.7 nJ/pulse are assigned to BR-570 alone.

4. References

1. Oesterhelt, D. & J. Tittor (1989) *TIBS*, **14**, 57.
2. Lanyi, J. (1992) *J. Bioenerg. Biomembr.* **24**, 169.
3. Atkinson, G.H. (1982) in *Advances in Infrared and Raman Spectroscopy*, eds. Clark, R.J. & R. Hester (North Holland Publishers, London), pp. 1-62.
4. Brack, T.L., & G. H. Atkinson (1989) *J. Molec. Structure* **214**, 289.
5. Van der Berg, B., J. Jang, & M. El-Sayed (1990) *Biophys. J.* **58**, 135.
6. Doig, S., P. Reid, & R. Mathies (1991) *J. Phys. Chem.* **95**, 6375.
7. Atkinson, G. (1992), in *XIII International Conference on Raman Spectroscopy*, eds. Kiefer, W., et. al. (J. Wiley & Sons, Chichester), pp. 192-193.
8. Ujj, L. & G.H. Atkinson (1993) *Israel J. Chem.* **33** (In press).
9. Ujj, L., B. Volodin, A. Popp, J. Delaney, & G.H. Atkinson *Chem. Phys.* (submitted.).
10. Ujj, L., A. Popp, B. Volodin, J. Delaney, & G.H. Atkinson, unpub. res.
11. Brack, T., & G.H. Atkinson (1991), *J. Phys. Chem.* **95** 2352.
12. Blanchard, D., D.A. Gilmore, T.L. Brack, H. Lemaire, D. Hughes, & G.H. Atkinson (1991) *Chem. Phys.* **154** 155.
13. Atkinson, G.H., D. Blanchard, & T.L. Brack (1991) *J. Lumines.* **48**, 410.

Programmable Phase and Amplitude Femtosecond Pulseshaping

M.W. Wefers and K.A. Nelson

Department of Chemistry, Massachusetts Institute of Technology, 77 Massachusetts Ave., Cambridge, MA 02139, USA

Abstract. We report programmable femtosecond pulseshaping using two commercially avaliable liquid crystal spatial light modulators (SLM). This allows generation of waveforms with arbitrary temporal phase and amplitude profiles within a 2.9 picosecond time window with features of less than 100 femtoseconds. Such waveforms suggest exciting prospects for the control of chemical behaviour.

1. Introduction

Pulses of 100 fs duration or shorter allow elememtary collective and molecular vibrations, chemical reactions, and structural rearrangements to be monitored directly in the time domain. These and other ultrafast events are usually monitored following their initiation by a single femtosecond pulse. However, in many cases more selective preparation of and more extensive control over a sample may be acheived with a sequence of femtosecond excitation pulses, or in general an ultrafast optical waveform, whose optical phase and amplitude profiles can be controlled.

A substantial theoretical literature already exists on the application of ultrafast optical waveforms to 'mode selective chemistry' [1]. This field expands the context of spectroscopy to emphasize detailed optical control over chemical events. Recent experiments have already demonstrated such ideas. Scherer et al [2] used a phased-locked pair of femtosecond pulses to control fluorescence and explore dynamics in molecular iodine. Weiner et al [3] demonstrated amplification and selectivity of lattice vibrations in perylene, using a timed sequence of femtosecond pulses, by non-resonant impulsive stimulated raman scattering (ISRS). The timing between pulses is resonant with the driven mode, hence only the resonant mode is driven and with lower peak intensity, thus circumventing sample damage. This technique made it possible for Dougherty et al [4] to explore low temperature polariton dynamics in lithium tantalate where low damage thresholds made signals from single pulse ISRS undetectable.

Such experimental success suggests interesting prospects for large amplitude displacement of lattice modes. Such displacemet could produce unusual nonequilibrium states and facilitate structural change that may not occur under ordinary conditions. Kosloff et al [5] and Cina and Smith [6] have recently proposed resonant multiple pulse ISRS schemes to generate large amplitude displacements. Both schemes involve definite phase relationships between the multiple excitation pulses. Furthermore Rabitz [7] has developed a general theoretical framework for determining appropriate temporal waveforms to achieve specific chemical objectives. It is clear that experimetal generation of ultrafast optical waveforms could provide a wealth of spectroscopic possibilities.

Springer Proceedings in Physics, Vol. 74 Time-Resolved Vibrational Spectroscopy VI
Editors: A. Lau, F. Siebert, W. Werncke

Femtosecond pulse-shaping methods have been developed by Weiner, Heritage, and coworkers at Bellcore [8-13]. The frequency components of an incident femtosecond pulse are dispersed and focused onto different regions of a spatially varying "mask" which blocks (or changes the phases of) some components and allows other components to pass unaffected. The transmitted components are recombined, producing the desired waveform. Originally, permanent mask patterns were etched onto glass substrates using microlithography techniques [8,9]. Recently, a homebuilt liquid crystal (LC) spatial light modulator (SLM) was used as a phase mask [10,11]. This was used to generate a temporal odd pulse as well as linear, quadratic, and cubic phase sweeps in the frequency domain.

Here we report results in which new waveforms with specified optical phase and amplitude profiles are computer-generated using only commercially available components. Using simulated annealing algorithms we have expanded the types of waveforms produced with a single mask to include pulse trains with various pre-specified intensity profiles. More importantly, we have reconfigured the optical set-up to incorporate a second SLM which can variably attenuate dispersed frequency components (figure 1). By compensating for retardation imparted by this attenuation with the first mask, complete manipulation of both spectral phase and amplitude is achieved. Furthermore, we have written algorithms that take into account the specific characteristics and limitations of the optical system including the mask(s). The important characteristics include the finite number and size of pixels, "dead" space between live pixel elements, the finite spot size for each optical frequency component at the mask(s), multiple diffraction arising from spectral filtering with two spatially separated masks, and (in contrast to earlier etched masks) the fixed pixel widths of the SLM. In practice, the user need not formulate the detailed mask pattern but need only specify to the computer the desired waveform. The computer designs the mask(s) through the algorithms and fabricates the mask(s) through its interface with the LC SLM's. The result is computer-controlled generation of arbitrary temporal waveforms with temporal range limited by the spatial resolution of the SLM and temporal resolution limited by the input pulse duration.

2. Experimental

A Ti:sapphire laser provided the incident 70-fs, 800-nm pulses at a 76 MHz repetition rate and with 400 mW of average power. These were directed into the optical network shown in Fig. 1, the critical element(s) of which is the LC SLM(s). The SLM (Meadowlark Optics, with CRI drivers) has an approximately 7-μm thick layer of nematic liquid crystal material sandwiched between two ITO-coated glass windows. One window acts as a ground plate while the second is patterned into 70 separate electrodes (pixels). The array of pixels is 2mm high and 8mm across. Each pixel is 115 μm wide, including a 15-μm "gap" between electrodes. The liquid crystals are driven by a 2 kHz bipolar square wave varying between 0 and 10 volts. The amplitude profiles of shaped waveforms were characterized by standard cross-correlation techniques.

An applied voltage causes the liquid crystals to twist along the plane perpendicular to the mask, varying the retardance for the **y** polarization while leaving the retardance for the **x** polarization constant. Thus if the beam is **y** (0°) polarized at the mask, the SLM acts as a phase mask, modulating only the relative phases of the different frequency components. If instead the incident beam is polarized at $+45^o$ and

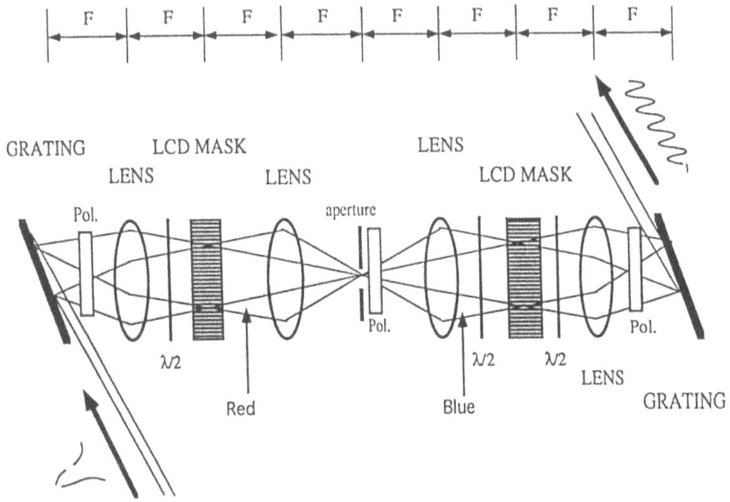

Figure 1
Schematic illustration of the pulse shaping apparatus using two masks. Two 1:1 telescopes are placed between the gratings so that the spatially dispersed frequency profile is cylindrically focused and filtered twice. A combination of polarizers and wave plates is used so that the first LCD mask attenuates the frequency profile and the second mask manipulates the relative phases of the frequency profile.

passed through a polarizer at -45° following the mask, then the SLM acts as an amplitude mask that variably attenuates the different frequency components. Note that this attenuation also imparts a phase shift, so the mask is not solely an 'amplitude' mask. The SLM's were calibrated by measuring the attenuation when a uniform voltage is applied across the masks [11].

Experiments using a single mask as a phase modulator were performed using 1800 line/mm gratings and 12 cm lenses. This provided spectral resolution sufficient for a 4.5 psec time window in which to pulseshape. Experiments involving two masks were performed with 1200 line/mm gratings and 12-cm lenses. In this case the time window was only 2.9 psec. In both case the 70-fs input pulse was not broadened. Careful (± 2-μm) alignment was required to ensure that the two masks were in register.

3. Results

A first test for phase-only filtering was to produce terahertz repetition-rate sequences of femtosecond pulses. This has been done using etched masks with binary retardations (e.g. 0 or pi) whose pixel widths were varied to vary the repetition rate of the pulse sequence [13]. In contrast , the LC SLM's have fixed pixel width but are capable of gray-scale retardation. Figure 2a shows the intensity profile of a 3.07 THz pulse train generated using the LC SLM with a grey-scale sequence. Similar pulse trains with repetition rates from 1.10 THz to 4.39 THz in 0.22 THz intervals were also easily produced.

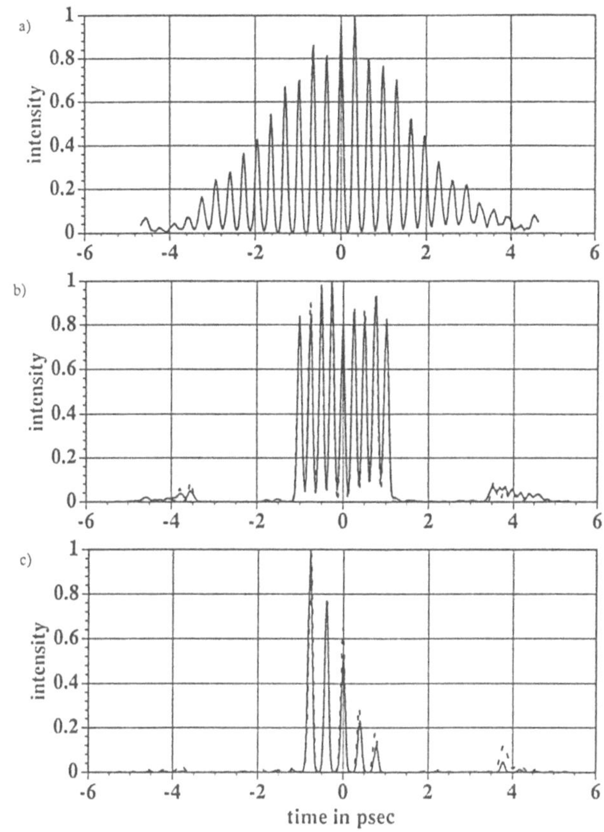

Figure 2.
Shaped waveforms produced using a single LCD phase mask. **(a)** 3.07 THz pulse train produced using a grey-scaled filtering array. Pulse trains with repetition rates from 1.10 to 4.39 THz were also generated. **(b), (c)** Shaped waveforms with specified intensity profiles along with the expected waveform (---). **(b)** Flat topped pulse train. **(c)** 'Staggered' pulse train.

Next we generated waveforms in which the intensity profile had been specified. Note that not all intensity profiles are possible with only phase modulation. Figure 2b shows a flat-topped pulse train. Figure 2c demonstrates a "staggered" pulse train. In both cases the fit (dotted curve) to the theoretical result is excellent.

The use of both phase and amplitude masks compounds the sensitivity of the setup to alignment, laser performance, and especially mask calibration (which was checked and reset each day). Although the algorithms converge easily on the desired waveform in simulation, small calibration errors lead to deviations between specified and experimentally achieved waveforms. In practice the algorithms were used with the computer-interfaced masks to generate an attempt at a desired waveform. Based on the result and on experience with the setup, a modified waveform was specified to help guide the algorithm toward generation of the desired

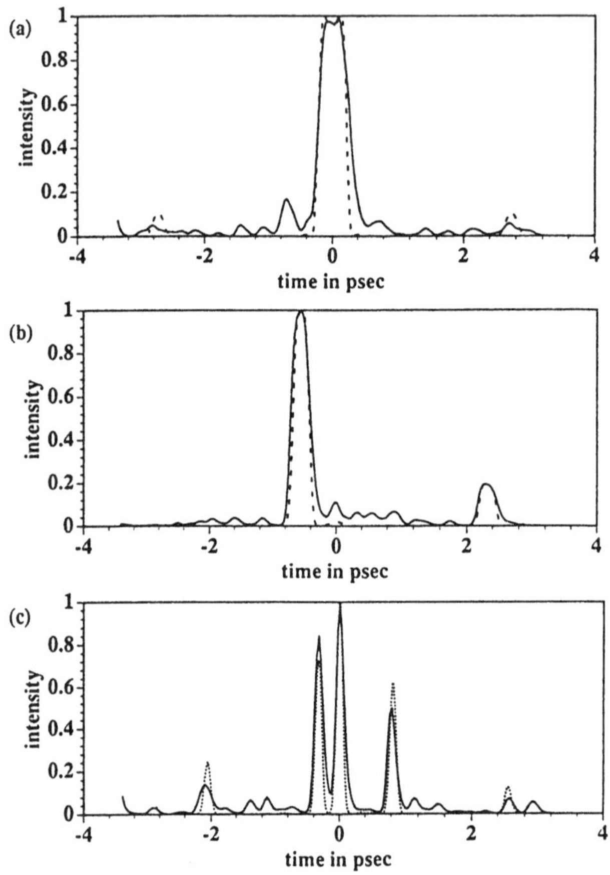

Figure 3.
Shaped waveforms produced using two LCD masks for manipulation of both spectral phase and amplitude profiles. The theoretical expectation (---) which assumes perfect calibration and alignment is also shown. **(a)** 550-fs optical square pulse. **(b)** 350-fs optical square pulse displaced in time. **(c)** Waveform involving three pulses with varied asymmetric timing. The waveform is designed so that the three pulses are all in phase with each other.

singular waveform. Improvements in the masks, discussed below, should alleviate this user-guided convergence and improve overall performance considerably.

"Square" pulses require both phase and amplitude manipulation. Figure 3a shows a 550-fs square pulse centered at t=0 with rise and fall times comparable to the unshaped 100-fs crosscorrelation width. The dotted curve gives the expected pulse, assuming accurate alignment and calibration. As a more demanding test, we tried displacing the square pulse away from t=0. Figure 3b shows a 300-fs square pulse displaced in time. Other than noise in the wings, both fits are excellent.

In anticipation of optical analogs of NMR spin-echo experiments, we tried to produce waveforms involving three ultrashort pulses with variable timing and well-defined phase relationships. Figure 3c shows such a waveform. This pulse shape is designed such that the three pulses are in phase with each other. Presently we have no way to test this since we only measure the cross correlation. However, creation of square pulses suggests that we are successfully manipulating the phase as desired.

Note the "replicas" in each of the waveforms generated using two masks. They are a consequence of each pixel sampling a finite range of frequency components. Increasing the spatial resolution of the masks, for example by increasing the number of pixels on each mask, will minimize this effect.

4. Prospects

Substantial improvements can be anticipated given current efforts to systematically improve the liquid crystal SLM. An SLM with 128 pixels and only 3 micron gaps (between 97 micron wide active regions) is under development. The greater number of pixels will improve the spectral resolution in the filtering plane and expand the temporal range of shaped waveforms to about 8 psec. The substantial reduction in gap size should reduce diffraction, produce a more robust calibration and simplify the pulse shaping algorithms.

In addition, two such masks will be incorporated into a single housing. Thus dispersed frequency components need only be filtered once within a single telescope. This should substantially simplify alignment of the optical apparatus, especially since the two masks will be permanently in register.

Furthermore, with an increasing number of detection techniques [14,15] that resolve both the amplitude and phase of ultrafast waveforms, iterative improvement of the shaped waveform is possible. Indeed the basis of our existing pulse shaping algorithm involves iterative improvement of the theoretically shaped pulse given an initial masking pattern. Rather than assuming accurate calibration and alignment to determine theoretically the shaped pulse, one could iterate based on the experimentally measured waveform. We have run simulations in which mask filters were based on inaccurate calibration curves, but still converged on the desired waveform when the iterative improvement was based on complete information of the shaped output waveform.

Iterative improvements of the shaped field need not only be focussed on improving the fidelty of the shaped pulse, but also on acheiving a particular chemical objective using the shaped pulse. Discussion involving such 'learning' algorithms have been presented [16].

In conclusion, we have demonstrated arbitrary programmamble waveform generation within a temporal range of 2.9 psec and resolution of less than 100 fsec. We foresee rapid improvement in this technology, and considerable spectroscopic utility in the emerging quest for experimental quantum control.

We acknowledge insightful discussions with Andy Weiner and J.R. Wullert. M.M.W. acknowledges a Canadian NSERC fellowship. This work was supported in part by NSF Grant No. CHE-8901722.

5. References

1. W. S. Warren, H. Rabitz, M. Dahleh, Science, **259** 1581 (1993)
2. N. F. Scherer, Rojer J. Carlson, Alexander Mato, Mei Du, Anthony J. Ruggiero, Victor Romero-Rochin, Jeffery A. Cina, Graham R. Fleming, Stuart A. Rice, J. Chem. Phys. **95**, 1487 (1991)
3. A. M. Weiner, D. E. Leaird, G. P. Wiederrecht, K. A. Nelson, Science **247**, 1317 (1990)
4. T. P. Dougherty, G. P. Wiederrecht, and K. A. Nelson, J. Opt. Soc. Am B. **9**, 2179 (1992)
5. R. Kosloff, A. D. Hammerich, D. Tannor, Phys. Rev. Lett. **69**, 2172 (1992)
6. J. A. Cina, T. J. Smith, submitted to J. Chem. Phys. (1992)
7. S. Shi, A. Woody, H. Rabitz, J. Cham. Phys. **88**, 6870 (1988)
8. J. P. Heritage, A. M. Weiner, R. N. Thurston, Opt. Lett. **10**, 609 (1985)
9. A. M. Weiner, J. P. Heritage, E. M. Kirschner, J. Opt. Soc. Am. B **5**, 1563 (1988)
10. A. M. Weiner, D. E. Leaird, J. S. Patel, J. R. Wullert, Opt. Lett. **15**, 326 (1990)
11. A. M. Weiner, D. E. Leaird, J. S. Patel, J. R. Wullert, IEEE J. Quan. Elec. **QE-28**, 908 (1992)
12. R. N. Thurston, J. P. Heritage, A. M. Weiner, W. J. Thomlinson, IEEE J. Quan. Elec. **QE-22**, 682 (1986)
13. A. M. Weiner, D. E. Leaird, Opt. Lett. **15**, 51 (1990)
14. D. J. Kane, R. Trebino, Opt. Lett. **18**, 823 (1993)
15. J. L. A. Chilla, O. E. Martinez, Opt. Lett. **16,** 39 (1991)
16. P. Gross, D. Neuhauser, H. Rabitz, J. Chem. Phys. **98**, 4557 (1993)

Time-Resolved Infrared Spectroscopy of Reaction Intermediates in Solution and Polymers

M.W. George and M. Poliakoff

Department of Chemistry, University of Nottingham,
Nottingham NG7 2RD, UK

Abstract. Time-resolved Infrared Spectroscopy is used to study the reactivity of organometallic intermediates using n-heptane and polyethylene solvents

Nanosecond Time-resolved Infrared Spectroscopy (TRIR) has proved a powerful tool for the characterisation of reaction intermediates and the elucidation of photochemical reaction mechanisms of organometallic molecules in solution[1]. The general reaction scheme for the generation of reaction intermediates and subsequent substitution, which has produced many previously unknown especially with H_2 and N_2, is as follows.

$$LnM(CO)_y \xrightarrow{h\nu} LnM(CO)_{y-1} \xrightarrow{<1ps} LnM(CO)_{y-1}\text{-Sol} \xrightarrow{X} LnM(CO)_{y-1}X$$

Sol = n-heptane, X = N_2 and H_2.

A typical example is shown below in figure 1, for the reactivity of $(\eta^5\text{-}C_5R_5)Mn(CO)_2$(n-heptane) with H_2 (R= H, Me and Et)[2]. The reactivity of $(\eta^5\text{-}C_5R_5)Mn(CO)_2$(n-heptane) depends significantly upon the nature R. The reactivity increases upon increasing the bulk of substituents around the C_5R_5 ring. This shows that steric and not electronic factors are responsible for governing the reactivity of these species.

Unsaturated organometallic intermediates are usually react with organic (C=C) double bonds to form stable species. We have chosen to study reactions in Hostalen

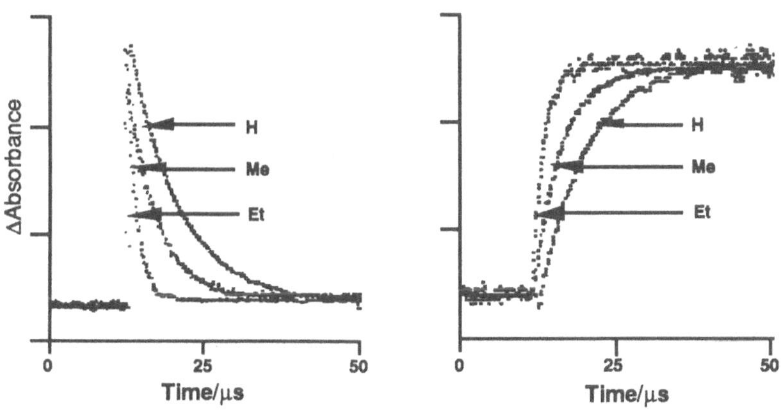

Figure 1: Variation in the rate of reaction of $(C_5R_5)Mn(CO)_2$ with H_2 depending on the nature of R (R = H, Me and Et).

Springer Proceedings in Physics, Vol. 74 Time-Resolved Vibrational Spectroscopy VI
Editors: A. Lau, F. Siebert, W. Werncke

Table1: The Second Order Rate of Reaction of $RM(CO)_x$(n-heptane) with CO.

$RM(CO)_x$	Rate / $dm^3mol^{-1}s^{-1}(x10^5)$
$(\eta^5C_5H_5)V(CO)_3$	130
$(\eta^6C_6H_6)Cr(CO)_2$	23
$(\eta^5C_5H_5)Mn(CO)_2$	8

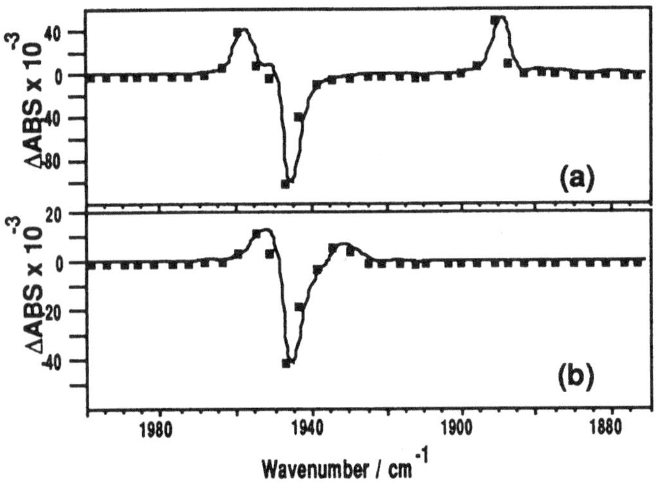

Figure 2: The TRIR spectrum of $CpMn(CO)_3$ in polyethylene (250 μm disc, $[CpMn(CO)_3] = 1.6x10^{-2}M$) in the presence of 800 PSI of CO: (a) 100 ns and (b) 50 μs after the UV flash.

GUR 415, a type of polyethylene which is very low in double bond content. We have also studied reactions under a high pressure of CO to make the reactions reversible and we expected only to be studying the reaction with the unsaturated intermediates with CO. The TRIR spectra obtained from photolysis of $(\eta^5C_5H_5)Mn(CO)_3$ in polyethylene are shown in figure 2.

The spectrum which is obtained after 100ns is identical to that recorded in n-heptane for $CpMn(CO)_2$-n-heptane, where n-heptane is weekly coordinating and is acting as a token ligand. The spectrum in polyethylene after 100 ns is assigned to be $CpMn(CO)_2$ weakly coordinated to polyethylene (probably through an agostic C-H bond). The spectrum after 50 μs shows that $CpMn(CO)_2$-polyethylene complex does not completely return to the starting $CpMn(CO)_3$ but forms a second species with bands which can be assigned to the dimeric species $Cp_2Mn_2(\mu CO)(CO)_2$ by comparison with the solution data from TRIR studies in n-heptane[3].

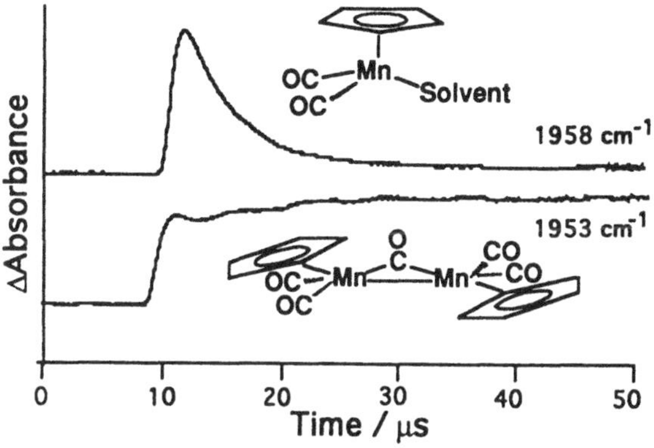

Figure 3: TRIR traces at 1953 and 1958 cm^{-1} after photolysis of $CpMn(CO)_3$ in polyethylene.

$Cp_2Mn_2(\mu CO)(CO)_2$ is formed by reaction of $CpMn(CO)_2$ with $CpMn(CO)_3$. Figure 3 shows TRIR traces assigned to $CpMn(CO)_2$ and $Cp_2Mn_2(\mu CO)(CO)_2$.

The band at 1958 cm^{-1} is due to the intermediate $CpMn(CO)_2$ while the band at 1953 cm^{-1} has overlapping contribution from the intermediate $CpMn(CO)_2$ and the dimer $Cp_2Mn_2(\mu CO)(CO)_2$. The growth of the dimer $Cp_2Mn_2(\mu CO)(CO)_2$ can be measured at a position where there is no overlap of bands (not shown). Using the starting concentration of $CpMn(CO)_3$ it is possible to estimate a second order reaction rate for $CpMn(CO)_2$ reacting with $CpMn(CO)_3$. Table 2 shows the rates of formation of $Cp_2Mn_2(\mu CO)(CO)_2$.

The rate in polyethylene is similar to that in n-heptane and cyclohexane. It might be expected that because polyethylene is more viscous than n-heptane or cyclohexane the rate obtained might be slower than the conventional alkane solvents. Actually the calculated rate constant appears to be significantly faster. This may be very informative about the polymer structure. The concentration of polyethylene was calculated using the infrared and UV extinction coefficients. The rate of formation of $Cp_2Mn_2(\mu CO)(CO)_2$ depends on the concentration of $CpMn(CO)_3$. On a microscale polyethylene is not homogeneous with a crystalline region that is not impregnated and an amorphous region which is impregnated. Although we have shown by scanning FTIR-microscopy that impregnation is uniform throughout the polymer the *local* concentration will be higher than calculated. Therefore the rate of formation of $Cp_2Mn_2(\mu CO)(CO)_2$ will depend on the crystallinity of the polymer

The use of polyethylene as a solvent for TRIR has produced new and interesting results. We have found that we could obtain higher concentrations of reactants, compared to conventional alkane solvents, such that reactions could be studied under conditions that were not normally accessible. In polyethylene, reactions can also be studied from 20 K to room temperature We have shown that polyethylene is a very convenient solvent for TRIR. As well as providing information on reactions in novel environments, the TRIR measurements in polyethylene may provide information about the polymer micro-structure.

Table 2: Showing the rate constants for the reaction of $CpMn(CO)_2$ with $CpMn(CO)_3$.

Solvent	Rate ($dm^3 mol^{-1} s^{-1}$) ($\times 10^6$)
Hostalen GUR 415	15.0
n-Heptane	3.4
Cyclohexane	1.3

We Drs. S.M.Howdle, F.P.A.Johnson, S.M.Nikiforov and Mr. M.J.Clarke for their help and advise. We thank the Royal Society Paul Instrument Fund, SERC and Nato for their support.

References:

1. M.Poliakoff, E.Weitz, Adv. Organomet. Chem. **25**, 277, (1986). F-W.Grevels, W.E.Klotzbücher, K. Schaffner Chem. Rev. in press.
2. F.P.A..Johnson, M.W.George, V.N.Bagratashvili, L.N.Vereshchagina, M.Poliakoff, Mendeleev Commun. **1**, 26, (1991).
3. B.S.Creaven, A.J.Dixon, J.M.Kelly, C.Long, M.Poliakoff, Organometallics **6**, 2600, (1987).

Vibrational IR-Multiphoton Excitation of Small Sulfoxides: Time-Resolved IR-Spectroscopic Detection of Dissociation Products

H. Groß, Y. He, M. Quack, and G. Seyfang

Laboratorium für Physikalische Chemie, ETH Zürich (Zentrum), CH-8092 Zürich, Switzerland

We report the time-resolved observation of the infrared laser induced dissociation of C_2H_4SO, $(CH_3)_2SO$ and $(CD_3)_2SO$ after IR-multiphoton excitation of the SO chromophore at wavenumbers around 1080 cm^{-1}. The CO_2 laser pulse lengths and time resolution of the detection scheme are both between 1 and 5 ns, also longer for some experiments. The collisionless IR-photochemical main primary products have been shown to be $SO(^3\Sigma^-)$ in all molecules together with C_2H_4 for C_2H_4SO and CH_3 for $(CH_3)_2SO$. For C_2H_4SO unimolecular decay constants of $k \cong 10^7$ to 10^8 s^{-1} after excitation above threshold are obtained. The dissociation process of $(CH_3)_2SO$ can be described by a two-step mechanism with the intermediate CH_3SO.

1. Introduction

The direct time-resolved detection of reactant and product quantum states with high resolution spectroscopic techniques is one of the central goals in infrared multiphoton excitation and laser chemistry [1]. We have over the years developed experimental schemes for this purpose and applied them to detection of HCl, CF_3I, CF_3 and I atoms among others [2 – 5]. Time-resolved IR-laser detection has also been successfully used by others in different contexts [6]. In the present work we have been interested in time-resolved spectroscopy of reactants and products of the infrared laser chemistry of sulfoxides. Apart from the relevance to organic and atmospheric chemistry [7, 8], the small organic sulfur compounds are of special interest due to their pyramidal configuration at the sulfur atom. This allows the construction of optically active chiral molecules and consequently, they have been proposed as candidates for investigating the influence of the parity-violating weak nuclear interaction on the IR-laser chemical racemization [9].
The SO chromophore in sulfoxides exhibits strong absorption near 1100 cm^{-1} in the higher frequency range of the CO_2 laser, with an integrated band strength of about 2 pm^2, thus providing the possibility to investigate IR-laser induced reactions on the basis of the chromophore principle [10]. The smaller sulfoxides are not too well characterized spectroscopically and kinetically. Therefore predictions of reaction channels and products for IR-laser induced processes can only be made on the basis of thermochemical data [11, 12] and quantum chemical calculations [13]. We have shown here by ns time-resolved vibrational spectroscopy of SO, CH_3 and C_2H_4 that the following reactions are the major primary laser induced chemical processes under collision free conditions:

$$C_2H_4SO \xrightarrow{nh\nu} C_2H_4 + SO(^3\Sigma^-) \tag{1}$$

$$(CH_3)_2SO \xrightarrow{nh\nu} CH_3SO + CH_3 \xrightarrow{nh\nu} 2CH_3 + SO(^3\Sigma^-) \tag{2}$$

2. Experimental

Figure 1 shows the experimental setup. The IR-excitation pulses are produced in a line tunable CO_2 oscillator-amplifier configuration allowing for quantitative pulse shaping control by means

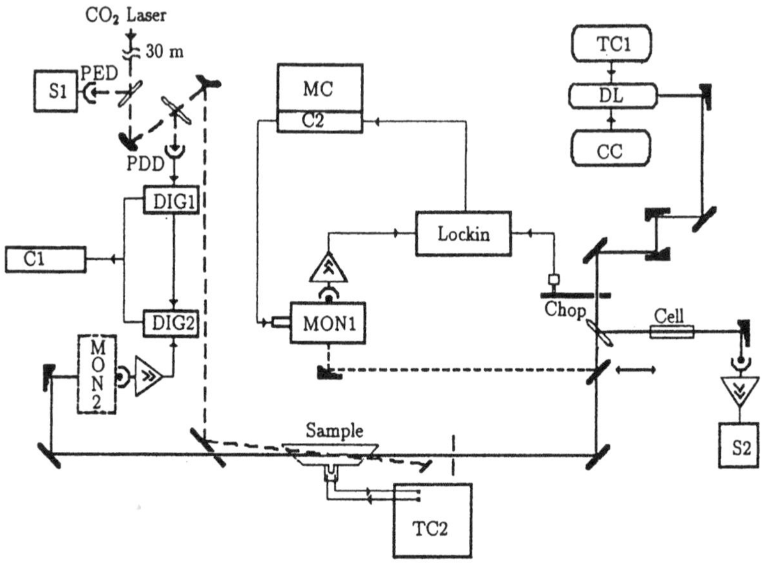

FIGURE 1: Experimental setup with PED: pyroelectric detector, S 1, 2: scope, PDD: photon-drag detector, DIG 1, 2: digitizer (Tektronix 7912AD), C 1, 2: computer, MON 1, 2: monochromator, MC: mode chart, TC 1, 2: temperature control, CC: current control, DL: diode laser (or alternatively a cw CO_2 laser). The pulsed CO_2 laser beam is shown as dashed line, the cw probe beam from the diode laser or cw CO_2 laser as full line. They overlap in the sample cell (Sample). A small part of the probe laser beam goes to a reference (Cell). A second dashed line shows the opportunity of controlling the frequency and intensity of the probe laser beam. A fast infrared detector (SAT) is placed directly after the second monochromator (MON 2). Collection and evaluation of the Measurements is done by the digitizers (DIG) and computers (C).

of unstable resonator optics, intracavity absorbers and a CdTe electrooptical switch [14]. This provides numerous pulse shaping options besides multimode (MM), single mode (SM) and mode-locked (ML) pulse trains with durations of pulses between 1 ns and 200 ns. We have used various lines of the 9R branch of the CO_2 laser with wavenumbers around 1080 cm^{-1}, near the strong SO-stretching fundamental absorption of the sulfoxides. A tunable diode laser or a cw CO_2 laser have been used to probe reactants and products about colinearly with the excitation beam in a cell of 30 cm lengths. The time resolution of 1 to 2 ns is determined by our combination of a fast IR detector, various preamplifiers and a transient digitizer. To avoid complications from wall adsorption effects the experiments have been performed at a constant vapour pressure over a thermostated liquid reservoir for the sulfoxides.

Further experimental details on our techniques for time-resolved vibrational spectroscopy during and after IR-multiphoton excitation are described in [2 – 5].

3. Results and Discussion

The dissociation process for C_2H_4SO has been followed by probing the nascent C_2H_4 on the ν_7 fundamental near 950 cm^{-1} and on different rovibrational transitions of $SO(^3\Sigma^-)$ and $SO(^1\Delta)$. Using CO_2 laser pulses with a duration of less than 5 ns the dissociation process could be separated from the excitation. A unimolecular decay constant of $(10 \pm 1) \times 10^7$ s^{-1} could be fitted to the experimental data (see figure 2).

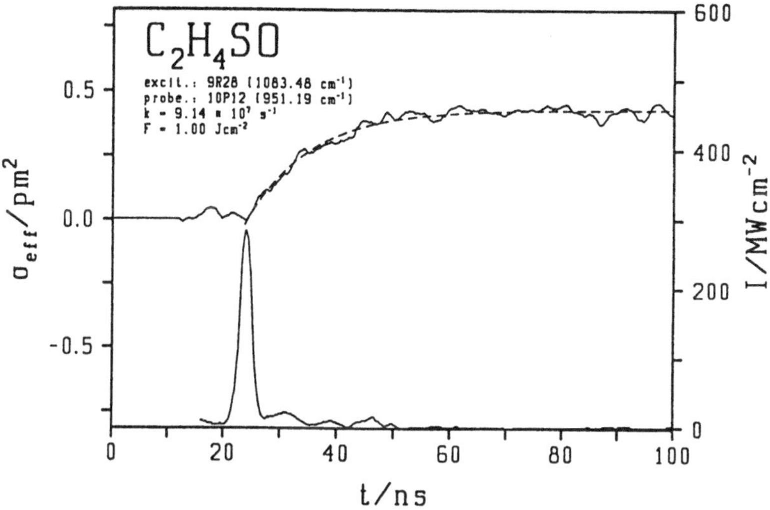

FIGURE 2: Time-dependent C_2H_4 absorption signal after excitation of C_2H_4SO with a pulsed CO_2 laser at $\tilde{\nu}= 1083.48\ cm^{-1}$ ($F = 1\ Jcm^{-2}$), probe line of cw CO_2 laser $\tilde{\nu}= 951.19\ cm^{-1}$. The lower trace shows the shape of the excitation pulse (FWHM ≈ 2.5 ns) with time dependent intensity I given by right hand side ordinate for a fluence of $F = 1\ Jcm^{-2}$.

For longer CO_2 laser pulses most of the dissociation takes place during the irradiation time. The measured effective absorption cross section σ_{eff} is defined by equation (3)

$$\sigma_{eff}(t > t_p) = (C \cdot l)^{-1} ln\,[I_0/I(t > t_p)] \tag{3}$$

where C is the reactant concentration before excitation (in molecules cm^{-3}) and l the length of the absorption path, $I(t>t_p)$ the transmitted intensity at some time after the pulse duration t_p, I_0 the incident intensity. The data for different SO rovibrational transitions are summarized in table 1. A rotational temperature $T_{rot} = (750 \pm 200)$ K could be deduced for the vibrational ground state. An approximate estimate for the vibrational temperature for $SO(^3\Sigma^-)$ is $T_{vib} = (850 \pm 100)$ K. The results in table 1 show that the spin allowed dissociation leading to $SO(^1\Delta)$ is of minor importance, if any.

Similar time dependent signals have been measured by following the $(CH_3)_2SO$ and $(CD_3)_2SO$ dissociation on various rovibrational transitions of $SO(^3\Sigma^-)$ and CD_3, using different laser diodes. Higher rotational levels are predominantly populated and no vibrational temperature could be determined, yet. Like in the dissociation of C_2H_4SO the contributions from the reaction channel leading to $SO(^1\Delta)$ is negligibly small.

The direct detection of CD_3 after multiphoton excitation of $(CD_3)_2SO$, together with results from static yield measurements [4] leads to the conclusion that $(CD_3)_2SO$ is dissociated in a two-step mechanism with CD_3SO as intermediate. The CD_3SO radical could not yet definitly be identified specroscopically. Further work on this question is in progress.

TABLE 1: Effective absorption cross section σ_{eff} (defined in equation (3)) on different rovibrational transitions of SO after multiphoton dissociation of C_2H_4SO (For nomenclature see [4], for pulse shapes see experimental section)

SO transition			$\tilde{\nu}_{SO}/cm^{-1}$ [15, 16]	pulse shape	F/Jcm^{-2}	σ_{eff}/pm^2
$^3\Sigma^-$	(v=0 → 1)	R(7,6)	1147.8004	MM	3.08	18.4
		R(9,10)	1152.9865	MM	2.97	20.4
		R(15,14)	1158.1544	MM	3.21	31.8
		R(15,14)	1158.1544	SM	1.89	22.4
		R(15,14)	1158.1544	ML	1.70	22.4
$^3\Sigma^-$	(v=1 → 2)	R(13,13)	1154.0552	MM	3.13	7.3
		R(15,16)	1147.7270	MM	3.10	8.5
		R(15,16)	1147.7270	SM	2.85	9.4
		R(15,16)	1147.7270	ML	2.85	9.8
		R(17,16)	1147.7422	MM	3.26	9.8
		R(25,25)	1158.2659	MM	2.95	6.1
$^3\Sigma^-$	(v=2 → 3)	R(26,27)	1147.5227	MM	3.47	≈ 1.7
		R(27,27)	1147.5475	MM	3.38	≈ 1.7
$^1\Delta$	(v=0 → 1)	R(38)	1147.1185	MM	3.40	≤ 1.5

References

[1] M. Quack: Adv. Chem. Phys. **50** 395 (1982)
[2] P. Dietrich, M. Quack and G. Seyfang: Faraday Disc. Chem. Soc. **82** 280 (1986), Chem. Phys. Lett. **167** 535 (1990)
[3] M. Quack, R. Schwarz and G. Seyfang: J. Chem. Phys. **96** 8727 (1992)
[4] H. Groß, Y. He, M. Quack, A. Schmid and G. Seyfang: Chem. Phys. Lett (1993) in press.
[5] Y. He, M. Quack, R. Ranz and G. Seyfang: to be published
[6] R.E. Weston and G. Flynn: Annu. Rev. Phys. Chem. **43** 559 (1992), J.I. Steinfeld: at this conference (1993)
[7] E.S. Saltzman and H.J. Cooper eds.: *Biogenic sulfur in the environment*, ACS Symposium Series **393** (Am. Chem. Soc. 1989)
[8] M.O. Andreae: Limnol. Oceanogr. **25** 1054 (1980)
[9] M. Quack: Angew. Chem. (Int. Ed. Engl.) **28** 571 (1989)
[10] M. Quack and H.J. Thöne: Ber. Bunsenges. Phys. Chem. **87** 582 (1983)
[11] G.E. Hartzell and J.N. Paige: J. Am. Chem. Soc. **88** 2616 (1966)
[12] S. Saito: Bull. Chem. Soc. Jap. **42** 667 (1969)
[13] H. Groß, T.K. Ha, M. Quack and G. Seyfang: to be published.
[14] M. Quack, C. Rüede and G.Seyfang: Spectrochim. Acta **46A** 523 (1990)
[15] J.B. Burkholder E.R. Lovejoy, P.D. Hammer, C.J. Howard and M. Mizushima: J. Mol. Spectrosc. **124** 379 (1987)
[16] M. Wong, T. Amano and P. Bernath: J. Chem. Phys. **77** 2211 (1982)

Energy-Transfer Processes in CF_2HCl and CF_2Cl_2 at Higher Pressures

W. Strube, J. Wollbrandt, M. Rossberg, and E. Linke

Forschungsgruppe Laser-Photochemie bei KAI e.V./WIP,
Rudower Chaussee 5, D-12484 Berlin, Germany

Abstract: The unimolecular decomposition of the halomethanes CF_2HCl and CF_2Cl_2 was studied under high IR laser fluence ($\Phi < 200$ J/cm^2) and high pressure conditions ($p < 2$ kPa). Under the given conditions considerable contributions of intermolecular energy transfer processes on dissociation behavour were found.
The rotational and vibrational excitation of the CF_2 dissociation fragments, as monitored by laser-induced fluorescence (LIF), was used as a probe for the energy content of the parent molecules. The observed strong post-pulse-dissociation was attributed to intermolecular vibrational energy transfer between undissociated, highly vibrationally excited parent molecules in the region also outside the focus core. This result supports the energy pooling model.

1. Introduction

The trend towards applications of IR-laser induced unimolecular reactions based on the multiphoton dissociation effect (IRMPD) requires studies of molecular processes occuring under conditions of high pressure and fluence (focussed radiation). As a result, irradiation geometry, inhomogeneous fluence field, and locally varying absorption make necessary measurements of the local reactant and/or product concentration at well defined fluences in space and time. To this end a special laser-induced fluorescence (LIF-)techniqe developed in our laboratory was employed for spatially resolved kinetic measurements of fragment concentrations in the laser focus and its neighbourhood. A description of this method is given elsewhere [01]. Apart from the usual concentration measurements, radial scans of a dye laser probe beam across the focus zone of the IR laser permit determination of the total amount $M(t) = 2\pi\int[M(r,t)]rdr$ of a species in the given cylindrically symmetrical configuration. Furthermore, for the first time fluence profiles in the focal region can be directly related to the dissociation yield distribution.

2. Results and Discussion

The most striking result was that for pressures above 100 Pa the maximum dissociation yield occurs spatially and temporally separate from the input of IR radiation. Especially for higher pressures the amount of CF_2 product observed after the end of the laser pulse can reach multiples of the CF_2 photolytically produced. (Fig: 1). Due to a favourable exchange rate between CF_2 and the freon molecules [02], the rotational and vibrational excitation of the CF_2 radicals, as monitored by LIF, was used as a probe to study the excitation of the parent molecules.

While the rotational excitation rapidly equilibrates with translation, such that R/T-temperature reaches room temperature values within about 40 µs ($p = 800$ Pa), the very slow decay of the vibrational temperature ($k_{VT} = 1.4 \cdot 10^{10}$ cm^3/(mol$\cdot$s)) reflects the high energy

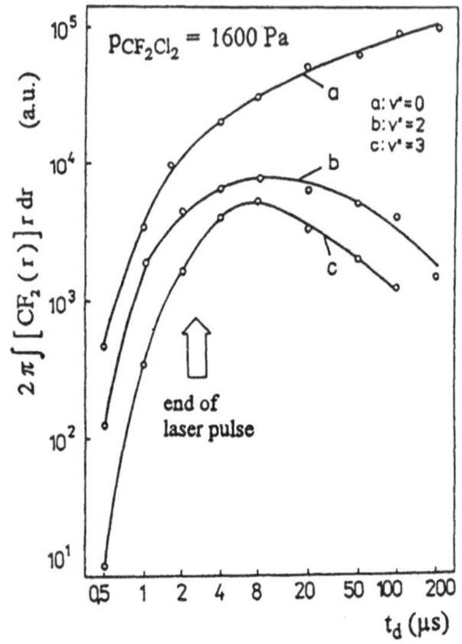

Fig. 1:
Total amount of $\tilde{X}^1A_1$ CF_2 (v" = 0, 2, 3) produced from CF_2Cl_2 in the center of the IR laser focus (t_d - delay time between IR laser pulse and probe laser pulse)

content of the vibrational energy reservoir, with which the CF_2 interacts. The measured maximum temperatures T_V exceed the vibrational excitation of the nascent CF_2 products observed under similar fluence conditions in the collisionless case [03]. (T_V = 2200 K vs. 1160 K for CF_2Cl_2 and T_V = 1600 K vs. 1050 K for CF_2HCl).

Spatially resolved CF_2 concentration measurements show that the absorption of the IR quanta leads to a complete dissociation (IRMPD) only in the relatively small focus core, while in the large double-cone region of the focus the freon molecules are left highly vibrationally excited (IRMPE).

Since the input of external energy has ended, the post-pulse dissociation must result from a special collision-controlled activation process

$$M^* + M^* \Rightarrow M^{**} + M$$

In this so-called energy pooling process [04][05], the vibrational energy stored in highly excited freon molecules M^* is redistributed in such a way that one of the collision partners becomes overexcited (M^{**}), i.e. exceeds the dissociation threshold, while the other becomes deactivated to some extent (M). The decomposition can continue until eventually vibrational cooling of the "reservoir", mostly by v-T exchange, stops this process.

Since CF_2Cl_2 has two energetically accessible decay channels as compared to the one-channel CF_2HCl, both substances exhibit different dissociation behaviour. A detailed discussion of the differences in dissociation behaviour between the two freons is given elsewhere [06].

References

[01] W. Strube, M. Rossberg, J. Wollbrandt, E. Linke, to be published in Rev. Sci. Instr.
[02] W. Strube, *Dissertation*, Humboldt-Universität Berlin, 1992, p. 111
[03] D.S. King, , Adv. Chem. Phys. **50**, 105(1982)
[04] I. Oref, J. Chem. Phys. 75, 131(1981)
[05] M. Ivanco, G.A. McRae, R.A. Back, J.W. Goodale, P.E. Lee, J. Chem. Phys. **96**, 5191(1992)
[06] W. Strube, J. Wollbrandt, M. Rossberg, E. Linke, to be published

Time-Resolved Infrared Spectroscopy of a CO_2 Reductant

J.R. Westwell, M.W. George, O. Ishitani, F.P.A. Johnson, and J.J. Turner*

Department of Chemistry, University of Nottingham,
Nottingham NG7 2RD, UK
*Permanent address: NIRE, 16-3 Onogawa, Tsukuba, Ibaraki 305, Japan

Abstract. Fast Time-resolved Infrared Spectroscopy is used to monitor the initial steps in the photocatalytic reduction of CO_2 using $[Re(CO)_2(bpy)\{P(OEt)_3\}_2]^+Br^-$.

Time-resolved Infrared Spectroscopy provides a new approach to the study of the excited states of coordination compounds [1, 2]. Flash photolysis from a UV/visible laser combined with a continuous wave IR laser is used to build a point-by-point spectrum of the ν(CO) bands of transition metal carbonyls.

The structure and dynamics of the excited states of coordination compounds have largely been the province of Electronic spectroscopy and of Resonance Raman spectroscopy, whether applied to the ground state or, as with TR^3, to the excited state. Each of these techniques has its own limitations and it is clear that the IR spectra of molecules in their excited states would provide additional or, in some cases, unique information. Positions of ν(CO) IR bands, for example, provide an excellent indication of the oxidation state of the metal centre, hence TRIR provides a powerful probe for electron transfer.

Recent studies have shown that the reduction of CO_2 can be achieved with a variety of transition metal complexes, with suitable electron donors in certain solvents. Examples of such compounds include $Re(CO)_3(bpy)Cl$ [3] and $[Re(CO)_2(bpy)\{P(OEt)_3\}_2]^+Br^-$ [4]. The proposed mechanism [4] by which the latter compound reduces CO_2 is shown in Fig. 1. Stage 3 summarises a series of complex chemical steps.

Recent work at Nottingham has focussed on monitoring stage 1, the generation of the excited state $[Re(CO)_2(bpy)\{P(OEt)_3\}_2]^{+*}$, and stage 2, its subsequent

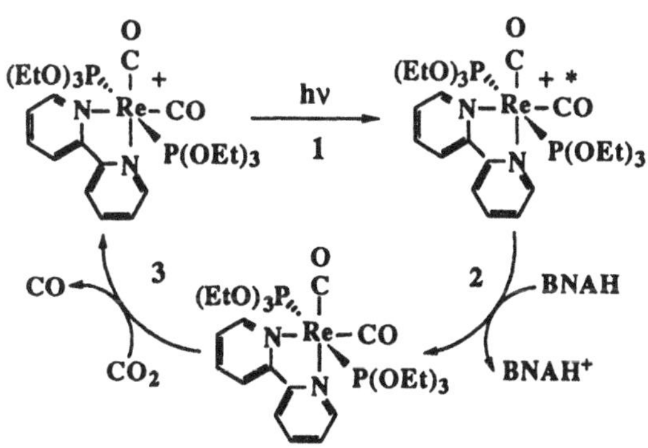

Figure 1: The catalytic reduction of CO_2 by $[Re(2,2'\text{-bpy})(CO)_2\{P(OEt)_3\}_2]^+$ in MeCN solution

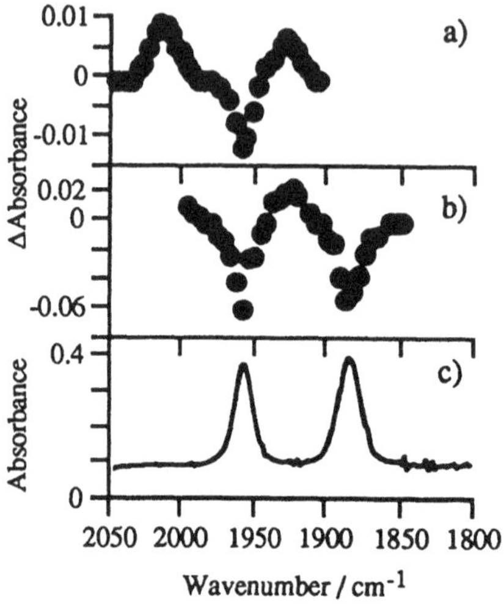

Figure 2: IR spectra in the ν(CO) region of $[Re(CO)_2(bpy)\{P(OEt)_3\}_2]^+Br^-$ in CH_2Cl_2 solution.
a) TRIR spectrum, using a diode laser, immediately after UV (308 nm, 20 ns) flash of $[Re(CO)_2(bpy)\{P(OEt)_3\}_2]^+Br^-$ ($5x10^{-4}$M) in CH_2Cl_2 under 2 atm Ar(g); path length 2 mm. Each point corresponds to a different IR laser wavelength.
b) As above, except a CO laser replaced the diode laser.
c) Fourier transform IR of $[Re(CO)_2(bpy)\{P(OEt)_3\}_2]^+Br^-$ ($5x10^{-4}$M) in CH_2Cl_2; path length 2mm.

reduction to the neutral species $Re(CO)_2(bpy)\{P(OEt)_3\}_2$. For example, the TRIR spectrum obtained for the generation of the excited state is shown in Fig. 2. Upward peaks represent generation of the excited state; downward peaks indicate depletion of the ground state. The increases in frequency in the excited state (1956 to 2014 cm^{-1} and 1883 to 1927 cm^{-1}) arise because in the MLCT excited state, the formal oxidation state of the metal increases. Future studies will concentrate on the reduction of CO_2 by the neutral species $Re(CO)_2(bpy)\{P(OEt)_3\}_2$ (stage 3).

[1] P. Glyn, M. W. George, P. M. Hodges, and J. J. Turner, J. Chem. Soc., Chem. Commun. 1655 (1989); P. Glyn, F. P. A. Johnson, M. W. George, A. J. Lees, and J. J. Turner, Inorg. Chem. **30**, 3543 (1991); J. J. Turner, M. W. George, F. P. A. Johnson, and J. R. Westwell, Coord. Chem. Rev. **125**, 101 (1993).
[2] S. K. Doorn, K. C. Gordon, R. B. Dyer, and W. H. Woodruff, Inorg. Chem. **31**, 2284 (1992).
[3] J. Hawecker, J.- M. Lehn, and R. Ziessel, J. Chem. Soc., Chem. Commun. 536 (1983).
[4] C. Pac, S. Kaseda, K. Ishii, S. Yanagida, and O. Ishitani, in "Photochemical Processes in Organized Molecular Systems", ed. K. Honda, Elsevier Science Publishers B.V (1991).

Short-Wavelength Flash-Photolytic Fragmentation of $Ru_3(CO)_{12}$ in the Presence of CO and Complementary Experiments with $Ru(CO)_5$: Transient IR-Spectroscopic Evidence for $Ru_3(CO)_{13}$

F.-W. Grevels, W.E. Klotzbücher, K. Schaffner, and J. Schrickel

Max-Planck-Institut für Strahlenchemie, Stiftstrasse 34–36,
D-45470 Mülheim a.d. Ruhr, Germany

Abstract. $Ru_3(CO)_{12}(\mu\text{-}CO)$ is shown to be a pivotal transient in the photofragmention of $Ru_3(CO)_{12}$ in the presence of added CO.

The photolytic behaviour of $Ru_3(CO)_{12}$ in the presence of CO or other prospective ligands **L** is known to depend on the wavelength of irradiation, the nature of **L**, the solvent and the temperature. In essence two types of processes were observed: cluster breakdown initiated by long–wavelength irradiation and dissociative CO–photosubstitution upon short–wavelength excitation. Our recent investigations establish a link between these two reactivity patterns.

Flash photolysis (308 nm) of $Ru_3(CO)_{12}$ in hydrocarbon solution at ambient temperature generates the $Ru_3(CO)_{11}$ fragment in the form of its CO–bridged isomer, the complete CO stretching vibrational pattern of which is recorded using our globar–based time–resolved IR instrumentation. Under pure argon or in the presence of CO in low concentration (up to 10^{-3} M) the parent $Ru_3(CO)_{12}$ is largely recovered. Further increase in CO concentration not only reduces the lifetime of $Ru_3(CO)_{11}$ below our apparatus limit of 2 μsec, but, quite unexpectedly, renders the system no longer reversible ! High CO–concentration favours, in competition with $Ru_3(CO)_{12}$ regeneration, the formation of a new transient species assigned as $Ru_3(CO)_{12}(\mu\text{-}CO)$. It decays with $k=2.2\cdot10^4$, ultimately yielding $Ru(CO)_5$ and $Ru_2(CO)_9$.

Complementary investigation into the photolytic behaviour of $Ru(CO)_5$ reveals that its known conversion to $Ru_3(CO)_{12}$ is strongly suppressed in the presence of added CO. In detail, flash photolysis (308 nm) of $Ru(CO)_5$ in hydrocarbon solution at 15–20 deg below ambient temperature generates the

$Ru(CO)_4$·solv fragment, with a lifetime two orders of magnitude shorter than the analogous $Fe(CO)_4$·solv species. In competition with $Ru(CO)_5$ regeneration, the dinuclear product $Ru_2(CO)_9$ is formed. Secondary excitation of this reaction mixture ('double–flash experiment') results in the transient appearance of the CO–stretching vibrational pattern associated with the above $Ru_3(CO)_{12}(\mu\text{–}CO)$ species, which in turn dissociates into $Ru(CO)_5$ and $Ru_2(CO)_9$. Both, the photo–induced depletion of $Ru_2(CO)_9$ and its formation from the $Ru(CO)_4$·solv reaction with $Ru(CO)_5$, are observable and clearly discernible by using C_6D_6–doped solvent – in this way the latter process is delayed by two orders of magnitude.

Microsecond Time-Resolved Infrared Study of the Electric-Field-Induced Reorientation of Nematic Liquid Crystals: Collective and Non-Collective Motions of nCB

T.I. Urano[1], *Y. Mori*[1], *and H. Hamaguchi*[2]

[1]Toshiba Corporation, Research and Development Center,
1, Komukai Toshiba-cho, Saiwai-ku, Kawasaki 210, Japan
[2]Kanagawa Academy of Science and Technology (KAST),
KSP East 301, 3-2-1 Sakato, Takatsu-ku, Kawasaki-shi 213, Japan

Under an applied electric field, liquid crystal (LC) molecules reorientate due to the interaction between their electric anisotropy and the electric field. In order to understand the mechanism of this LC response to an applied electric field, which is not well known at the molecular level, information on the time course of the reorientation is necessary. Polarized time-resolved infrared spectroscopy is most suitable for obtaining this class of information.

We selected 4-n-alkyl-4'-cyanobiphenyl (hereafter abbreviated as nCB. n : carbon number in the alkyl chain) as the sample and investigated the time response of this LC molecule to pulsed applied electric fields. Generally speaking, a LC molecule is composed of two distinct sub units, the rigid core part and the flexible part. In the case of nCB, the cyanobiphenyl group corresponds to the rigid core part and the n-alkyl group corresponds to the flexible part. The CN stretch and methylene anti-symmetric CH stretch modes were used as the probe to monitor the response of the rigid core part and the flexible part of nCB, respectively. By using a submicrosecond time-resolved infrared spectrometer based on a dispersive infrared monochromator (time-resolution : 500 nanosecond, sensitivity : 10^{-6} in the absorbance change), we observed a marked difference in the time response between the rigid core part and the flexible part when the absorbance change was smaller than 10^{-4} [1]. Contrary to the expectation that the flexible part is inert to the electric field, the flexible part appeared to move earlier both when applying and removing the electric field. The response of the flexible part consisted of two components. One was the same as that of the rigid part which corresponds to the reorientation of the director. The other was due to the internal rotation around the C-C bond in the alkyl chain.

In Figure 1, the time response of (A) the flexible part, (B) the rigid core part of 5CB are shown for a pulsed rectangular electric field (pulse width : 2 ms, voltage : 14 V, repetition rate : 5 Hz). Both response curves are normalized at t = 5 ms. The difference (C) corresponds to the extra motion of the flexible part which is due to the internal rotation of the C-C bond of the n-alkyl group. The curve B shows a delay in the rise (an inertia effect) and represents a collective motion of the bulk LC molecules. The curve C shows exponential-like rise and decay and is most likely to correspond to some non-collective motion of the pentyl group through the internal rotation. This observation together with other experimental results such as the dependence of the time response of each part on the repetition rate of the electric field

Editors: A. Lau, F. Siebert, W. Werncke

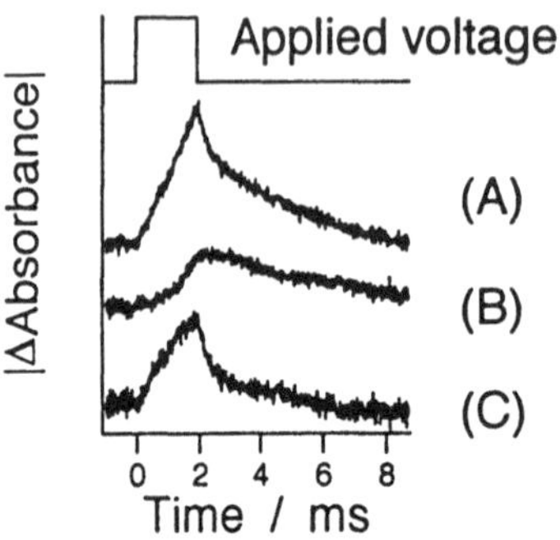

Figure 1. Time response of (A) the flexible part, (B) the rigid core part, and (C) the difference (A) - (B) of 5CB to a pulsed rectangular electric field. (A) and (B) are normalized at t = 5 ms.

[2], temperature, and the carbon number in the alkyl chain, are well interpreted in the following way. In the reorientation process, the application of the electric field induces first the perturbation of the LC molecules near the orienting membrane which is too small to be observed by the absorbance change in the CN stretch. This movement induces the absorbance change in the CH stretch through the internal rotation. Note that the internal rotation can be change the direction of the transition moment greatly without changing the direction of the molecular axis (director). After the induction period in which the environment of the bulk LC molecules is changing gradually, the bulk LC molecules start to reorientate in a cooperative manner. The internal rotation of the flexible part is also induced by this large movement. In this way, the time response of the LC molecules near the orienting membrane to an applied electric field can be observed with the aid of the time response of the flexible part which is effectively amplified by the internal rotation. The mechnism of the LC reorientation will be further studied on the basis of these experimental results.

[1] T. I. Urano and H. Hamaguchi, Chem. Phys. Lett. 195(4), 287 (1992).
[2] T. I. Urano and H. Hamaguchi, Appl. Spectrosc. in press.

Photoreduction of 2,2'-Bipyrimidine by Aliphatic Amines: Indirect Evidence of the Formation of a Very Short-Lived Ion Pair

O. Poizat[1] *and G. Buntinx*[2]

[1]Laboratoire de Spectrochimie Infrarouge et Raman, CNRS, 2, rue Henri Dunant, F-94320 Thiais, France
[2]Laboratoire de Spectrochimie Infrarouge et Raman, CNRS, USTLFA, Bât. C5, F-59655 Villeneuve d'Ascq, France

As a part of a general investigation of the photoreactivity of the imine function in aza-aromatic molecules [1-3], we present here a discussion of the photoreduction of 2,2'-Bipyrimidine (22BPMY) by amines in acetonitrile as studied by nanosecond transient absorption and resonance Raman scattering.

Excitation of 22BPMY at 248nm in acetonitrile or water yields the formation of a transient species characterized by a strong absorption at 340nm, the lifetime of which (ca. 20μs) is reduced notably by addition of oxygen. We ascribe it to the lowest triplet state T_1.

In the presence of 1,4-diazabicyclo(2.2.2)octane in acetonitrile, transient absorption measurements show the quenching of the T_1 state by the amine ($k_q = 2x10^{10}M^{-1}s^{-1}$) and the appearance of a new transient species identified by two absorptions at 380nm and in the 480-550nm range. We ascribe it to the radical anion $22BPMY^{-\cdot}$ produced by electron transfer from the amine. The corresponding Raman spectra probed at 370 and 532nm show specific resonance enhancements which are comparable to those reported for the biphenyl anion in resonance with its 405 and 630nm absorptions [4] and for the 4,4'-bipyridine anion in resonance with its 385 and 570/605nm transitions [2]. The observed Raman lines can be ascribed to the totally symmetric modes expected for a planar, D_{2h} conformation.

In the presence of triethylamine (TEA), T_1 22BPMY is also quenched by electron transfer ($k_q \sim 10^{10}M^{-1}s^{-1}$) but in this case the transient absorption spectrum which appears as the triplet spectrum decays is different from the anion spectrum and exhibits two bands at 355nm and in the 430-460nm range. This spectrum is assigned to the N-protonated radical 22BPMYH· for two reasons : (i) A similar spectrum is observed upon photolysis of 22BPMY in hydrogen atom donor solvents such as 2-propanol or cyclohexane and (ii) it is also observed, instead of the anion spectrum, on photoreduction of 22BPMY by DABCO in proton donor solvents (water, alcohols). On reduction by TEA, this radical results probably from protonation of the anion $22BPMY^{-\cdot}$ by the cation $TEA^{+\cdot}$, as it has been observed in the photo-reduction of ketones and of bipyridines [2] by TEA.

Raman spectra probed at 355nm after excitation at 248nm of 22BPMY in acetonitrile in the presence of TEA show exclusively bands due to vibrations of the 22BPMYH· radical, in agreement with the analysis by absorption. However, spectra probed at 370nm, i.e. in sharp resonance for the anion $22BPMY^{-\cdot}$ but in lower resonance for the radical, show the superposition of the anion and radical peaks.

More detailed information is thus obtained by Raman than by absorption because of the higher spectral resolution of the former technique (the anion is not detected by absorption as its spectrum is too similar to the radical spectrum and is probably hidden by it). The kinetic analysis of the anion and radical Raman lines indicates that the growth of the 22BPMYH· radical is much faster than the decay of the anion. This observation reveals that (i) the anion species observed by Raman is not the precursor of the radical and (ii) this radical is not produced via a diffusional process of ions recombination. These results suggest that the protonation of 22BPMY$^{-\cdot}$ by TEA$^{+\cdot}$ arises directly after the electron transfert, before the separation of the ions. This mechanism is thus assuming the existence of a short-lived ion pair which decays via two competiting processes : dissociation into free ions and intra-pair proton transfer. The former process prevails in the case of DABCO whereas the second is predominant in the presence of TEA, the cation of which is a better proton donor. This photoreduction process is summarized in the following scheme :

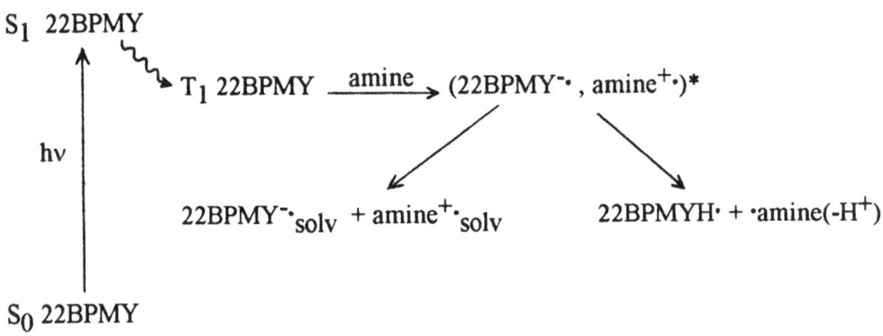

[1] O. Poizat, G. Buntinx, M. Ventura and M.F. Lautié, J. Phys. Chem. 95, 1245 (1991)
[2] G. Buntinx, P. Valat, V. Wintgens and O. Poizat, J. Phys. Chem. 95, 9347 (1991)
[3] O. Poizat, G. Buntinx, P. Valat, V. Wintgens and M. Bridoux, J. Phys. Chem. 97, 5905 (1993)
[4] V. Aleksandrov, Y.S. Bobovitch, Y.G. Maslov and A.N. Sidorov, Opt. Spectrosc. 38, 387 (1975)

The Benzyl Radical: Resonance-Raman Spectra and Quantum-Chemical Calculations: Evidence for Vibronic Coupling

R. Wilbrandt[1], *F.W. Langkilde*[1,*], *K. Bajdor*[2], *F. Negri*[3], *F. Zerbetto*[3], *and G. Orlandi*[3]

[1]Department of Environmental Science and Technology, Risø National Laboratory, DK-4000 Roskilde, Denmark
[2]Industrial Chemistry Research Institute, ul. Rydygiera 8, PL-01-793 Warsaw, Poland
[3]Dipartimento di Chimica "G. Ciamician", University of Bologna, I-40126 Bologna, Italy
*Present address: Astra Hässle AB, S-43183 Mölndal, Sweden

The benzyl radical $C_7H_7^{\cdot}$ is one of the simplest and most basic aromatic free radicals and is of importance in many chemical reactions in atmospheric chemistry and combustion processes. A characterization of the potential energy surfaces in its ground and excited states is therefore of great interest.

Time-resolved resonance Raman (RR) spectra of the benzyl radical and its perdeutero isotopomer are presented[1]. The radicals are created by laser flash photolysis (λ = 266 nm) of

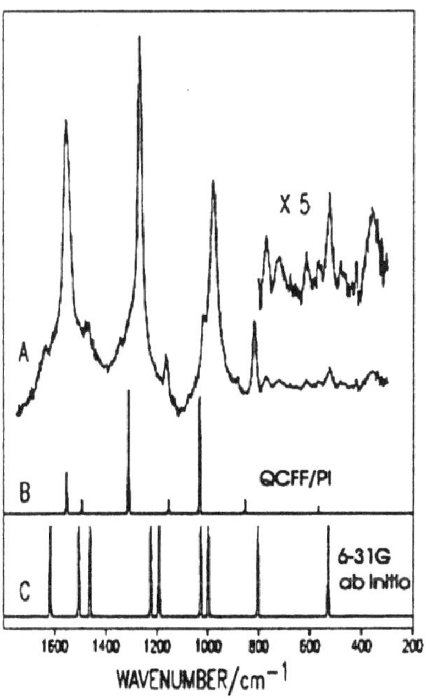

Figure 1 A) Resonance Raman spectrum of the benzyl radical produced by photolysis of 0.009 M benzylchloride in methanol. Pump and probe wavelengths are 266 and 315 nm. B) and C) Calculated spectra.

benzylchloride in solution. The spectra are excited at a wavelength of 315 nm, in resonance with the intense D_3 (2^2A_2) $\leftarrow$ D_0 (1^2B_2) transition. Twenty Raman bands, both polarized and depolarized, are observed. The RR spectra are analyzed through quantum chemical (QCFF/PI and *ab initio*) calculations of equilibrium geometries, vibrational frequencies, Franck-Condon factors, and vibronic coupling. Polarized intense bands are assigned to totally symmetric a_1 modes and depolarized bands to fundamentals of b_1 modes. The observed activity of b_1 modes suggests vibronic coupling, which is confirmed theoretically by calculations of vibronic interactions between the D_3 (2^2A_2) and D_5 (4^2B_2) states. The results from semiempirical and *ab initio* calculations are compared with experiment. The contributions of the different internal coordinates to the normal modes are calculated together with the changes in the molecular structure upon electronic excitation from D_0 to D_3 and D_5.

[1]A preliminary report of this work has been published: F.W. Langkilde, K. Bajdor and R. Wilbrandt, "Resonance Raman study of the benzyl radical", Chem. Phys. Letters, 193, 169-175, (1992)

Time-Resolved FTIR-Studies of H-Ras P21 GTPase Activity

K. Gerwert[1], *V. Cepus*[1], *A. Scheidig*[2], *and R.S. Goody*[2]

[1]Institut für Biophysik, Ruhruniversität Bochum,
ND 04.596, D-44780 Bochum, Germany
[2]Max-Planck Institut für molekulare Physiologie,
Postfach 102664, D-44026 Dortmund, Germany

The small GTP-binding protein H-ras p21 plays a central role in signalling leading to cell-growth (1). P21 appears to act as a switch: in the GTP bound form a signal is given to an effector molecule which is part of a cascade leading to cell proliferation and differentiation. The intrinsic GTPase activity is accelerated by GAP (GTPase activating protein). The transition from the active GTP bound to the inactive GDP bound form is accompanied by a conformational change which has been characterized by time-resolved x-ray structure analysis (2). In order to specify in more detail the structural changes occurring in the GTP to GDP reaction, we have begun complementary investigations by time resolved FTIR difference spectroscopy.

The reaction is studied using caged GTP complexed to p21 instead of GTP to prevent hydrolysis. Using an intense UV flash (308nm, 100mJ) the cage group is photolysed, GTP released and GTPase activity initiated. In fig. 1, a difference spectrum between p21.caged GTP and p21.GTP is shown. Characteristic bands for the caged GTP to GTP reaction are at 1526 cm^{-1} and 1343 cm^{-1} representing disappearance of the NO_2 of the caged group during photolysis. The region between 1300 and 900 cm^{-1} is dominated by phosphate bands.

In fig. 2, a 3-D graph of the difference spectra betweeen spectra of p21.GTP, taken directly after photolysis of caged GTP, and spectra appearing during GTP hydrolysis are shown.

The difference band at 1262/1236 cm^{-1} is indicative of the GTP to GDP reaction as our experiments with model systems and isotopically labelled compounds show (data not shown). In fig. 3 the control experiment with p21.caged GDP, is shown. After photolysis, no reaction is seen. Having now determined reliably the IR

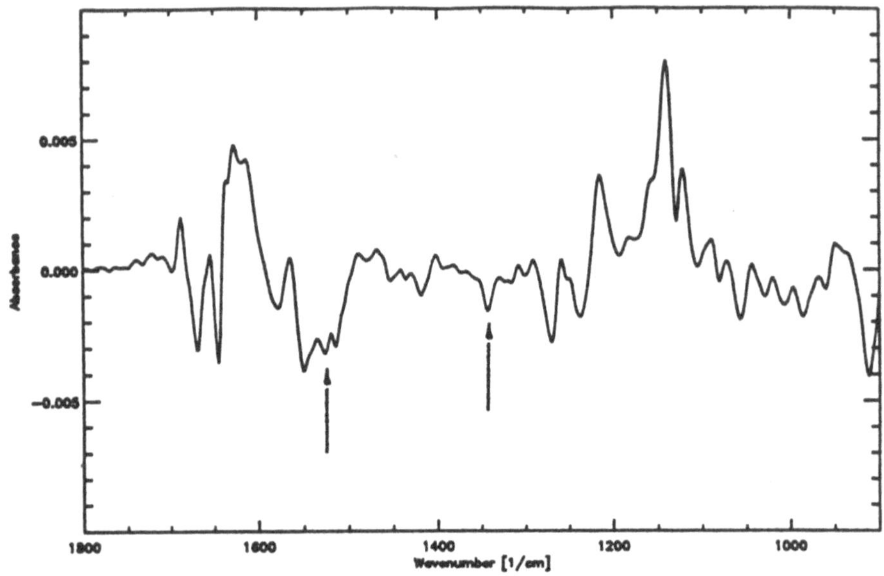

FIG. 1: The IR absorbance changes accompanying the p21.caged GTP (negative bands) to p21.GTP (positive bands) photolysis reaction are shown

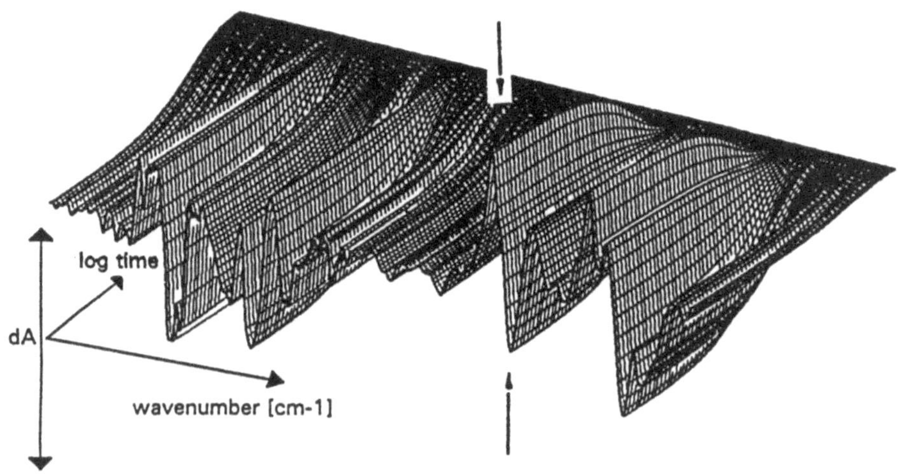

FIG. 2: 3-D graph of IR absorbance changes between 1800 and 900 cm^{-1} during the p21.GTP to p21.GDP reaction. The time axis is logarithmic and the overall time 30 min. Positive disappearing bands indicate disappearance of p21.GTP and negative disappearing bands rising of p21.GDP-bands

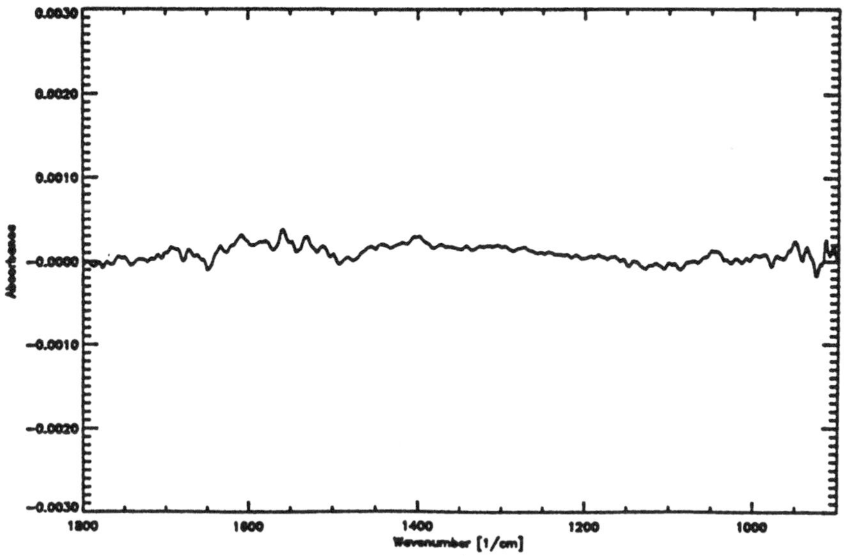

FIG. 3: Control experiment starting with caged GDP: after photolysis of p21.caged GDP, as expected no reaction is observed; the control indicates the baseline quality

absorbance changes, in further experiments assignments of bands to specific molecular groups will be performed, which will allow specification of the structural changes at the atomic level.

References:

1 Wittinghofer, A. and Pai, E. , TIBS, 1991, 16, 382-387

2 Schlichting, I. , Almo S. , Rapp, G. , Wilson, K. , Petratos, K. , Lentfer, A. , Wittinghofer, A. , Kabsch, W. , Pai, E. , Petsko, G. , Goody, R. S. , Nature, 1990, 345, 309-315

Part V

Photobiology

Vibrational Dynamics in Condensed Phases and Proteins

R.M. Hochstrasser, B.R. Cowen, P.L. Dutton, C. Galli, S. LeCours, S. Maiti, C.C. Moser, D. Raftery, M. Therien, G. Walker, and K. Wynne

Department of Chemistry and Department of Biochemistry and Biophysics, University of Pennsylvania, Philadelphia, PA 19104, USA

Abstract. Ultrafast processes in condensed phase are explored with transient vibrational spectroscopy. Examples include chemical reactions, protein dynamics and energy transfer.

Direct interrogation of molecular vibrations by infrared or Raman scattering is complementary to optical spectroscopic studies of ultrafast processes. The vibrational spectroscopies have some advantage in as much as the structurally important modes can be observed directly even when electronic transitions may be inaccessible. Although there are as yet no table top ultrashort pulse lasers in the structurally useful region of the infrared spectrum, say from 500 to 4000 cm^{-1} the recent advances in ultrafast technology in combination with advances in nonlinear materials, permit effective generation of IR pulses and shorter time resolution. Free electron lasers represent another approach with promise. At time of writing the shortest IR pulse used in an experiment was 160 fs [1]. There appears to be no technological barrier in obtaining useable pulses of about one-half this width. It is therefore evident that ultrafast processes can be studied directly via the vibrations and the purpose of this talk is to summarize some of the recent results obtained in this laboratory using a variety of IR and finally Raman methods.

First a very brief summary of the current capabilities of transient IR methods will be given. The methods themselves have been adequately described in recent review articles [2, 3]. There then follows an account of various areas of research in which these methods have been employed by us. For this talk, the following topics will be considered in more detail:

(1) Bimolecular abstraction reactions producing HCl or HCN in solution.
(2) Protein dynamics: Reaction center electron transfer reaction
(3) Partial Dephasing in the Raman scattering of porphyrins related to the

reaction center.

The use of IR in probing small molecules whose electronic transitions are not accessible in solutions is the first example. The remainder of the talk is concerned with protein phenomena, the identification of both electronic and vibrational transitions in the IR and with dephasing dynamics of porphyrin-like molecules that occur in the reaction center. The topic of population relaxation of simple ions is discussed in the paper by Owrutsky et al. in this volume.

Transient infrared spectroscopy

Important characteristics of ultrafast technologies include such features as the available time resolution, the repetition rate which is a critical feature in determining the ultimate signal-to-noise, the range of accessible IR frequencies and the available energy in the pump and probe pulses. Table I shows some of these parameters that are typical for the experiments reported here.

Table I Parameters of IR Spectroscopy	
Pulsewidth :	≥ 150 fs
Time resolution :	≥ 100 fs
Repetition rate :	≤ 5 kHz
Energy/IR pulse :	ca. 0.5 μJ (pump)
Probe range :	700-4000 cm^{-1}

Experiments have been performed with significantly more intense pulses than those in Table I but only at the expense of repetition rate. A very useful summary of work on developing higher intensity, mainly slower repetition rate infrared sources is given the in article by Laubereau [4]. Our goal has been to minimize the incident pulse energy and maximize the rep rate. In this laboratory transient IR spectroscopy was used to study dynamics in hemoproteins [5, 6], photochemistry of metal carbonyls [7], orientational dynamics in proteins [8, 9], relaxation of ions in water [10], heat transduction in proteins [11], free induction decay of molecules on metal surfaces [1], dynamics and spectra associated with electron transfer in reaction centers [12, 13] and with Bacteriorhodopsin [14] and bimolecular chemical reactions in solutions [15]. These experiments have involved a variety of different laser configurations and methods of generating IR pulses for pumping and probing samples. The first method used by us, and one that is particularly useful for protein studies, involved gated detection of quasi-CW IR beams [5]. In other experiments ultrashort IR pulses were generated in a conventional manner by means of difference frequency mixing.

Bimolecular hydrogen abstraction reactions (with D. Raftery)

Here is an area where IR methods allow detection in the condensed phase of transiently generated small molecules that could not be sensed with current laser optical methods.

Experimentally, gas phase studies have relied on collisionless conditions that allow evaluation of product states long after the reaction has taken place. Recent experiments have shown that the vibrational relaxation in condensed phase is often slow (10's to 100's of picoseconds), so that evaluation of the vibrational states should

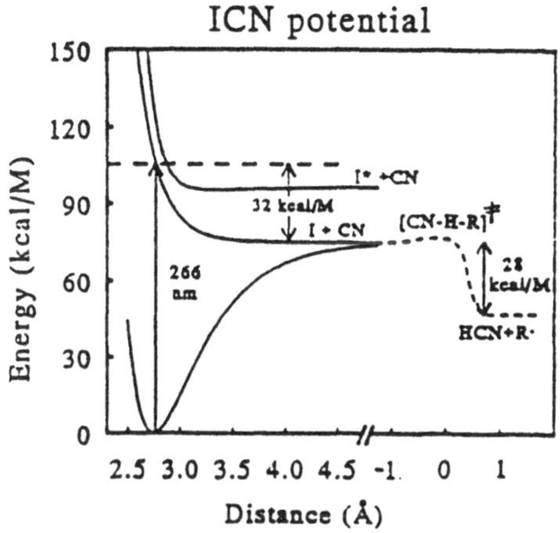

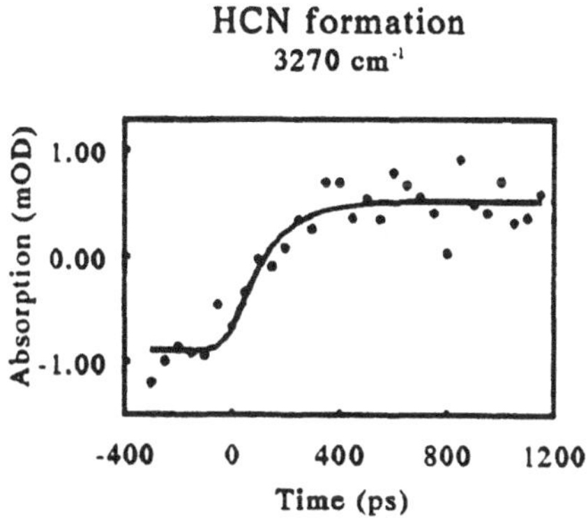

Figure1: (Left): Combined potential energy diagram of the ICN molecule and the reaction coordinate for hydrogen abstraction. The 266 nm transition is shown from the ground state to the dissociative state with excess energy 27 kcal/M per CN radical. The potential energy along the reaction coordinate (hydrogen asymmetric stretch) is indicated by the dotted line. (Right): IR transient of the formation of HCN in its ground vibrational state at 3265 cm^{-1}. The formation time is 120 ps ± 30.

indeed still be possible. Over the past several years, methods in transient IR spectroscopy have progressed to the point where the sensitivity and time resolution are sufficient to probe these distributions before they thermalize and to measure reaction rates.

A prototype experiment for a bimolecular reaction, hydrogen abstraction by CN, is shown in the Figure1. A near UV photon excites an ICN molecule from its ground state to a dissociative state. In the gas phase, photolyzing ICN molecules at 266 nm would result in ~ 27 kcal/M of translational energy per CN radical, which would easily be enough to surmount the small activation barrier to hydrogen abstraction from the reactant. However, the results of the experiment indicate that most of the reaction is thermal, and thus the radicals lose their kinetic energy before they react with the solvent.

Similar considerations apply to H abstraction from cyclohexane by Cl atoms [15]. Our results show that the majority of chlorine atoms, like the CN radicals in solution, do not react during their first encounter with a solvent molecule. From kinematic considerations, after 3 collisions, occurring within a time of 1 ps, chlorine atoms would lose the 11 kcal/M of initial translational energy (imparted by the 355 nm photolysis pulse) and thus become thermalized quickly.

It is expected that the mean force potential for the bimolecular process will be modified from the gas phase potential by solvation. The resulting change in activation free energy will cause the rates in the gas and solution phase to be different. Therefore, the slow solution phase reaction rates could simply be caused by an increased activation barrier in solution. Although no direct measurement of the barrier has been reported for the Cl and C_6H_{12} reaction, there are several gas phase studies of the reaction of chlorine with cyclopentane that give an activation barrier of about 0.5 kcal/M [16, 17]. A change in activation free energy of approximately 1.1 kcal/M would be required to account for the factor of 14 observed in the rates, assuming that chlorine is reacting thermally. One might expect the increased barrier height to be due to a larger (negative) free energy change of solvating the Cl atom compared to the solvation of the transition state. There is also information which could imply that part of the reduction in the rate is due to entropic factors caused by the solvent caging effect on Cl [18].

In general, the influence of vibrational excitation on simple gas phase bimolecular reactions has been frequently considered both theoretically and experimentally [19]. In the condensed phase the effect of vibrational relaxation needs to be considered. In many cases the vibrational energy relaxation induced by the solvent forces will occur on timescales that are very long compared with the time for activation and passage through a transition state region. In this limit the medium induced relaxation should have little effect on a reaction product distribution. For example, we have observed transitions generated in the reaction of CN with $CHCl_3$, that appear to signal the presence of HCN in its v=0, 1 and 2 states in solution [20]. It should be possible in this limit to prepare specific reactant vibrational states in solution that will live long enough to undergo reaction. For example, if the vibrational relaxation time of a mode is 100 ps, the molecule, having a radius of 3Å will most probably diffuse a distance of about 10Å before the vibrational excitation is lost in a liquid having a viscosity of 0.1 cp. This situation leaves the vibrationally excited molecule with plenty of opportunity to undergo a bimolecular reaction. However much work, both theoretical and experimental, remains to be accomplished before a broad understanding of the solvent effects on transition state dynamics is achieved.

Protein dynamics (with B.R. Cowen, P.L. Dutton, G. Walker, S. Maiti, C.C. Moser)

Proteins exhibit IR transitions that are characteristic of structure in the amide regions where vibrations of the peptide backbone occur. Transitions in this region (1400-1800 cm^{-1}) have relatively small cross sections so high sensitivity methods are needed to detect changes. Cofactors also have characteristic IR transitions such as the strong FeCO absorption in heme proteins and the carbonyl sidechains of chlorophylls.

It is well-known from optical studies that the early steps of electron transfer in photosynthetic reaction centers are very fast [21]: An electron moves from the excited chlorophyll dimer, $BChl_2^*$, to the bacteriopheophytin PPh in about 3 ps. The first quinone, Q, is reduced by BPh^- in about 200 ps. Thus the protein states $BChl_2^*$:BPh:Q , $BChl_2^+$:BPh^-:Q and $BChl_2^+$:BPh:Q^- are all interceptable by the current IR technology. If the protein or cofactor sidechains undergo changes in their IR spectra during these transformations it will be possible to characterize them and relate them to structural alterations.

We first studied the changes in the vibrational transitions of the protein and redox cofactors of the photosynthetic reaction center by picosecond infrared spectroscopy [12]. The spectra in the vibrational mid-infrared region (1800-1550 cm^{-1}) of

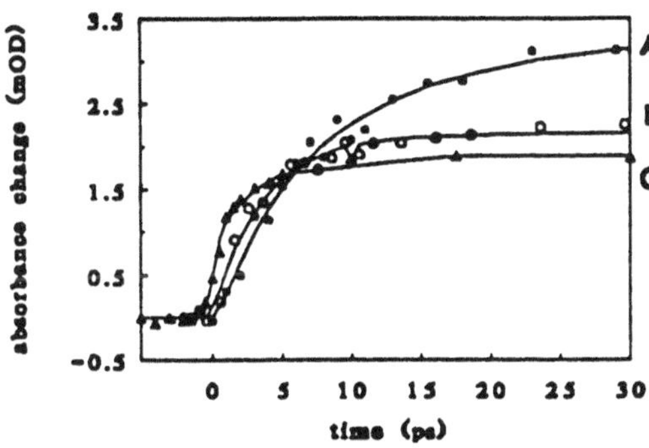

Figure 2: Some ultrafast IR transients in reaction centers. (a) 1702 cm^{-1}; (b) 1710 cm^{-1}; (c) 1960 cm^{-1}

hydrated and partially dehydrated reaction centers were investigated from 50 ps to 4 ns after photoinitiation of the electron transfer. Features in the infrared difference spectra were identified with both protein and redox cofactor vibrational modes and correlated with electron transfer events whose kinetics were measured in the infrared and visible regions. The observed protein response is confined to a few amide I transitions (1644 cm^{-1}, 1661 cm^{-1}, 1665 cm^{-1}) and carboxylic residues (1727 cm^{-1}). About 85% of the observed signal corresponded to alterations in the cofactor associated ester and keto- carbonyls. The amide I and carboxylic transitions appeared prior to 50 ps suggesting that the primary electron transfer event is coupled with a specific piece of the protein backbone and to Glu or Asp residues nearby to the special pair. Infrared absorption changes accompanying bacteriochlorophyll dimer cation formation, dominated the signal at all times investigated.

IR spectral changes observed in hydrated and partially dehydrated reaction centers were distinctly different; a band at 1665 cm^{-1} with a spectral width of 6 cm^{-1} in the hydrated protein corresponding to a protein amide I bleach was not present in the dehydrated film. These differences point to different electron-transfer kinetics observed in the presence of water.

Femtosecond IR was used recently to investigate the earliest steps in electron transfer [22]. Some data are shown in Figure 2. The 9-ketocarbonyl region around 1702 cm^{-1}, curve (a) in Figure 2 shows non-single-exponential growth kinetics which is influenced by sample heterogeneity. Besides the expected 3.5 ps response we have seen in this region even stronger signals exhibiting growth times of ca. 11 ps. At 1708 cm^{-1}, curve (b) only the 3.5 ps rise has been seen. These results are not yet understood but they clearly present another perspective on this electron transfer step.

Trace (c) in Figure 2 shows a response typical for the 1800-2000 cm^{-1} region where we have seen transitions in the IR due to electronic excitation of $BChl_2^*$ and $BChl_2^+$. In the case of $BChl_2^*$ we observe an interexciton transition which is between states that are symmetric and antisymmetric to interchange of the excitation. For $BChl_2^+$ we confirm the existence of a transition between states that are symmetric and antisymmetric to the transfer of a positive hole. Such a state was already conjectured to be present [23, 24].

These new results, taken together with our measurements on hemoglobin and bacteriorhodopsin suggest a wide variety of new insights can be obtained on protein dynamics using transient IR spectroscopy.

Partial dephasing (with C. Galli, S. LeCours, M. Therien and K. Wynne)

The occurrence of rapid electron transfer in reaction centers suggest that careful consideration be given to dephasing effects involving the various states involved. Some aspects of this were brought out in the paper by Jean and coworkers [25]. We have noted [26, 27] the similarity in the expected dephasing processes in porphyrin-like electronic states and dimers: the reaction center contains both such structures. Although this part of the research does not involve IR spectroscopy it does concern Raman transitions and the results are very different for the various vibrational symmetries that occur in porphyrins and chlorins.

The notion of partial dephasing is related to Raman dephasing and arises from consideration of a pair (or multiplet) of states that are nearly degenerate by symmetry or by accident, in contact with a bath. These could be the Q_x and Q_y states of a porphyrin or the accidentally degenerate states Q_{yL} and Q_{yM} of $BChl_2$*. When driven by a short enough linearly polarized light pulse a molecule having two levels $|a\rangle$ and $|b\rangle$ acquires the density matrix:

$$\sigma = (\varepsilon_x^2 + \varepsilon_y^2)\sigma_0 + (\varepsilon_x^2 - \varepsilon_y^2)\sigma_z + 2\varepsilon_x\varepsilon_y\sigma_x$$

where σ_0, σ_z and σ_x are the Pauli spin operators and ε_i is the projection of the polarized pump field on the ith molecular dipole allowed state. In a pump probe experiment all three components of σ can be probed but the signals have different anisotropy. We have shown that the pump-probe anisotropy signal acts as if there were a three state kinetic system of the form:

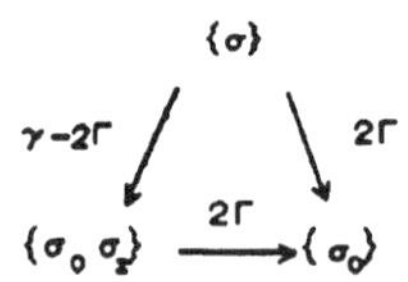

where $\{\sigma_i\}$ means σ_i is determining the ensemble averaged pump-probe signal. The "rate" coefficients γ and 2Γ satisfy a three state anisotropy equation:

$$r(t) = 0.7e^{-\gamma t} + 0.4(e^{-2\Gamma t} - e^{-\gamma t}) + 0.1(1 - e^{-2\Gamma t})$$

The evolution of σ under the influence of bath fluctuations corresponds to the transition to the statistically averaged state $\{\sigma_0\}$ through the partially coherent state characterized by the loss of σ_x only. In the experiment with MgTPP [27] the coefficients γ and 2Γ were measured to be 210 fs^{-1} and 1.6 ps^{-1} respectively indicating that the partial dephasing rate γ-2Γ is 240 fs^{-1}, quite similar to the rate 2Γ of creating of a statistical mixture. The application of this idea to higher symmetry cases (e.g. A→T transitions) and to dimers and trimers is straightforward.

Furthermore the influence of such effects on determining pathways of electron transfer from such structures to nearby donors is evident. In recent experiments with porphyrin-quinones [28] the rate of electron transfer is demonstrably limited by the partial dephasing rate. These effects are dramatically different for different vibrational states involved in the process [26].

Acknowledgements:

This work was supported by NIH and NSF. Thanks are due to my colleagues, post-Doctorals and students who contributed to the research projects described in this lecture.

References:

1. J. C. Owrutsky, Y.-R. Kim. M. Li, M.J. Sarisky, R.M. Hochstrasser, J. Chem. Phys. 97, 4421 (1992).
2. P.A. Anfinrud, C.K. Johnson, R. Sension, R.M. Hochstrasser, "Ultrafast Spectroscopic Methods," in Applied Laser Spectroscopy; Techniques, Instrumentation and Applications, ed. D.L. Andrews (VCH Publishers, 1992).
3. B.R. Locke, R. Diller, R.M. Hochstrasser, "Ultrafast Infrared Spectroscopy and Protein Dynamics," in Biomolecular Spectroscopy, vol. 20, eds. R.J. Clark and R.E. Hester (John Wiley & Sons, Ltd. 1993) in press .
4. A. Laubereau, in Topics in Applied Physics, ed. W. Kaiser (Springer-Verlag, Berlin, 1992) vol. 60, p. 35.
5. J.N. Moore, P.A. Hansen, R.M. Hochstrasser, Chem. Phys. Lett. 138, 110 (1987).
6. P.A. Anfinrud, C. Han, R.M. Hochstrasser, Proc. Natl. Acad. Sci. 86, 8387 (1989).
7. P.A. Anfinrud, C. Han, T. Lian, R.M. Hochstrasser, J. Phys. Chem. 94, 180 (1990).
8. P.A. Hansen, J.N. Moore, R.M. Hochstrasser, Chem. Phys. 131, 49 (1989).
9. B.R. Locke, T. Lian, R.M. Hochstrasser, Chem. Phys. 158, 409 (1991).
10. M. Li, J.C. Owrutsky, M. J. Sarisky, J.P. Culver, A.G. Yodh, R.M. Hochstrasser, J.Chem. Phys. 98, 7 (1993).
11. T. Lian, B.R. Locke, R.M. Hochstrasser, Biophys. J. 61, A53 (1992).
12. S. Maiti, B.R. Cowen, R. Diller, M. Iannone, C.C. Moser, P.L. Dutton, R.M. Hochstrasser, Proc. Natl. Acad. Sci. 90, 5247 (1993).
13. R. Sension, S.T. Repinec, A.Z. Szarka, R.M. Hochstrasser, "Ultrafast Photoreactions of Cis-stilbene," in Ultrafast Phenomena VIII, eds. J.L. Martin, A. Migus, G. Mourou, A.H. Zewail (Springer Series in Chemical Physics, 1993) vol. 55.
14. R. Diller, M. Iannone, B.R. Cowen, S. Maiti, R.A. Bogomolni, R.M. Hochstrasser, Biochemistry 31, 5567 (1992).

15. D. Raftery, M. Iannone, C.M. Phillips, R.M. Hochstrasser, Chem. Phys. Lett. 201, 513 (1993).
16. V.N. Kondratiev, Constants of Gas Phase Reactions Reference Book, ed. R.M. Fristrom (NBS, Washington, 1972).
17. D.D. Davis, W. Braum, A.M. Bass, Intern. J. Chem. Kinetics 2, 101 (1970).
18. F. Galiba, J.M. Tedder, J.C. Walton, J. Chem. Soc. B., 604 (1966).
19. J.I. Steinfeld, J.S. Francisco, W.L. Hase, Chemical Kinetics and Dynamics (Prentice Hall, Englewood Cliffs, N.J.,1989).
20. D. Raftery and R.M. Hochstrasser, in process of publication (1993).
21. J. Breton, J.-L. Martin, A. Migus, A. Antonetti, A. Orszag, Proc. Natl. Acad. Sci., 83, 5121 (1986).
22. B.R. Cowen, G.C. Walker, S. Maiti, C.C. Moser, P.L. Dutton, R.M. Hochstrasser, Biophys. J. 64, A213 (1993); G.C. Walker, S. Maiti, B.R. Cowen, P.L. Dutton, R.M. Hochstrasser, "Ultrafast Steps in Reaction Center Electron Transfer Studied by IR Spectroscopy," in process of publication.
23. W.W. Parson, Z.-T. Chu, A. Warshel, Biochim. Biophys. Acta. 1017, 251 (1990).
24. J. Breton, E. Nabedryk, W.W. Parson, Biochemistry 31, 7503 (1992).
25. J.M. Jean, R.A. Friesner, G.R. Fleming, J. Chem. Phys. 76, 5632 (1992).
26. K. Wynne and R.M. Hochstrasser, Chem. Phys. 171, 179 (1993).
27. C. Galli, K. Wynne, S. LeCours, M.J. Therien, R.M. Hochstrasser, Chem. Phys. Lett. 206, 493 (1993).
28. K. Wynne, C. Galli, S. LeCours, M.J. Therien, R.M. Hochstrasser, in process of publication (1993).

A Picosecond Time-Resolved Resonance-Raman Comparison of Deoxy Hemoglobin and Carbonmonoxy Hemoglobin

S. Franzen[1], B. Bohn[2], C. Poyart[2], and J.-L. Martin[1]

[1]Laboratoire d'Optique Appliquée, ENSTA-Ecole Polytechnique, F-91128 Palaiseau, France
[2]INSERM U299, Kremlin-Bicetre, France

A sub-picosecond time resolved resonance Raman experiment is used to measure the time scale of conformational changes in hemoglobin following ligand photolysis. Comparison with an unligated species allows the heme photophysical contribution to the signal to be isolated.

1. Introduction

Hemoglobin is a four-subunit protein consisting of two homodimers whose function is oxygen transport. Each subunit contains a heme molecule where a central iron atom serves as the binding site for oxygen or other ligand. The protein exists in two stable conformations, the deoxy and ligated structures, which have been characterized by X-ray crystallography [1,2]. In the allosteric model proposed by Monod et al. [3], the ligated structure corresponds to the R or relaxed state which has a high affinity for ligand binding and the deoxy structure corresponds to the T or tense state which has a low affinity for ligand binding. The transition between these two states is cooperative. The mechanism of cooperativity has been studied extensively by spectroscopy. Flash photolysis techniques have been used to study the time dependence of both ligand rebinding and protein structural changes. Time-resolved resonance Raman spectroscopy TRRR has been extensively used in recent years to obtain information on the conformational changes in the protein associated with ligand photolysis. In hemoglobin, the goal of these studies was to understand the details of how one heme communicates ligand departure to the other hemes in the protein. It has been suggested by severals investigators that the recoil of the iron out of the heme plane, and therefore heme doming, is the first configurational change to follow ligand photolysis. As part of the present TRRR study we have observed what we believe is the first direct evidence that the initial heme doming occurs on the femtosecond time scale.

The extension of the TRRR experiment to the sub picosecond (ps) time scale is the route to determine the time scale for heme doming, as well as to determine whether there are changes in protein tertiary structure on the time scale of 1 to 100 ps. In order to separate protein structural changes (including heme doming) from heme photophysics, we have used ps TRRR spectroscopy to compare deoxy Hb and HbCO. In deoxy Hb, we assume that there are no protein structural changes and all contributions to the TRRR signal are due to either vibrational cooling or excited states of the heme. In the ligated HbCO species the heme is surrounded by a conformationally unrelaxed protein during the first μs after photolysis. The contribution of the protein relaxation to TRRR signals has been studied on time scales ranging from 25 ps to μs [4]. In addition, the departure of the CO ligand means that there is less energy on the heme subsequent to photolysis for a photon energy identical to that used to photoexcite deoxy Hb. This difference could be manifested in a different initial vibrational temperature of the heme. Despite possible differences in temperature and excited state dynamics which can affect the comparison of deoxy Hb and HbCO, we have established that heme doming occurs on the sub ps time scale and that there are no subsequent protein structural changes on the 100 ps time scale.

2. Experimental

The experimental apparatus used was a two-color pump/probe of hemoglobin samples in the 90° scattering geometry as described by Petrich et al. [6]. The pump pulse had a duration of 100 fs and sufficient energy to photoexcite 50% of the hemes in the sample and therefore to dissociate 50 % of the CO ligands (the quantum yield for CO ligand photodissociation is unity). The wavelength of the pump pulse was 570 nm which is in the Q-band region of the absorption spectrum, as seen in the absorption spectrum of both HbCO and deoxy Hb. The probe pulse at 435 nm was weak enough not to detectably photodissociate HbCO. The 435 nm Raman pulse was in resonance with the Soret band of both HbCO and deoxy Hb. The Raman pulse had duration of 700 fs which is the at the limit of time resolution in TRRR because the spectral bandwidth of the pulse, determined by the uncertainty principle, is as large as typical Raman bands. In order to calibrate time dependent intensity changes, ammonium sulfate was used as an internal standard in some of the measurements. The extent of photolysis was monitored by measuring the intensity change in Fe-ligand stretching mode at 505 cm^{-1} (data not shown).

3. Results

The equilibrium resonance Raman spectrum for HbCO and deoxy Hb was determined by allowing the Raman scattering pulse to arrive at the sample 5 ps before the pump pulse. The RR spectra over the range from 900 to 1700 cm^{-1} is shown in Fig. 1A for HbCO which has a six coordinate low spin heme iron. The most significant feature in this RR spectrum is the electron density marker mode or ν_4 at 1372 cm^{-1}. The core size marker region between 1450 and 1700 cm^{-1} is spectrally congested and no attempt will be made to analyze the well-known time dependent frequency shifts in this region, which have been characterized by Spiro and co-workers [5]. The equilibrium resonance Raman spectrum for deoxy Hb is

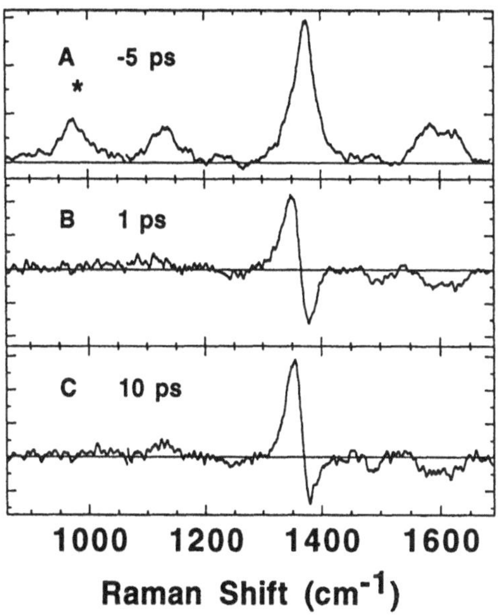

Figure 1 The time dependent resonance Raman spectrum of carbonmonoxy hemoglobin from 860 to 1700 cm^{-1}. **A.** At a time delay of -5 ps the equilibrium resonance Raman spectrum is observed. The asterisk marks the sulfate peak of the internal standard (1M $(NH_4)_2SO_4$). **B.** At 1 ps the largc band shift of the ν_4 mode at 1372 cm^{-1} is apparent in this Raman difference spectrum, but no relaxation toward the equilibrium deoxy spectrum has taken place. **C.** In the Raman difference spectrum at a delay of 10 ps, the intensity of modes has reached the same value as observed in deoxy Hb giving rise to the assymmetry of the ν_4 band.

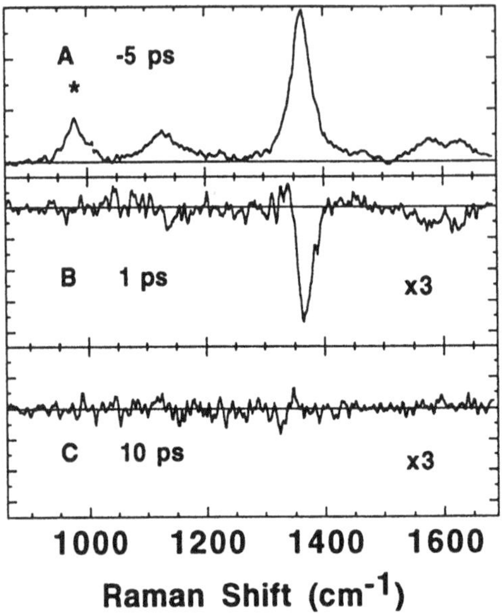

Figure 2 The time dependent resonance Raman spectrum of deoxy hemoglobin from 860 to 1700 cm^{-1}. **A.** At a time delay of -5 ps the equilibrium resonance Raman spectrum is observed. The asterisk marks the sulfate peak of the internal standard (1M $(NH_4)_2SO_4$). **B.** At a delay of 1 ps a decrease in intensity of all modes is apparent in this Raman difference spectrum. In addition, there is a shift of the ν_4 mode whose equilibrium position is 1357 cm^{-1} in deoxy Hb. **C.** The Raman difference spectrum at 10 ps shows complete relaxation of all intensity changes and frequency shifts back to their equilibrium values.

shown in Fig. 2A. The electron density marker band is shifted by around 15 cm^{-1} to lower energy. In general all of the modes show increased scattering intensity of about 1.5 in deoxy Hb compared to HbCO. This is due solely to differences in the extinction coefficients between the two species. The bands in the core size marker region are shifted and less intense in deoxy Hb. The heme iron is five coordinate and high spin in this species.

The time-dependent changes observed in deoxy Hb consist of an overall reduction in intensity and a band shift of the ν_4 band as seen in Fig. 2B. At 1 ps the intensity of the RR bands is 59 (5) % of the equilibrium value and the ν_4 band shift is 6.8 (0.6) cm^{-1} to lower energy than the equilibrium value. The return of these two changes signal to the equilibrium

Table 1 Comparison of shift of ν_4 mode and relative amplitude change of 674 cm^{-1} mode in deoxy Hb and HbCO

Time (ps)	Shift (cm^{-1}) HbCO	DeoxyHb	Relative Amplitude[a] HbCO	Deoxy Hb
1.0	6.2 (0.6)[b]	6.8 (0.6)	0.73 (0.16)	0.59 (0.045)
2.0	6.3 (0.2)	6.2 (0.8)	0.83 (0.12)	0.73 (0.100)
3.0	5.5 (0.4)	5.0 (1.2)	1.05 (0.11)	0.75 (0.095)
5.0	4.4 (0.4)	2.4 (0.7)	0.97 (0.12)	0.80 (0.056)
10.0	3.1 (0.5)	0.1 (0.3)	0.97 (0.22)	0.98 (0.100)
60.0	2.5 (0.4)	- 0.2 (0.5)	0.87 (0.05)	0.96 (0.064)

a. The amplitude at the indicated time delays was normalized to the value measured at -5 ps.
b. Quantities in parentheses are root-mean-square errors based on the average of five to eight data sets.

values occurs with a time constant of 3.5 (1.0) ps. The values of the intensity changes and ν_4 frequency shift are given in Table 1. Similar changes in intensity to those given in the Table are observed in all modes of deoxy Hb. The return to equilibrium values for deoxy Hb is shown in Fig. 2C and is clearly complete by 10 ps.

The TRRR spectra of HbCO after photolysis must be compared to the equilibrium deoxy Hb-HbCO difference spectrum should be identical to that in Fig. 3. Since the time scale for the protein relaxation associated with the quaternary structural change is known to occur on the μs time scale, the ps spectra reported here will not be identical to that reported in Fig. 3. Note that in addition to the band shift of the ν_4 RR band, there are overall changes in RR intensity which are due to the change in the absorption cross section in the photolyzed species at the Raman scattering wavelength. The time-dependent changes in the HbCO TRRR signal are shown in Fig. 1B and 1C. After 1 ps the intensity of all Raman bands is still similar to that in HbCO and there is a large band shift in the ν_4 band. The intensity of Raman bands increases with time approaching the equilibrium deoxy Hb value with a time constant of 3.5 (1.0) ps. The change in frequency of the ν_4 band has a different behavior from that seen in deoxy Hb. In Table 1 the ν_4 band shift as a function of time is compared for the two species. Although the band shifts are similar at early times, the ν_4 band shift in

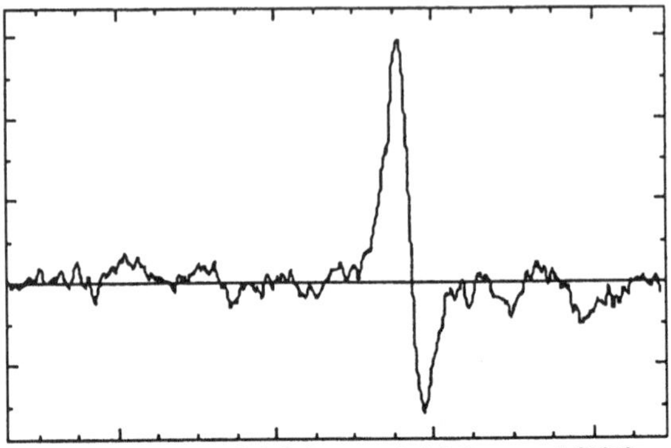

Figure 3 The equilibrium difference spectrum of deoxy and carbonmonoxy hemoglobin.

photolyzed HbCO clearly approaches a value which shifted 2 to 3 cm^{-1} from the equilibrium deoxy Hb value as has been observed by several investigators.

In the low frequency region, at the earliest delay time of 1 ps, we have observed the appearance of the well-known Fe-His stretching mode [7]. This mode is indicative of the iron out-of-plane displacement and is therefore strong evidence that the heme is at least partially domed within 1 ps of photolysis. This mode appears at 228 cm^{-1}, which is the position observed by other investigators on the time scale from 25 ps to 1 μs. In addition to the Fe-His stretching mode, there is a mode at 303 cm^{-1} which is observed to appear and change intensity on exactly the same time scale. The intensity changes in the low frequency modes are similar to those already discussed for the high frequency modes, although they may be slightly larger (data not shown).

4. Discussion

There are three major contributions to the TRRR signal, conformational changes, vibrational cooling, and heme excited states. Following photodissociation, the heme molecule is in an excited state which may decay to other excited states or directly to the ground state. In addition, the molecule may have excess vibrational energy due to the absorption of a photon which requires some time to dissipate. Finally, structural changes such as motion of the iron out of the plane of the heme result in an unrelaxed (strained) heme-protein configuration which relaxes on time scales ranging from ps to μs. Ultimately, the protein conformational changes lead to the R -> T transition on the time scale of tens of μs. Prior to this quaternary (or intersubunit) transition there are tertiary structure changes which include the rotation and tilt of the proximal His and the large 1 Å movement of the F-helix. The effect of these relaxations on the RR signal is known in the sense that several phases of relaxation of shifted RR bands have been identified on the ns and μs time scale. One color experiments using 25 ps pulses have given the same result as those using 10 ns pulses suggesting that there is little activity on the ps time scale. However, it is clear from the ps experiments that heme iron is at least partially domed at 25 ps [4] and the time scale for this doming has been inferred to be rapid based on the transient absorption spectroscopy results of Petrich and Martin [8].

The Raman signal can be affected by heme photophysics in three ways, depletion of the ground state (and its signal), RR scattering from a populated excited state, and repopulation of a hot ground state. Comparison of Stokes and anti-Stokes TRRR in deoxy Hb on the 10 ps time scale by Lingle and Hopkins has been used to suggest that the main contribution is from a hot ground state [9]. Moreover, the ν_4 band shift resembles a band shift one would expect from anharmonic coupling to low frequency modes in a hot molecule [6,10]. However, this interpretation has been questioned by Champion and co-workers based on saturation RR spectroscopy [11]. The results presented here for deoxy Hb show that the relaxation of the TRRR signal occurs on the same time scale as an excited state (Hb_{II}^*) known from transient absorption spectroscopy [8]. Calculations show that a uniform reduction in intensity across the RR spectrum is not consistent with the effect of a hot heme on the RR spectrum [7,11]. Our Stokes data agree qualitatively with the spectrum presented by Lingle and Hopkins in that we see a reduction in the RR intensity when compared to an internal standard. However, we also see a band shift in the ν_4 band which they were unable to time resolve. A similar band shift is reported by Li, Sage, and Champion and assigned to an excited state based on the RR saturation experiment [11]. Thus, we find the most resonable explanation for the time-dependent changes in deoxy Hb to be a combination of ground state depletion (leading to an overall decrease in intensity of all Raman bands) combined with an excited state

Raman spectrum having a reduced RR cross section and a shifted ν4 band.

These heme photophysical changes are also observed in photolyzed HbCO. The time constant for relaxation is nearly identical to that observed in deoxy Hb. However, after 10 ps there are residual band shifts in HbCO which are due to protein conformational effects on the heme and Fe-His molecules. These changes are clearly separated from the heme photophysical events by the comparison presented here. The insight gained from this comparison is that the heme doming event is extremely rapid and it occurs without participation of the surrounding globin. Rather, this conformational event is a means of storing the information that the ligand has departed in order to trigger further conformational changes on a much longer (μs) time scale in agreement with the results of ab initio and molecular dynamics calculations [12,13,14]. The ultrafast trigger event stores the information required to give rise to structure changes on the μs time scale which are ultimately responsible for communication between the hemes and the release of the free energy of cooperativity in hemoglobin.

References

1. Baldwin, J.; Chothia C. J. Mol. Biol. **1979**, 129, 175-220
2. Perutz, M. F. Ann. Rev. Phys. Chem. **1979**, 48: 327-386
3. Monod, J.; Wyman, J.; Changeux; J.-P. J. Mol. Biol. **1965**, 12, 88-118
4. Scott, T. W. ; Friedman, J. M. J. Am. Chem. Soc. **1984**, 106, 5677-5687
5. Spiro, T.G.; Smulevich, G.; Su, C. Biochemistry **1990**, 29, 4497-4508
6. Petrich, J. W.; Martin, J. L.; Houde, D.; Poyart, C.; Orszag, A. Biochemistry **1987**, 26, 7914-7923
7. Franzen, S. F.; Bohn, B.; Poyart, C.; Martin, J. L. Manuscript in preparation
8. Petrich, J. W.; Poyart, C.; Martin, J. L. Biochemistry **1988**, 27, 4049-4060
9. Lingle, R. Xu, X. Huiping, Z.; Yu S.-C.; Hopkins, J. B.; McClellan, J. L. J. Am. Chem. Soc. **1991**, 95, 9320-9331
10. Asher, S. A.; Murtaugh, J. J. Am. Chem. Soc. **1983**, 105, 7244-7251
11. Li, P.; Sage, J. T.; Champion, P. M. J. Chem. Phys. **1992**, 97, 3214-3227

12. Henry, E. R.; Levitt, M.; Eaton, W. A. Proc. Natl. Acad. Sci. USA **1985**, 82, 2034-2038
13. Kuczera, K.; Kuriyan, J.; Karplus, M. J. Mol. Biol. **1990**, 213, 351-373
14. Olafson, B. D.; Goddard, W. A. Proc. Natl. Acad. Sci. USA **1977**, 74, 1315-1319

Light-Induced Reaction Sequence of the Chromophore in Bacteriorhodopsin Studied by Time-Resolved RR Spectroscopy

R. Lohrmann, T. Althaus, W. Eisfeld, and M. Stockburger
Max-Planck-Institut für biologische Chemie, D-37018 Göttingen, Germany

Retinylidene Schiff–Base Chromophore in Bacteriorhodopsin

Bacteriorhodopsin (bR) is a retinal-binding protein in the so-called purple membrane of *Halobacterium halobium* where it acts as a light-driven proton pump. The protein backbone consists of 7 α-helical segments in cylindrical shape which span the membrane. In the interior part of the protein a retinal molecule is bound to a lysine residue via a protonated Schiff base linkage, thus forming the chromophore (Figure 1). In its parent state (BR_{570}) the chromophore is held in a fixed position in the binding pocket by electrostatic and van der Waals forces. A key role is played by the carboxylate side chain of the residue Asp-85 which forms a negative counterion to the positively charged Schiff base (for reviews, see [1,2]).

The resonance Raman (RR) spectrum of BR_{570} is shown in Figure 2. The strongest band at 1528 cm^{-1} can be assigned to the delocalized C=C stretching vibration of the retinal chain. The frequency of this band is sensitive to electrostatic interactions with the counterion and other charged residues in the binding pocket. The $C{=}NH^+$ stretching vibration of the Schiff base group depends on its direct environment, in particular on its distance from the negative counterion Asp-85. The bands in the "fingerprint" region indicate trans-to-cis configurational changes of the retinal chain. Hydrogen out-of-plane (hoop) bending vibrations which occur between 700 and 1000 cm^{-1} reflect distorted conformations along the molecular axis [3].

On light absorption BR_{570} runs through a completely reversible cyclic reaction (photocycle) which under physiological conditions (20°C, pH 7) has a period of ~6 ms. This reaction sequence controls bR's unique function of a light-induced proton pump. During the photocycle the chromophore takes in a series of intermediate states

$$K_{590}\ (6\,\text{ps}),\ L_{550}\ (1.2\,\mu\text{s}),\ M_{412}\ (80\,\mu\text{s}),\ N_{560}\ (0.5\,\text{ms}),\ O_{640}\ (3\,\text{ms})$$

which are listed in alphabetic order according to their appearance time and are further classified by their optical absorption maxima in nm. The primary transition in the photocycle from BR_{570} to K_{590} involves the light-induced rotation around the 13–14 double

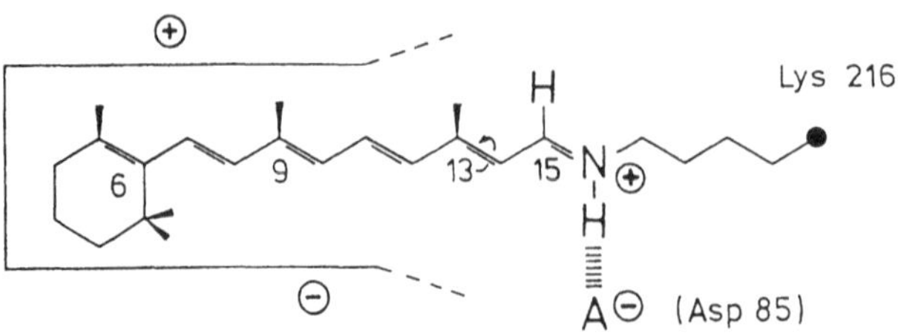

Fig. 1: Chromophore in the binding pocket of bR in its parent state BR_{570}.

Springer Proceedings in Physics, Vol. 74 Time-Resolved Vibrational Spectroscopy VI
Editors: A. Lau, F. Siebert, W. Werncke

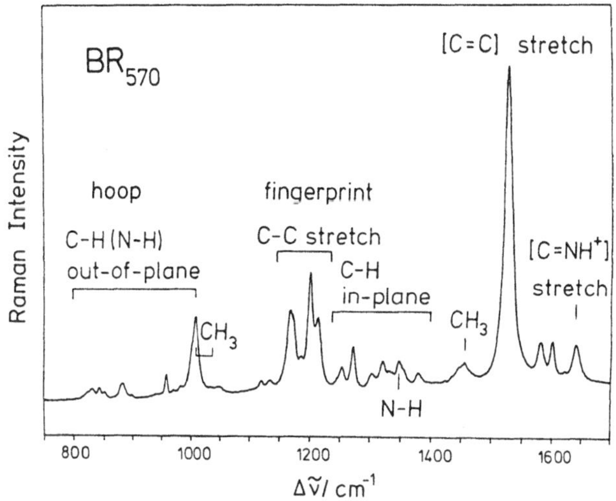

Fig. 2: RR spectrum of the parent chromophore BR_{570}. For details, see ref. [3].

bond. All subsequent reaction steps are thermally controlled. During the transition from L to M the Schiff base becomes deprotonated and at the stage of N is reprotonated [1].

For an understanding of the molecular mechanism of the proton pump it is necessary to study the structural changes of the chromophore and its interactions with the protein during the photocycle. For this purpose we have performed a variety of laser photolysis experiments with flowing aqueous suspensions of purple membranes which enabled us to record the RR spectra of the intermediates.

Evolution of the Intermediate K: 10 ps – 2 μs

The RR spectra of the short-lived intermediate K were obtained by combining a laser beam with a free liquid jet of a purple membrane suspension as described in Figure 3. In this experiment the laser beam serves both for photolysis of bR and for RR excitation of the products. For probing the "early" K two different laser systems with pulse durations of 10 ps (synchronously pumped dye laser with cavity dumper: 585 nm, 15 nJ per pulse, 3.5 cm^{-1} bandwidth) and 6 ns (cw argon laser with cavity dumper: 515 nm, 45 nJ per pulse) were used. In the liquid jet the sample element which is photolyzed in the beam waist by a single laser pulse is replaced in $\approx 2\,\mu s$ by fresh sample (Figure 3). Thus, a repetition rate of the pulses as high as 500 kHz could be used. – The short dwell time of a sample element in the beam waist of the laser ($\approx 2\,\mu s$) made it possible to obtain the RR spectra of the "late" K intermediate, i.e. the direct precursor of L, by the application of cw lasers [4]. To obtain spectra of good quality the RR signals were collected over several hours.

The RR spectra of K for the three different time domains are depicted in Figure 4. For comparison the spectra of BR_{570} and L_{550} were added. It can be seen that the spectra of K obtained from 10 ps- and 6 ns-pulses closely resemble each other, whereas distinctive spectral changes occur when the observation time is prolonged from 6 ns to 2 μs. This indicates that during this period a late K intermediate is formed which can be

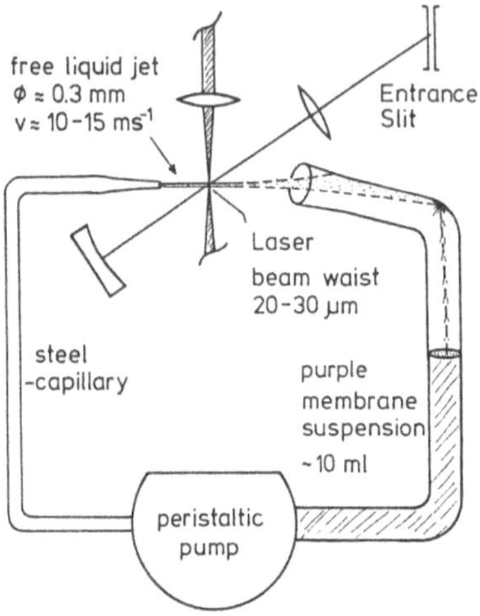

Fig. 3: Photolysis and RR excitation of an aqueous purple membrane suspension is performed by a single laser beam in a free liquid jet. The dwell time of a sample element in the beam is $\approx 2\,\mu s$. Scattered RR radiation is monitored perpendicularly to the laser beam by a conventional Raman spectrometer with photomultiplier detection.

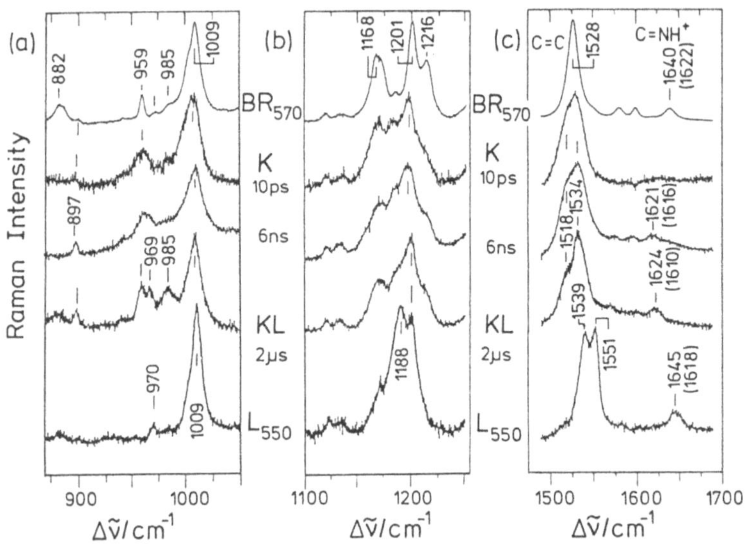

Fig. 4: RR spectra of K (10 ps, 6 ns) and KL ($2\,\mu s$), compared with BR_{570} and L_{550}. All spectra recorded at 22°C and pH 7. (a): hoop–; (b): fingerprint–; (c): (C=C/C=N) stretch region, frequencies for $C{=}ND^+$ in brackets. The "pure" spectra of K, KL and L_{550} were obtained by subtracting the contribution of BR_{570}.

distinguished from the early K. Spectroscopic evidences for a transition from the primary K to a secondary form in the low temperature range 80 K – 135 K had already been found in static FTIR experiments [5]. This secondary K intermediate could also be identified in time–resolved FTIR measurements at ambient temperature by probing the sample in the time range 200 – 500 ns after the photolysis event [6,7]. These findings justify to design the late K by a special name. Here we call it KL, following an earlier proposal [8].

Taking into account that KL is only accumulated for illumination times > 6 ns but on the other hand decays to L_{550} with a time constant of 1.2 μs (22°C, own measurements), the transition from K to KL at ambient temperature should be in the range 50 - 250 ns.

A characteristic feature in the spectrum of KL is the band at 985 cm^{-1}. On the basis of isotopic labeling we could assign it to a C_{15}–H hoop mode. From quantum chemical model calculations it has been concluded that single bond twists in the polyene chain of the chromophore induce a vibrational coupling of hoop modes with deformation modes of the carbon skeleton [9]. Here we propose that this coupling enables a hoop mode, which per se is not RR active, to "borrow" significant intensity from an allowed RR transition in the same frequency range. In this sense the significant intensity of the 985–cm^{-1} band in KL, which in BR_{570} appears only as a weak feature, could be explained by a twisted C_{14}–C_{15} single bond in KL which induces a vibrational coupling with deformation modes of the carbon skeleton. Below we shall specify one of these "lending" modes for the case of L_{550}.

The spectral changes between BR_{570} and K(10 ps) in the fingerprint region (Figure 4b) reflect the rotation around the 13–14 double bond of retinal during the primary photochemical event. The fact that between K(10 ps) and KL no significant changes in the fingerprint region are observed indicates that the geometry of the retinal moiety is largely preserved during this period.

In the C=C stretching region (Figure 4c) K is characterized by a broad spectral feature which can be reasonably fitted by the superposition of two bands at 1518 and 1534 cm^{-1}. This also holds for KL. The existence of the low-frequency peak is in line with the observed red-shift of the absorption maximum of K with respect to BR_{570}[4]. - The appearance of two bands in K is not yet fully understood.

In K(6 ns) and KL the $C{=}NH^+$ stretches could be identified at 1621 and 1624 cm^{-1} on the basis of isotopic shifts in D_2O [10]. The parallel frequency down-shifts of the C=C and C=N stretches in the K intermediates can be qualitatively explained by a model in which the positive charge at the Schiff base is removed from the negative counterion [4] (one-counterion model [11]). This would imply that during the primary photochemical event the Schiff base counterion complex is desintegrated and the two charges are separated.

In K(10 ps) no single peak for the $C{=}NH^+$ stretch could be identified. It thus appears that at this early stage the structure around the Schiff base is fairly inhomogeneous, and only in KL a more homogeneous structure is established.

Transition from KL to L_{550}

This transition is accompanied by drastic spectral changes (Figure 4). Thus, the hoop band at 985 cm^{-1} disappears, the intensity distribution in the fingerprint region is modified and the C=C as well as the $C{=}NH^+$ stretches are up-shifted by about 20 cm^{-1}. This latter effect indicates that a new complex around the Schiff base is established in which the counterion has approached the positive charge. The fact that the shape of the band at 1645 cm^{-1} is narrow demonstrates that the structure of this new complex in L is fairly homogeneous.

If the electrostatic interactions between the chromophore and the protein backbone would be exclusively governed by the Schiff base counterion complex one would expect that the frequencies of the C=C and C=N stretches change in a parallel way for the various chromophores [4]. For a comparison one has to consider the pure C=N stretches which are found in D_2O where the nitrogen is deuterated and the coupling between the C=N stretching and N-H bending motion is removed (the C=N frequencies in D_2O are given in Figure 4c in brackets). It can be seen that during the transition from KL to L the pure C=N stretch is upshifted by 8 cm^{-1} whereas the upshift for the C=C stretch is 21 cm^{-1} (compare the low-frequency components). It is suggested that this big shift of the C=C stretch is caused by changes of the protein environment at the retinal moiety of the chromophore. It has been argued that the observed shifts of the C=C stretches in the KL-to-L transition can be modeled by the approach of a positive charge to the retinal moiety [4]. This view is in line with the conclusion from FTIR experiments that the formation of L is accompanied by conformational changes of the protein backbone [12].

Structure of the Intermediate L_{550}

At the stage of K a certain amount of excess energy which is left after the primary photochemical event is stored at the chromophoric site [13]. The finding that the intermediate L_{550} takes in a fairly homogeneous and stable structure suggests that during the formation of L this excess energy is transferred to the protein backbone where it is available for functionally important conformational changes.

It is interesting in this respect to ask for the detailed structure of L. It has been proposed that in this state the chromophore is in a 13-*cis*, 15-*anti* configuration [14]. This, however, can only be a crude description of the structure of L. - Thus a peculiar isotope effect is observed in the hoop region in D_2O where the hydrogen at the Schiff base nitrogen is exchanged for deuterium [Figure 5]. The fairly strong band at 1009 cm^{-1} which involves the bending vibrations (rocks) of the two methyl groups of the retinal chain at C_9 and C_{13} [3], looses nearly half of its intensity and a new band appears at $\sim$ 985 cm^{-1} which could be assigned to the C_{15}-H hoop mode [15]. Previously these effects were explained in terms of a hoop mode in which the two hydrogens at the Schiff base undergo a coupled motion giving rise to a band at 1009 cm^{-1} which accidentally coincides with the

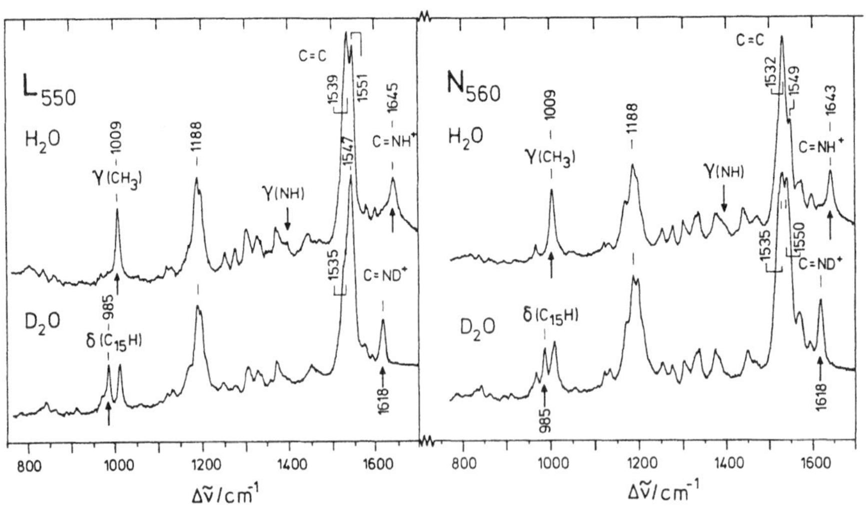

Fig. 5: RR spectra of L_{550} and N_{560} from purple membrane suspension in H_2O and D_2O obtained in cw pump-probe experiments with a spinning cell [18]

D_2O [15]. In this model the observation of RR intensity in such hoop bands would require a distortion around the C=N bond. This, however is unlikely since the C_{15}–H hoop mode is found at the same position (985 cm^{-1}) in BR_{570}, KL and L_{550}.

Here we therefore propose that the same mechanism we had invoked for the RR intensity of the 985-cm^{-1} hoop mode in KL might be active in L_{550}. The intensity decrease of the band at 1009 cm^{-1} in D_2O [Figure 5] suggests that the C_{13} methyl rock might be coupled via a twisted 14–15 single bond to the isolated C_{15}–H mode at 985 cm^{-1} and in this way may lend RR intensity to this mode. – The finding that the 985-cm^{-1} band disappears in H_2O might be explained by a preferential coupling of the C_{13} methyl rock with the isolated N–H hoop which is expected near 1000 cm^{-1} [15] and therefore is in near resonance with the methyl rock at 1009 cm^{-1}.

The fact that in KL the 985-cm^{-1} band could be observed in H_2O as well as in D_2O [10], whereas in L it is only found in D_2O, indicates a different configuration of the Schiff base for the two species. A possibility could be that in KL this configuration is 15-*anti* and during the transition to L is converted to 15-*syn* by a rotation around the C=N bond. This view is corroborated by the fact that for the 13-*cis* component in dark-adapted bR (BR_{555}), which is known to exist in the *syn* form [16], the same coupling effects between the C_{13} methyl rock and the C_{15}–H hoop were found as for L_{550} [15]. This behavior strongly suggests that the configuration of the Schiff base in L_{550} is identical with that in BR_{555}, namely 15-*syn*.

The Intermediate N_{560}

It could be shown that the appearance of N is associated with a conformational transition of the protein backbone [17]. A comparative analysis of the RR spectra of L and N in Figure 5 reveals that the main spectral change occurs for the C=C stretches whereas the characteristic bands of the Schiff base and their respective isotope effects are completely identical for L and N. Only small changes are observed in the fingerprint region. This behavior implies that the conformational transition of the protein backbone leaves the Schiff base and its immediate environment unperturbed and mainly modifies the π-electron distribution in the retinal moiety.

We have shown that in the manifold of bR molecules the intermediate N_{560} can be formed either as a direct product of L (L→N) or in the sequence L→M→N [18] where the formation of M is accompanied by the transfer of the Schiff base proton to the counterion Asp-85. The fact that the Schiff base counterion complex is identical in L and N implies that Asp-85 is also deprotonated in N. From this it follows that during the M→N step Asp-85 becomes deprotonated. The most plausible interpretation of this reaction step thus would be that the Schiff base is directly reprotonated from Asp-85. – This would imply that during the photocycle the Schiff base proton would not escape the Schiff-base counterion complex. This conclusion is in contradiction with a widely accepted model in which the Schiff base acts as an active transfer switch for proton pumping [1]. – This is one of the reasons why we arrived at an alternative description of the pump mechanism which is presented in ref. [18].

Conclusions

In this paper we have obtained the RR spectra of the chromophore of bR in various intermediate states of its ligth-induced reaction cycle on a time scale from 10 ps to several milliseconds. The vibrational analysis revealed detailed structural changes of the chromophore and its immediate environment during this period. About 2 μs after the primary photochemical event the chromophore has reached a fairly stable state (L_{550}). The subsequent reactions of L (de- and reprotonation of the Schiff base) are confined to the immediate environment, or they are induced by conformational changes of the protein backbone (L→N, cf. ref. [18]). It thus appears that at the stage of L the primary excess energy is transferred from the chromophore to the protein backbone. This energy may be used for chromophore-independent structural changes of the protein which control the biological function of a proton pump.

REFERENCES

[1] Mathies, R. A., Lin, S. W., Ames, J. B., & Pollard, W. T. (1991) *Ann. Rev. Biophys. Biophys. Chem 20*, 491 - 518.
[2] Henderson, R., Baldwin, J. M, Ceska, T. A., Zemlin, F., Beckmann, E., & Downing, K. H. (1990) *J. Mol. Biol. 213*, 899 - 929.
[3] Smith, S. O., Braiman, M. S., Myers, A. B., Pardoen, J. M., Courtin, M. L., Winkel, C., Lugtenburg, J., & Mathies, R. A. (1987) *J. Am. Chem. Soc.109*, 3108-3125.
[4] Lohrmann, R., & Stockburger, M. (1992) *J. Raman Spec. 23*, 575 - 583.
[5] Rothschild, K. J., Roepe, P., & Gillespie, J. (1985) *Biochim. Biophys. Acta 808*, 140-148.
[6] Noelker, K., Weidlich, O., & Siebert, F. (1992) in *Time Resolved Vibrational Spectroscopy V*, Springer : Berlin, Takahashi, H., Ed., p., 57-59.
[7] Sasaki, J., Maeda, A., Kato, C., & Hamaguchi, H. (1993) *Biochemistry 32*, 867-871.
[8] Shichida, S., Matuoka, S., Hidaka, Y., & Toshizawa, T. (1983) *Biochim. Biophys. Acta 723*, 240-246.
[9] Fahmy, K., Siebert, F., & Tavan, P. (1991) *Biophys. J. 60*, 989-1001
[10] Lohrmann, R. (1992) *Ph. D. Thesis, University of Göttingen, Germany*
[11] Grossjean, M. F., Tavan, P., & Schulten, K. (1990) *J. Phys. Chem. 94*, 8059 - 8069.
[12] Gerwert, K., Hess, B., Soppa, J., & Oesterhelt, D. (1989) *Proc. Natl. Acad. Sci. USA 86*, 4943 - 4947.
[13] Rohr, M., Gärtner, W., Schweitzer, G., Holzwarth, A. R., & Braslavsky, S. E. (1992) *J. Phys. Chem. 96*, 6055 - 6061.
[14] Smith, S. O., Myers, A. B., Pardoen, J. A., Winkel, C., Mulder, P. J., Lugtenburg, J., & Mathies, R. (1984) *Proc. Natl. Acad. Sci. USA 81*, 2055-2059.
[15] Stockburger, M., Alshuth, T., Oesterhelt, D., & Gärtner, W. (1986) in *Spectroscopy of Biological Systems*, 483 - 535, John Wiley & Sons, Ed. Clark, R. J. H., & Hester, R. E.
[16] Harbison, G. S., Smith, S. O., Pardoen, J. A., Winkel, C., Lugtenburg, J., Herzfeld, J., Mathies, R., & Griffin, R. G. (1984) *Proc. Natl. Acad. Sci. USA 81*, 1706 - 1709.
[17] Ormos, P., Chu, K., & Mourant, J. (1992) *Biochemistry 31*, 6933-6937.
[18] Eisfeld, W., Pusch, C., Diller, R., Lohrmann, R., & Stockburger, M. (1993) *Biochemistry*, issue July.

Solvent Effects and Vibronic Couplings of a_g-Type Stretching Vibrations in T_1-Bacteriochlorophyll a and S_0- and S_1-Spheroidene: Environment of these Pigments in the LH2 Complex of *Rhodobacter sphaeroides*

Y. Koyama[1], *E. Nishizawa*[1], *M. Kuki*[1], *and H. Nagae*[2]

[1]Faculty of Science, Kwansei Gakuin University, Uegahara, Nishinomiya 662, Japan

[2]Kobe City University of Foreign Studies, Gakuen Higashi-Machi, Nishi-ku, Kobe 651-21, Japan

Theories of vibronic coupling predict that its strength is proportional to the square of coupling constants, and is inversely proportional to the energy gap of the electronic states involved; a coupling constant is proportional to the L_x matrix of the normal mode and the bond transition-density matrix between the pair of electronic states [1, 2, 3, 4]. Thus, Raman frequencies can reflect changes in the electronic energies through vibronic couplings.

1. The ring-breathing frequency of T_1-state bacteriochlorophyll a [3]

Figure 1 shows the ring-breaching (ν_r) frequencies (mainly Ca-Cm stretching) of bacteriochlorophyll (BChl) and bacteriopheophytin (BPhe) in the S_0 and T_1 states. In the S_0 state, the ν_r frequency has been used as a coordination marker [5]; It appears in the region of 1603 - 1610 cm^{-1} (1594 - 1600 cm^{-1}) in the penta-(hexa-)coordinated state. Depending on the strength of ligation, we classify solvents into (i) non-ligating solvents (dichloromethane and carbon tetrachloride), (ii) singly-ligating solvents and (iii) doubly-ligating solvents (the latter two kinds of solvents are defined to form the penta- and hexa-coordinated states in the S_0 state).

In the T_1 state, no discrete changes in the ν_r frequencies were found, although the frequency is in the order of non-ligating > singly-ligating > doubly-ligating solvents. Strikingly, the ν_r frequency decreases linearly as a function of $(n^2-1)/(n^2+2)$ in the region of 0.21 - 0.26. In the case of solvents which exhibit the linear relation, the oxygen or nitrogen atom of each solvent is expected to *axially* ligate the Mg atom in the center of the macrocycle. For the rest of the solvents, much stronger and direct van der Waals interaction between their large aliphatic or aromatic part and the aromatic macrocycle of BChl most probably modifies the above way of axial ligation. No solvent effects were observed for the ν_r Raman lines of BPhe in the S_0 and T_1 states.

In order to explain the linear relation, we built a bacteriochlorine model in the ligated complex, in which some electronic charge is transferred from the solvent oxygen or nitrogen to the Mg of BChl. Further, this axial dipole was assumed to be under the influence of the reaction field proportional to $(n^2-1)/$

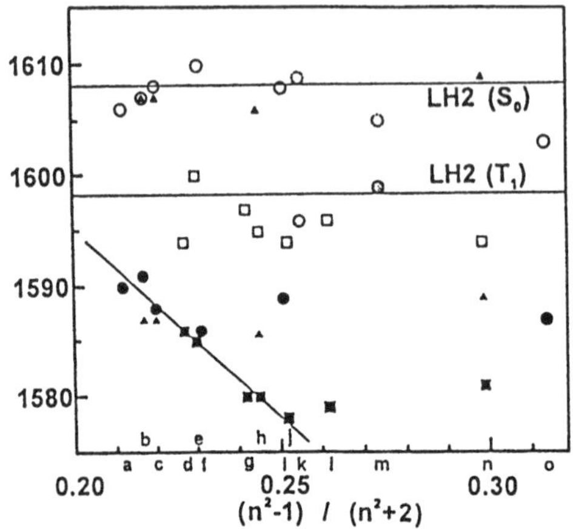

Figure 1. The ring-breathing (ν_r) frequencies of BChl and BPhe in the S_0 and T_1 states shown as a function of $(n^2-1)/(n^2+2)$. The S_0 frequencies of BChl in the non-ligating (⁘), singly-ligating (○) and doubly-ligating (□) solvents and those of BPhe(△); and T_1 frequencies of BChl in non-ligating (◎), singly-ligating (●) and doubly-ligating (■) solvents and those of BPhe (▲). Solvents: (a) acetonitrile, (b) diethyl ether, (c) acetone, (d) ethyl acetate, (e) 2-propanol, (f) 2-butanone, (g) 1-butanol, (h) tetrahydrofuran, (i) 2-octanone, (j) 1-hexanol, (k) dichloromethane, (l) 1-decanol, (m) carbon tetrachloride, (n) pyridine, and (o) benzyl alcohol.

(n^2+2). Thus, the effect of higher reaction field was assessed in model calculations by increasing the electronic charges on the dipole.

A PPP-CI calculation for the model showed that the lowest triplet state (T_1) is the $^3B_{2u}$ state, and that the nearest electronic state which can vibronically couple through the a_g-type vibration is the 2^3B_{2u} state. When the electronic charge on the dipole was increased, the 2^3B_{2u} - $^3B_{2u}$ energy gap decreased linearly. The same trend was seen for the optically-allowed transitions, which was evidenced by recording the $T_n \leftarrow T_1$ absorptions. The bond transition-density matrix calculated for the $2^3B_{2u} \leftarrow {}^3B_{2u}$ transition predicted a large coupling constant. All the above results support the idea that the linear decrease in the ν_r frequency can be explained in terms of the 2^3B_{2u} - $^3B_{2u}$ vibronic coupling.

2. The C=C stretching frequencies of S_0- and S_1-spheroidene [4]

Figure 2 shows the frequencies of the C=C stretching (a_g-type) of (a) S_0- and (b) S_1-spheroidene. The considerably low S_0-state frequencies and the abnormally high S_1-state frequencies indicate the presence of a vibronic coupling between these states through the particular vibration, as in the case of β-carotene [6]. Since the S_0 state of the carotenoid is the $^1A_g^-$ state, the S_1 state which can vibronically couple with it through the a_g-type vibration is the

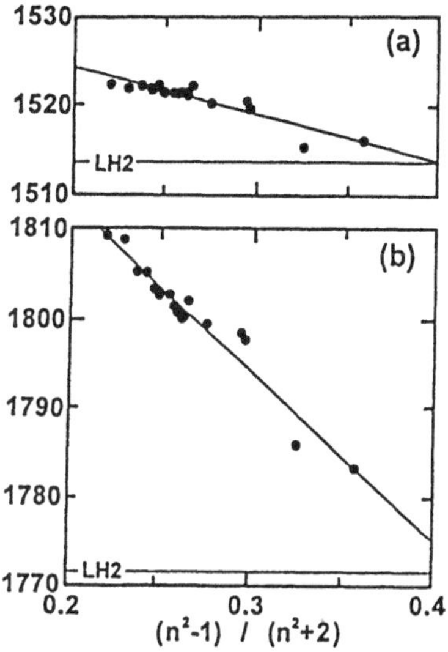

Figure 2. The C=C stretching (a_g) frequencies of spheroidene in the (a) S_0 ($^1A_g^-$) and (b) S_1 ($2^1A_g^-$) states shown as a function of $(n^2-1)/(n^2+2)$. Solvents [in the order of $(n^2-1)/(n^2+2)$]: n-pentane, n-hexane, n-octane, n-nonane, n-decane, n-dodecane, n-tridecane, n-tetradecane, n-pentadecane, n-hexadecane, n-hexane / CS_2 (3 / 1), n-undecane, carbon tetrachloride, n-hexane / CS_2 (1 / 1), benzene, n-hexane / CS_2=1 / 3, and carbon disulfide.

$2^1A_g^-$state. The $^1A_g^-$- and $2^1A_g^-$ -state frequencies decrease as a linear function of $(n^2-1)/(n^2+2)$. The slope in the $2^1A_g^-$ state is steeper than that in the $^1A_g^-$ state, a fact which indicates that an additional vibronic coupling with the $3^1A_g^-$ state takes place.

Fluorescence spectroscopy of spheroidene at 170 K in n-hexane, CS_2 and their mixtures determined the energies of the $^1A_g^-$, $2^1A_g^-$ and $3^1A_g^-$ states to be $E_1 = 0$, $E_2 = 15403 - 2124\ (n^2-1)/(n^2+2)$, and $E_3 = 25532 - 5983\ (n^2-1)/(n^2+2)$ [7]. Calculations of bond transition-density matrices showed that the coupling constants with the $2^1A_g^-$ state are in the order $^1A_g^- > 3^1A_g^- >> 4^1A_g^-$ states. Thus, we took into account a pair of vibronic couplings, i.e. one between the $^1A_g^-$ and $2^1A_g^-$ states and the other between the $2^1A_g^-$ and $3^1A_g^-$ states, and we calculated the diabatic frequencies and the coupling constants to be ν_1^0=1914 cm^{-1}, ν_2^0=2103 cm^{-1}, V_{12}=2417 cm^{-1} and V_{23}=2530 cm^{-1}, by the use of the above set of energies and the relations between the C=C stretching frequencies and $(n^2-1)/(n^2+2)$ shown in the figure.

3. Environment of BChl and spheroidene in the LH2 complex [3, 4]

Figure 1 shows also the ν_r frequencies in the S_0 and T_1 states of BChl which is bound to the light-harvesting (LH2) complex of *Rhodobacter sphaeroides* 2.4.1. In the S_0 state, the bound BChl is in the penta-(or tetra-)coordinated state. In the T_1 state, it is not ligated at all or very weakly ligated, since its frequency corresponds to those in dichloromethane and carbon tetrachloride.

Figure 2 shows the C=C stretching frequencies in the $^1A_g^-$(S_0) and $2^1A_g^-$ (S_1) states of spheroidene bound to the same LH2 complex. The frequency for the bound spheroidene is considerably (by far) lower than those for spheroidene in solutions in the $^1A_g^-$ ($2^1A_g^-$) state. By using the energies of the

$2^1A_g^-$ and $3^1A_g^-$ states of the bound spheroidene (14500 and 23200 cm^{-1}, respectively) which were determined by fluorescence spectroscopy at 170 K [8], and by using also the set of diabatic frequencies and coupling constants obtained above, the frequency for the bound spheroidene was calculated to be 1511 (1770) cm^{-1} in the $^1A_g^-$($2^1A_g^-$) state in good agreement with the observed value of 1514 (1772) cm^{-1}. The results indicate that not the coupling constants but the state energies of the carotenoid are changed upon binding.

References

[1]F. Zerbetto, M.Z. Zgierski, G. Orland and G. Marconi, J. Chem. Phys. **87**, 2505 (1987).

[2]H. Nagae, T. Kakitani, T. Katoh and M. Mimuro, J. Chem. Phys. **98**, 8012 (1993).

[3]E. Nishizawa, L. Limantara, N. Nanjou, H. Nagae, T. Kakuno and Y. Koyama, submitted to Photochem. Photobiol.

[4]M. Kuki, H. Nagae, R.J. Cogdell, K. Shimada and Y. Koyama, submitted to Photochem. Photobiol.

[5]P.M. Callahan and T.M. Cotton, J. Am. Chem. Soc. **109**, 7001 (1987).

[6]H. Hashimoto and Y. Koyama, Chem. Phys. Lett. **154**, 321 (1989).

[7]Y. Watanabe, T. Kameyama, Y. Miki, M. Kuki and Y. Koyama, Chem. Phys. Lett. **206**, 62 (1993).

[8]Y. Koyama, Y. Miki, T. Kameyama, R.J. Cogdell and Y. Watanabe, Chem. Phys. Lett. **208**, 479 (1993).

Probing Low-Frequency Vibrational Motion in Bacterial Reaction Centers by Femtosecond Absorption Spectroscopy

M.H. Vos[1], *F. Rappaport*[1], *J.-Ch. Lambry*[1], *J. Breton*[2], *and J.-L. Martin*[1]

[1]Laboratoire d'Optique Appliquée, INSERM U275,
Ecole Polytechnique-ENSTA, F-91120 Palaiseau, France

[2]Service de Biophysique, Département de Biologie,
Centre d'Etudes Nucléaires de Saclay, F-91191 Gif-sur-Yvette Cedex, France

Abstract. Oscillatory features in the stimulated emission kinetics of chromatophores containing ***Rhodobacter capsulatus*** D_{LL} reaction centers reflect vibrational motion on the excited state potential energy surface. The dependence of the features on the probe wavelength is discussed with an emphasis on the detection of overtones.

1. Introduction

Vibrational motion in simple alkali molecules in molecular jets can be visualized using femtosecond optical spectroscopy, as has been shown by Zewail and coworkers [1]. Recently these studies have been extended to condensed phase environments by several groups [2-5]. In small molecules, such as iodine, the analysis is facilitated by the restricted number of vibrational modes and the known potential energy surfaces. However, the same concepts may apply for large molecules where the method in principle allows to specifically study vibrational modes which are coupled to an optical transition.

Low frequency (period > 400 fs) features were observed in bacterial reaction centers [6], which are multiple chromophore membrane proteins. We have recently shown that these oscillations reflect low-frequency (< 100 cm^{-1}) vibrational motion on the excited state potential energy surface [7]. Thus a physiological perturbation invokes specific global (because of the low frequency [8]) protein motions and the phase of these motions is preserved for a period of a few picoseconds. As the excited state is the precursor of electron transport in functional reaction centers and as primary electron transport also takes place in a few picoseconds [9,10], this finding may have important consequences for our understanding of the mechanism of primary electron transport in reaction centers.

The main evidence for the assignment of the observed vibrational modes to excited state rather than ground state modes came from the dependence of the features on the probe wavelength when scanning the stimulated emission band of a mutant in which electron transfer is blocked and in which the excited state is long-lived. At both sides of the stimulated emission band the fundamental frequencies are best detected (with opposite phases) and at the center of the band the first overtone, as well as sum- and difference frequencies of two vibrational modes, are observed with a much lower amplitude. In this paper we will discuss the origin of the latter features in detail.

2. Materials and Methods

Reaction centers of the D_{LL} mutant of *Rb. capsulatus* do not contain the bacteriopheophytin primary electron acceptor H_L [11]. Chromatophores containing no other pigmented proteins than D_{LL} reaction centers were prepared by S.J. Robles and D.C. Youvan as described [7,12]. Stimulated emission kinetics were recorded at 10K as described [7], using a 40 fs pump pulse (full width half maximum 25 nm centred at 870 nm) and a 30 fs continuum probe pulse.

3. Results

Fundamental frequency (ν) features of 15 cm^{-1} (ν_1) and 77 cm^{-1} (ν_2) are observed at both sides of the stimulated emission band (Fig. 1), with a diminishing relative amplitude towards the peak of the band. One expects to observe the first overtone component (see Discussion) at the peak of the band, i.e. near 920 nm. We concentrated on the 77 cm^{-1} feature, as its overtone component should be well separated from its fundamental component, in contrast to the 15 cm^{-1} feature. At λ_{probe} = 921.5 nm weak but significant irregular oscillatory features are observed. The Fourier Transform (FT) of the oscillatory part shows no oscillator strength at 77 cm^{-1} and the peak at 15 cm^{-1} is also strongly diminished compared to λ_{probe} = 945 nm. The main feature appears at 153 cm^{-1} and additional peaks are visible at 62 and 91 cm^{-1}. These frequencies correspond to, respectively, $2\nu_2$, $\nu_2 - \nu_1$ and $\nu_2 + \nu_1$.

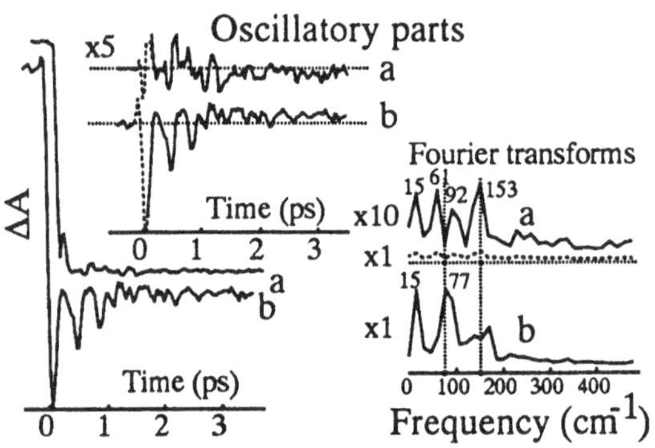

Fig. 1 Kinetics of stimulated emission taken at probe wavelengths of 921.5 nm (a) and 945 nm (b) and normalized to the same total signal. Oscillatory parts were obtained by subtraction of a Heaviside-function convoluted with the instrument-response function from the data. Fourier transforms were taken from the oscillatory parts after the pump-probe overlap time.

4. Discussion

We will first discuss the observation of the 2ν component at the bottom of the well in a scheme of one-dimensional wave packet motion. In the case of displaced harmonic potential energy surfaces for the ground and excited states, a linear relation holds between the (probe) transition energy and the nuclear coordinate probed, for each site. A wave packet which is created off-equilibrium by the pump pulse will move on the excited state potential energy surface with the frequency of the well (Fig. 2). A narrow and non-dephasing wave packet dwells once per vibrational period at each turning point and twice at all coordinates between the turning points. In the latter case, the time delays between the passages of the wave packets are unequally spaced, with the exception of the bottom of the well. Fourier transformation of the packet passage patterns leads to frequency components at ν, 2ν, 3ν, etc.. The relative strengths of the different components depends on the probed coordinate and on the width of the wave packet. The fundamental (ν) frequency component diminishes when tuning from the turning points towards the bottom of the well, where it is absent. Widening the wave

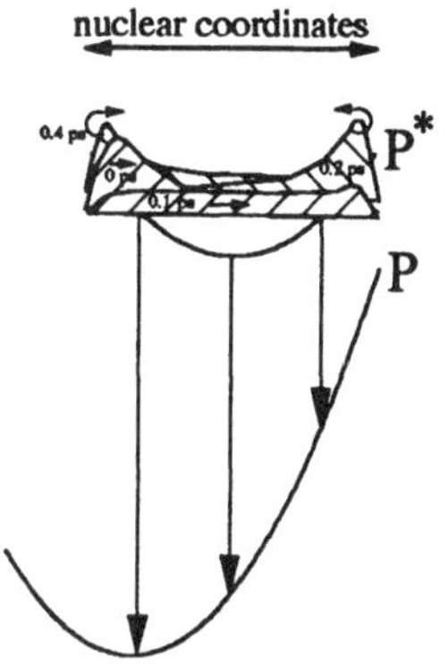

Fig. 2 One-dimensional scheme of potential energy surfaces of the ground state (P) and the excited state (P) of bacterial reaction centers. The product state (P^+) is not shown, because electron transfer is blocked in the D_{LL} mutant of R. capsulatus. Wave packets were simulated for a well frequency of 75 cm^{-1}, at different times after population of the excited state with an 80 fs pump pulse. Dephasing was modelled by diffusion of the wave packet in phase coordinates, using a model inspired by the frozen Gaussian approximation [15]. Vertical arrows indicate the different transition energies to the ground state, corresponding to different stimulated emission probe wavelengths.*

packets makes the passage pattern more sinusoidal and therefore relatively diminishes, at all probing coordinates, the overtone components. Moreover, near-spaced double passages of the wave-packet congest to one passage and thus extend the virtual "turning point" coordinate area. Therefore, widening of wave packets leads to a restriction of the area in which the 2ν component can be clearly observed. In general, the momentum of the wave packet is highest near the bottom of the well, which leads to both short dwell times and relative broadening of the wave packets in nuclear coordinates as compared to the turning points. For the case of reaction centers two more phenomena are of importance in understanding the low relative amplitude of the 2ν component:

- site inhomogeneous broadening is quite substantial (~170 cm^{-1} of a total 500 cm^{-1} fwhm band [13,14]). This implies that, at a single transition energy, one probes a distribution of sites and thus a distribution of nuclear coordinates. This especially masks the 2ν features, which are sensitive to the probe coordinate.
- dephasing mechanisms (quasi-elastic and non-elastic collisions and anharmonicity) progressively broaden the wave packet and eventually destroy its dynamics. This is illustrated in Figure 2.

The wavelength dependence of the fundamental frequency components clearly identified both the 77 and the 15 cm^{-1} modes as excited state vibrational modes [7]. In the most simple, harmonic, picture one would expect the disappearance of the fundamental (ν) components and appearance of the 2ν components for both modes at the same probe wavelength. At the probe wavelength (921.5 nm) where the 77 cm^{-1} feature disappears, a minor 15 cm^{-1} feature is still observed (Fig. 1). However, that the analysis of the 15 cm^{-1} feature is complicated by its strong damping within one vibrational period (~2 ps). This gives a relatively large incertitude (± 8 cm^{-1}) in its frequency and therefore the fundamental and first overtone features may be congested. Another complication is that the probe wavelengths for the apparent minimum of the well may be slightly displaced for the two vibrational modes, due to non-harmonic potential energy surfaces. A more detailed investigation of the wavelength region near 921.5 nm is needed to clarify this point.

Finally we note that the one-dimensional potential energy surface picture is a simplification which cannot reproduce all details of the experimentally observed features. It is clear that at least two low-frequency modes are coupled to the optical transition. In a two-dimensional potential energy surface picture center of the wave packet follows a Lissajous-like pathway which spirals towards the center. At each probe wavelength one probes a line cross-section of the plane of motion. Semi-classical simulations (not shown) predict the presence of mixed frequency components ($\nu_2 - \nu_1$ and $\nu_2 + \nu_1$) in the probe kinetics, as observed at λ_{probe} = 921.5 nm (Fig. 1). As for the 2ν components, they should be best observable near the bottom of the well, where the dominating ν components vanish.

Acknowledgments. D_{LL} mutant bacteria's were kindly supplied by S.J. Robles and D.C. Youvan. M.H.V. is recipient of an EMBO long term fellowship during the course of this work. Part of this work has been supported by the Human Frontiers Science Organization.

References

1 L.R. Khundkar and A.H Zewail, Annu. Rev. Phys. Chem. **41**, 15 (1990).
2 N.F. Scherer, L.D. Ziegler, and G.R. Fleming, J. Chem. Phys. **96**, 5544 (1992).
3 U. Banin, A. Waldman and S. Ruhman, in *Ultrafast Phenomena VIII*, edited by J.-L. Martin, A. Migus, G.A. Mourou and A.H. Zewail (Springer, Berlin, 1993), p. 68.
4 K. Wynne, C. Galli and R.M. Hochstrasser Chem. Phys. Lett. **193**, 17 (1992).
5 Y. Yan, Y., R.M. Whitnell, K.R. Wilson and A.H. Zewail, Chem. Phys. Lett. **193**, 402 (1992).
6 M.H. Vos, J.-C. Lambry, S.J. Robles, D.C. Youvan, J. Breton, and J.-L. Martin, Proc. Natl. Acad. Sci. USA **88**, 8885 (1991).
7 M.H. Vos, F. Rappaport, J.-C. Lambry, J. Breton, and J.-L. Martin, Nature **363**, 320 (1993).
8 Gō, N., Noguti, T. & Nishikawa, T., Proc. Natl. Acad. Sci. USA **80**, 3696 (1983).
9 J.-L. Martin, J. Breton, A.J. Hoff, A. Migus, and A. Antonetti, Proc. Natl. Acad. Sci. USA **83**, 957 (1986).
10 J.-L. Martin and M.H. Vos, Annu. Rev. Biophys. Biomol. Struct **21**, 199 (1992).
11 S.J. Robles, J. Breton and D.C. Youvan, Science **248**, 1402 (1990).
12 E.J. Bylina, E.J., S.J. Robles and D.C. Youvan, Isr. J. Chem. **28**, 73 (1988).
13 S.G. Johnson, D. Tang, R. Jankowiak, J.M. Hayes, G.J. Small, and D.M. Tiede J. Phys. Chem. **94**, 5849 (1990).
14 T.R. Middendorf, T.R., L.T. Mazzola, D.F. Gaul, C.C. Schenck, and S.G. Boxer, J. Phys. Chem. **95**, 10142 (1991).
15 E.J. Heller, J. Chem. Phys. **75**, 2923 (1981).

Picosecond Infrared Spectroscopy of the Photosynthetic Reaction Center

S. Maiti, B.R. Cowen, R. Diller, M. Iannone, C.C. Moser, P.L. Dutton, and R.M. Hochstrasser

Department of Chemistry and Department of Biophysics and Biochemistry, University of Pennsylvania, Philadelphia, PA 19104-6323, USA

Abstract. We report the first picosecond infrared spectroscopic study on the reaction center of bacterial photosynthesis. Infrared difference spectra in the range from 1560 to 1800 cm^{-1} were recorded and investigated from 50 ps to 4 ns after photooxidation of the primary donor. Bands at 1755, 1707 and 1683 cm^{-1} can be assigned to vibrations of cofactor carbonyl groups. Protein contributions at 1644, 1661 and 1665 cm^{-1} (amide I) and at 1727 cm^{-1} (carboxylic residues) and their kinetics reflect the fast protein response to intermolecular charge transfer.

The study of protein dynamics is of substantial importance for the understanding of protein function. A prerequisite for time resolved experiments on this subject is a fast molecular trigger as provided by a photoreactive chromophore. In the past, protein dynamics on the picosecond (ps) and femtosecond (fs) time scale have been addressed on an experimental basis mostly by means of optical spectroscopic methods (cofactor response to the perturbation). Though these methods are very useful for the characterization of chromophore dynamics, information concerning the protein moiety can only be derived indirectly. Time resolved infrared (IR) spectroscopy, however, is a method to monitor protein dynamics directly on the basis of spatially more or less localized vibrational modes. Recently we have applied this technique to Bacteriorhodpsin [1,2] and now to the bacterial reaction center (RC) [3]. The electron transfer kinetics of the reaction center within the first nanosecond is characterized by a ~3 ps step for the transfer from the primary donor, the dimeric bacteriochlorophyll ($BChl_2$), to the intermediate acceptor bacteriopheophytin (BPh), and by a 200 ps step for the succeeding transfer to the Quinone-A (Q_A) [4-7]. The molecular structure is known from X - ray diffraction studies [8]. Fourier Transform Infrared (FTIR) (for a review see [9]) and μs time resolved IR studies (e. g. [10]) already provide a wealth of structural information concerning the charge separated states. The transient IR absorbance changes presented here yield dynamic information about how the protein medium responds to the electron transfer induced change of the electric field on the ps time scale and how it becomes involved in the electron transfer process.

The experiments were performed at ambient temperature. Partially dehydrated films and suspensions of rhodobacter sphaeroides reaction centers embedded in lipid vesicles were used. Two dye lasers at 760 and 810 nm were pumped by the second harmonic of a Q-switched mode-locked Nd-YAG laser operating at 500 Hz. The 810 nm beam (200 nJ/pulse) was chopped at 250 Hz and used for excitation of the bacteriochlorophyll monomer (BChl). This leads to a subpicosecond energy transfer to the $BChl_2$ that initiates electron transfer [6]. The

Springer Proceedings in Physics, Vol. 74 Time-Resolved Vibrational Spectroscopy VI
Editors: A. Lau, F. Siebert, W. Werncke

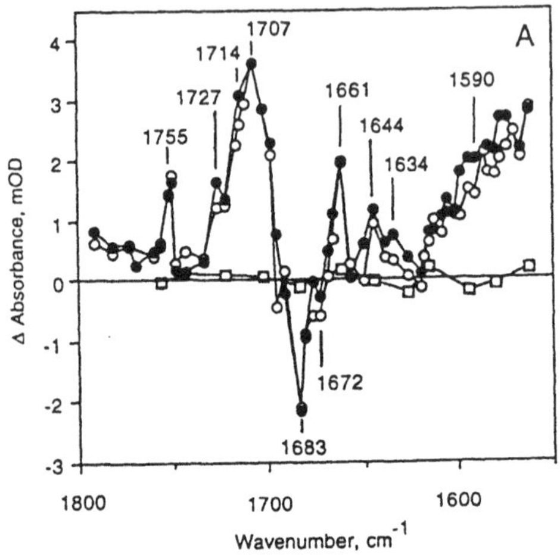

Figure 1. Infrared difference spectra (photoexcited minus ground state) at 70 (●) and 600 ps (O) of a partially dehydrated reaction center film. Control spectrum obtained from a lipid film free of RC (□). Errors in measurements are 0.3 mOD or smaller. Neighboring points are joined by straight lines in order to clarify the Figure - these lines do not necessarily reproduce the correct spectral shape.

760 nm beam (2μJ/pulse) was passed through a delay line, and used to upconvert the probe infrared CO laser radiation in an $AgGaS_2$ nonlinear crystal. Further processing of the sum frequency pulses via a photomultiplier, lock-in amplifiers and a computer yielded the photoinduced IR absorbance changes. The instrument function was about 60 ps, permitting the identification of transients in the 50 ps range. More details and principles of the instrument are provided in reference [5]. In order to check the activity of the film preparations, the bleach and the recovery of the 860 nm $BChl_2$ band was measured on a millisecond time scale.

Fig. 1 shows IR difference spectra (absorbance of excited minus unexcited reaction center film) at delay times of 70 and 600 ps. It is known [11] that at 70 ps the excited sample contains mainly $BChl_2^+BPh^-Q_A$, whereas at 600 ps the dominant species is $BChl_2^+BPhQ_A^-$.

Our assignments of IR difference absorption bands to reaction center vibrational modes are based on IR group frequencies and the results of FTIR studies ([9] and references therein). We assign the bands at 1755, 1707, 1683 and 1590 cm^{-1} to the $BChl_2^+$ 10a ester C=O stretch, the $BChl_2^+$ 9keto C=O stretch, the $BChl_2$ 9keto C=O stretch and C=C stretch vibrations of BPh^-, respectively. The broad positive feature between 1600 and 1562 cm^{-1} is probably due to various C=C stretches of $BChl_2^+$. The bands at 1727, 1672 and 1661 cm^{-1} are most likely protein contributions from an amino acid side chain carboxylic C=O stretch (1727 cm^{-1}) and amide I backbone C=O stretches. A superposition of amide I and BPh^- contributions leads probably to the doublet at 1644 cm^{-1}. The 1714 cm^{-1} band is assigned to a BPh^- ester carbonyl or a C=O stretch of a protein carboxylic group near BPh^-. The 70 ps spectrum demonstrates that not only the cofactors but also parts of the protein adjust on this time scale to the new charge distribution.

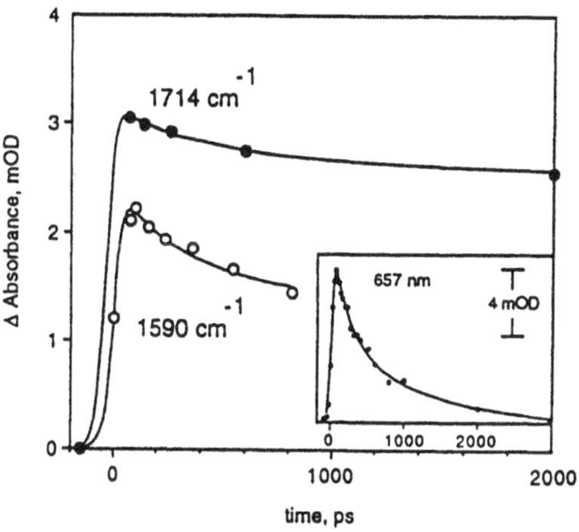

Figure 2. Evolution of IR absorbance changes at 1714 (●) and 1590 cm^{-1} (O). *Inset.* Absorbance change of RC film at 657 nm monitoring electron transfer by probing the BPh^- population. The lines are fit to a double exponential $A_0\Theta(t) + A_1e^{-t/\tau_1} + A_2e^{-t/\tau_2}$, convoluted with the instrument response function, with $A_1/A_2 = 1.0(0.2)$, $\tau_1 = 220$ ps and $\tau_2 = 1200(200)$ ps, where $\Theta(t)$ represents a step function corresponding to creation of $BChl_2^-$. For 657 nm, $A_0 = 0$; for 1714 cm^{-1}, $A_0/A_1 = 7(1)$; for 1590 cm^{-1}, $A_0/A_1 = 1.7(0.3)$.

Changes between the 70 and the 600 ps spectrum can be seen at 1714, 1672, 1644 and 1590 cm^{-1}. In order to compare the time evolution of IR and optical absorbance changes, kinetics were taken at 657 nm and at 1714 and 1590 cm^{-1} (Fig.2). They can be fit well with the same set of two time constants (see caption) showing the strong time correlation of vibrational and electronic response at least for these two vibrational modes.

IR difference spectra and kinetics of reaction centers suspended in D_2O (data not shown) demonstrate the impact of water on reaction center dynamics. The appearance of a narrow band at 1665 cm^{-1} (not seen in the spectra of the film sample) which we assign to an amide I mode near BPh^- or Q_A^- indicates a higher flexibility of a specific part of the backbone. Its kinetics compare well with that of electron transfer probed by the BPh^- absorption at 657 nm. Here the electron transfer from BPh^- to Q_A is fully determined by a 220 ps time constant.

We have shown that the dynamic changes following electron transfer taking place in a protein can be measured on picosecond time scales by infrared difference spectroscopy. While no global conformational change could be identified during the primary steps of charge separation in the reaction center, specific regions in the protein are seen to be selectively perturbed. Contrary to this result a molecular dynamics simulation on reaction center [12] predicts a rather uniform structural reorganization with the protein medium behaving akin to a dielectric continuum. We find the first protein response to be faster than 50 ps, later relaxation processes most likely take longer than 4 ns. This behaviour is also found in Bacteriorhodopsin although here the initial perturbation is of steric and not exclusively of electronic nature. Meanwhile similar experiments were performed on the femtosecond timescale and are subject of another publication.

References

1. R. Diller, M. Iannone, R. Bogomolni and R. M. Hochstrasser, Biophys. J. **60**, 286 (1991).
2. R. Diller, M. Iannone, B. R. Cowen, S. Maiti, R. Bogomolni and R. M. Hochstrasser, Biochem. **31**, 5567 (1992).
3. S. Maiti, B. R. Cowen, R. Diller, M. Iannone, C. C. Moser, P. L. Dutton and R. M. Hochstrasser, Proc. Natl. Acad. Sci. USA **90**, 5247 (1993).
4. M. H. Vos, J. Lambry, S. J. Robles, D. C. Youvan, J. Breton and J. L. Martin, Proc. Natl. Acad. Sci. USA **89**, 613 (1992).
5. M. Du, S. J. Rosenthal, X. Xie, T. J. DiMagno, M. Schmidt, D. K. Hansen, M. Schiffer, J. R. Norris and G. R. Fleming, Proc. Natl. Acad. Sci. USA **89**, 5817 (1992).
6. C. Kirmaier and D. Holten, Proc. Natl. Acad. Sci. USA **87**, 3552 (1990).
7. W. Holzapfel, U. Finkele, W. Kaiser, D. Oesterhelt, H. Scheer, H. U. Stilz and W. Zinth, Proc. Natl. Acad. Sci. USA **87**, 5168 (1990).
8. J. Deisenhofer and H. Michel, Ann. Rev. Biophys. Biophys. Chem. **20**, 247 (1991).
9. W. Mäntele, *The Photosynthetic Reaction Center*, J. Deisenhofer and J. Norris, Eds. (Academic Press, New York, 1992) vol.2, in press.
10. R. Hienerwadel, D. Thibodeau, F. Lenz, E. Nabedryck, J. Breton, W. Kreutz and W. Mäntele, Biochem. **31**, 5799 (1992).
11. C. Kirmaier and D. Holten, Photosyn. Res. **13**, 225 (1987).
12. H. Treutlein, K. Schulten, A. T. Brunger, M. Karplus, J. Deisenhofer and H. Michel, Proc. Natl. Acad. Sci. USA **89**, 75 (1992).

Time-Resolved FTIR Spectroscopy of the Rebinding of Photodissociated Carboxy-Myoglobin

C. Rödig, O. Weidlich, and F. Siebert

Institut für Biophysik und Strahlenbiologie, Albert-Ludwigs-Universität, Albertstraße 23, D-79104 Freiburg, Germany

Abstract. The recombination of Myoglobin (Mb) and CO after laser flash induced photolysis of Carboxymyoglobin (MbCO) in aqueous solution has been studied by Time-resolved Fourier-Transform Infrared Spectroscopy (FTIR) using a Step-Scan-Interferometer. Difference specra in the spectral range between 1100 cm^{-1} and 2100 cm^{-1} were obtained at different times after photodissociation. The kinetics of the CO stretching band and the amide-II band show effects of photoselection.

Myoglobin is a dioxygen storage protein showing some characteristics that make it ideal as a model system for investigations about interactions between protein structure and function. Firstly, it is a relatively simple protein so that the knowledge of its features could be helpful in elucidating those of more complex heme proteins. Secondly, Mb is capable of reversible binding of CO which can easily be photodissociated with a quantum yield of nearly 1. This reversible ligand binding capacity must depend on specific interactions of the heme with the protein, and therefore the direct coupling between biological function and protein dynamics can be observed in the ligand binding process of Mb.

The method of time-resolved vibrational spectroscopy is appropriate to provide information about molecular relaxation processes as stated before. Until now, Mb has been investigated by dispersive infrared spectroscopy in the spectral range of about 1900 cm^{-1} to 2200 cm^{-1} with time resolutions from the fs- up to ks-range.[1,2,3] The advantages of the time-resolved FTIR-spectroscopy (i.e. multiplex advantage) enabled us to take measurements in the whole spectral range between 1100 cm^{-1} and 2100 cm^{-1} with a time resolution of 100 ns. The time resolution in our experiments was obtained by using the step-scan technique which requires the implementation of a special interferometer into the spectromter as described earlier.[4]

The principle of our measurements can be described in the following way: the interferometer mirror is held at a fixed postion. Subsequently a flash of an excimer-laser-pumped dye-laser (wavelength 540 nm) is applied to the Mb-sample, the data acquisition is optically triggered and the time course of the change of the interferogram is recorded. By moving the mirror stepwise, this procedure is repeated at each sampling point. From these data the spectral changes at each time of interest can be computed by the Fourier-Transformation.

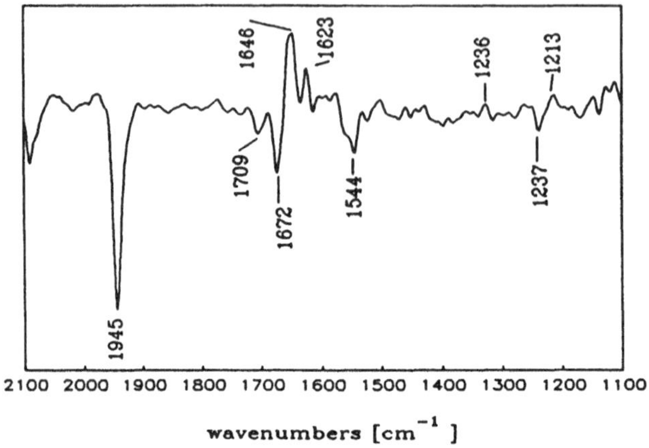

Figure 1. Difference spectrum of a 5 mM solution of MbCO in 0.1 M phosphate buffer (pH 7) on CaF_2 windows, pathlength 5 μm at ambient temperature. The spectrum was obtained at a time 700 ns after flash with a spectral resolution of 4 cm^{-1}. In order to reduce the noise level two spectra were added.

Figure 1 shows an IR difference spectrum of photodissociated MbCO in H_2O measured 700 ns after the flash with a spectral resolution of 4 cm^{-1}.The largest absorbance change at 1945 cm^{-1} is due to the CO stretching mode of the liganded CO. Further important changes can be seen in the amide-I region between 1700 cm^{-1} and 1620 cm^{-1}. Bands in this spectral range are generally assigned to the C=O stretching mode of the peptide bonds. The signal-noise ratio in this region is decreased because of the lower transmission due to the water molecules of the solvent. Another large band appears in the amide-II region at 1544 cm^{-1}. It corresponds mainly to the NH in-plane bending mode, but also other parts of the peptide groups like C-N, C^α-C (bending modes) and CO (in-plane bending mode) are contributing to the amide-II band. Both amide bands are therefore indicators for changes in the backbone of the protein. Especially shifts of these bands in deuterated samples could help to assign these bands. In the range below 1500 cm^{-1} several bands, due to amino acid side-chaines, can be observed (significant bands are marked in Fig.1). To make more detailed assumptions about the spectral range below 1500 cm^{-1}, it is desireable to improve the signal-noise ratio.

The kinetics of the bands at 1945 cm^{-1} and 1544 cm^{-1} are shown in Figure 2. The fast rise time of the two kinetics is due to the detector. At least three processes with different rate constants are occuring in the signal of the CO-stretching mode of the ligand (Fig. 2b). There is a slight increase of the signal until a time of about 10 microseconds before the signal is relaxing within a time of about 1 ms. Corresponding to the increase of the signal in fig. 2b there is a rise in the kinetic of the amide-II band (Fig. 2a) within the same time. The relaxation of the amide-II

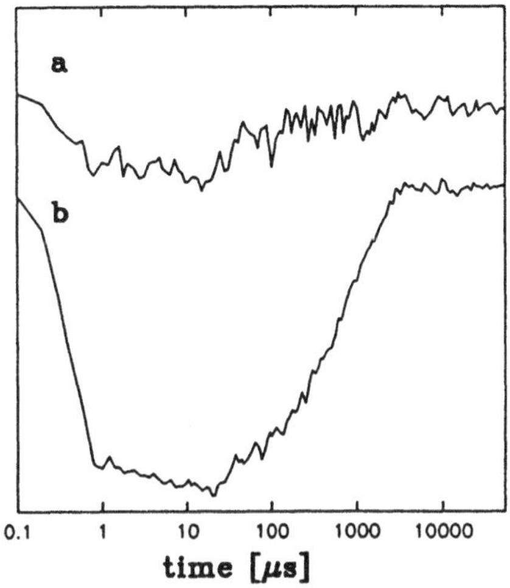

Figure 2. Time evolution of the IR absorbance changes at 1945 cm^{-1} and 1545 cm^{-1}.
(a) The kinetic data at 1945 cm^{-1} are representing rebinding of CO.
(b) The signal at 1544 cm^{-1} is due to changes in the amide-II band and, thus, reflects protein structural changes.

signal is much faster than the overall relaxation of the rebinding signal of the ligand. In both signals the relaxation appears to be dominated by two processes. The relative amplitude of the fast process is larger in the signal caused by the amide-II mode. The occurance of processes with the same kinetics in these different parts of the molecule show that there is a close relationship between structural changes and the rebinding process.

The slow increase of the signal in the first microseconds after the flash can be explained by assuming a photoselection process during the excitation with the laser pulse. An anisotropic distribution of liganded and unliganded molecules is created by photoselection (depending on the orientation of the transition moments of the heme-CO bonds). The decraese of this anisotropy is monitored by the infrared probe beam, which is orientated nearly parallel to the excitation pulse, when the molecules randomize by rotational diffusion.[5]

Our measurements show that the step-scan technique is a suitable method to provide information about backbone conformational processes during the CO rebinding to Mb after photodissociation. In order to assign the bands in the amide-I, amide-II and amide-III region in more detail, it is desireable to carry out experiments with both deuterated Mb and mutants of Mb. Also a better time resolution together with low temperature experiments could help to get more insights into the kinetics of the rebinding process.

References

1. R.H.Austin, K.W.Beeson, L.Eisenstein, H.Frauenfelder Biochemistry, **14**, 5355 (1975)
2. A.J.Dixon, P.Glyn, M.Healy, P.Hodges, T.Jenkins, M.Poliakoff, J.T.Turner, Spectrochimica Acta, **44A**, 1309 (1988)
3. P.A.Anfinrud, C.Han, R.M.Hochstrasser Proc.Natl.Acad.Sci., **86**, 8387 (1989)
4. W.Uhmann, A.Becker, C.Taran, F.Siebert Applied Spectroscopy, **45**, 390 (1991)
5. A.Ansari, C.M.Jones, E.R.Henry, J.Hofrichter, W.A.Eaton Biophys. J., **64**, 852 (1993)

Time-Resolved Infrared Spectroscopy of Proton Transfer in Reaction Centers of Photosynthetic Bacteria

R. Hienerwadel[1], *S. Grzybek*[1], *C. Fogel*[1], *W. Mäntele*[1], *J. Breton*[2], *E. Nabedryk*[2], *and M.Y. Okamura*[3]

[1]Institut für Biophysik und Strahlenbiologie der Universität Freiburg, Albertstraße 23, D-79104 Freiburg, Germany
[2]SBE/DBCM, CEN Saclay, F-91191 Gif-sur-Yvette Cédex, France
[3]Department of Physics, UCSD, La Jolla, CA 92093-0319, USA

Abstract. Electron transfer in photosynthetic reaction centers from *Rhodobacter sphaeroides* was investigated by time-resolved infrared spectroscopy using tunable lead salt diode lasers. In the spectral region between approximately 1690 and 1770 cm^{-1}, transient µsec to msec signals could be characterized upon electron transfer from the primary quinone acceptor Q_A to the secondary acceptor Q_B. The dominant transient absorbance changes are three positive signals at 1725, 1735, and 1706 cm^{-1} and two negative signals at ca. 1716 and at 1698 cm^{-1}. They are interpreted in terms of the C=O frequencies of aspartic or glutamic acid side chain residues affected by electron transfer. The major signal at 1725 cm^{-1} exhibits a fast kinetic component of $t_{1/2}$ = 0.2 ms and a slow one of $t_{1/2}$ = 1 ms at pH 7. Analysis of its time course and amplitude under different conditions (pH, $^1H/^2H$ echange, mutations) allows its assignment to glutamic acid L212 near Q_B and indicates the uptake of a significant amount of a proton upon Q_B^- formation. We propose that this proton transfer leads to a favorable electrostatic interaction in the site since it lowers the interaction energy of Glu L212 with Q_B^-.

1. Introduction

In photosynthetic reaction centers (RC), excitation of the bacteriochlorophyll dimer (P) results in transfer of an electron to the primary electron acceptor quinone (Q_A) in ≈ 200 ps, and from there to the secondary electron acceptor Q_B in ≈ 0.2 ms. Q_B^- is stable for several hundred ms and can proceed to the quinol Q_BH_2 with a second turnover of this reaction. The coupling of electron transfer to proton transfer occurs on the level of the secondary electron acceptor quinone, and involves proton binding to the protein with the formation of Q_B^- and proton transfer to Q_B as a consequence of the second turnover (for a review on these processes see [1]). The X-ray crystal structure of the photosynthetic RC of *Rhodopseudomonas viridis* [2] and *Rhodobacter* (*Rb.*) *sphaeroides* [3,4] revealed that Q_B is occluded within the protein out of contact with bulk water, and suggested a number of protonable amino acid side chain residues which could be responsible for proton transfer to Q_B. The role of some of these residues has been probed by site-directed mutagenesis (reviewed in [1]) revealing sites that can block (i.e. slow down rates by 2-3 orders of magnitude) both electron and proton transfer, such as Ser L223 and Asp L213, and sites that block only proton transfer (such as Glu L212) upon mutation. The measurement of electron transfer rates ($Q_A \rightarrow Q_B$, $Q_B \rightarrow P$) as a function of pH has indicated that acidic residues near Q_B affect electron transfer rates via electrostatic interactions.

Infrared spectroscopy has previously been used to identify the changes in the vibrational spectrum of the RC due to light-induced electron transfer (for a review, see [5]). In steady-state FTIR Q_A^-/Q_A or Q_B^-/Q_B difference spectra [6-9] most of the bands can be attributed to vibrational modes of protein residues perturbed by quinone reduction. In the 1700- 1750 cm^{-1} range, where clearly no contributions from the quinone molecule itself appear (the highest frequency mode of the quinone in the 2000 - 1000 cm^{-1} range is the C=O peaking at 1670 - 1630 cm^{-1} [10]) absorption bands of carboxylic groups can contribute, indicating proton uptake (formation of a COOH form), release (decay of a COOH form), or changes in the environment of a C=O bond. In the work presented here, we analyze the transient µs to ms IR spectrum in this range.

2. Experimental

IR samples from reaction centers from *Rhodobacter sphaeroides* R26 were prepared as described in [11]. Time-resolved IR signals (µs-s) were recorded with the kinetic IR laser photometer described by Mäntele et al. [12]. Laser modes in the 1690 - 1760 cm^{-1} range were obtained every 3-6 cm^{-1} by temperature tuning. Stabilization of temperature and laser current were optimized to achieve a good S/N ratio even for single-shot measurements of transients as small as $\Delta I/I \approx 10^{-4}$. Excitation was done with a 590 nm, 5 mJ, 15 ns pulse from an excimer/dye laser setup. Signals were subjected to a global fit analysis using the ***mexfit*** program [13].

3. Results and Discussion

Figure 1 (upper trace) shows a time-resolved IR signal at 1725 cm^{-1} for a native RC sample with the Q_B pocket fully occupied at pH 7 and 5°C. A rapid absorbance increase is observed after the flash, which cannot be time-resolved with our present instrument resolution (~500 ns). This immediate rise is followed by a slow transient with a half time of around one millisecond. The entire signal decays with a half time of around one second. Treatment of the sample with the herbicide terbutryn leads to the loss of Q_B. In this case, when Q_A is the terminal electron acceptor, the transient ms signal is lacking (Fig. 1, lower trace), and the initial non-resolved rise decays with a half time of around 60 ms. The rapid, non-resolved phase of the signal can thus be associated with charge separation between P and Q_A.

The time course of the ms transient present only with Q_B as a terminal electron acceptor does not correlate with those observed for an electronic transition of the pigments or quinones. IR signals of this type were already reported for the spectral range around 1620 cm^{-1}, and tentatively attributed to the antisymmetric COO^- stretching mode of Asp or Glu residues affected by $Q_A \rightarrow Q_B$ electron transfer [11]. The ms transient at 1725 cm^{-1} represents the

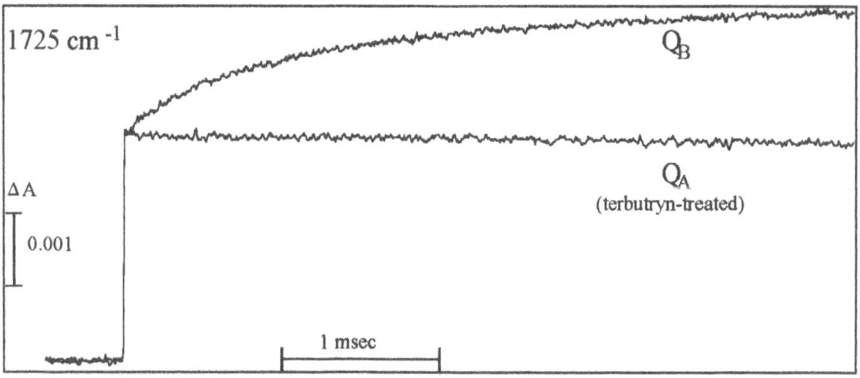

Fig. 1 Time-resolved IR-signal from native reaction centers with Q_B (upper trace) and Q_A (lower trace) as terminal electron acceptor.

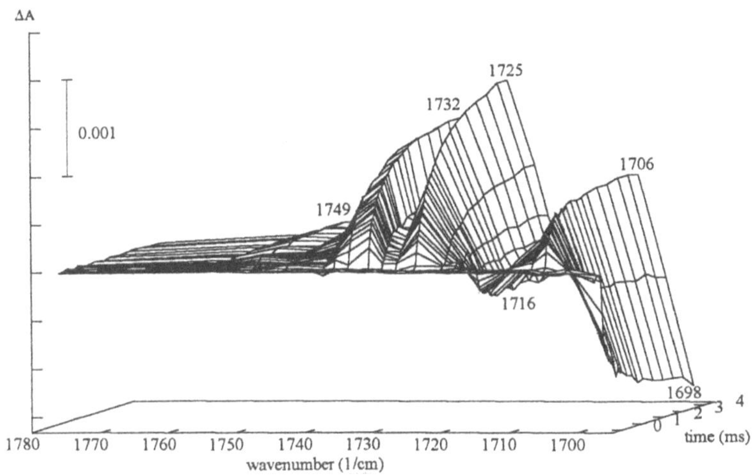

Fig. 2 3-D representation of time-resolved IR-signals from native reaction centers

COOH counterpart for these signals, and supports the view of Asp and Glu residues in the vicinity of Q_B responding to $Q_A \rightarrow Q_B$ electron transfer by protolytic reactions.

Figure 2 shows a three-dimensional representation of the signals obtained in the spectral range from 1690 to 1760 cm^{-1}. For better display, the strong instantaneously rising signal (Fig. 1) was removed by setting the baseline to the absorbance value for the state $P^+Q_A^-$ at t=5 μs. Three distinct positive peaks are detected at 1732, 1725, and 1706 cm^{-1}, and two negative signals appear at ca. 1716 cm^{-1} and below 1700 cm^{-1}. In case that all signals arise from COOH ↔ COO$^-$ transitions, several Asp or Glu residues undergo (partial) protonation or deprotonation as a consequence of the first electron transfer between Q_A and Q_B. Note that deprotonation occurs from a group at <1700 cm^{-1} which should be highly solvated, and that protonation occurs for groups absorbing at higher frequencies, which could be more secluded. All signals were subjected to a global fit analysis: a sum of two exponentials was sufficient for a satisfying fit at all wavelengths, yielding half times of 1 ms and 100 μs.

The transient difference spectrum shown in Fig. 2 was analyzed in detail for the pH range between 4 and 10, for $^1H/^2H$ substituted samples, and for several mutants. The details of this analysis will be reported elsewhere [14]. There is convincing evidence that the transient signal at 1725 cm^{-1} can be attributed to Glu L212 in the Q_B site. The sign and size of the signal corresponds to the uptake of a significant amount of a proton upon Q_B^- formation. Whether this proton is transferred from one of the groups which are deprotonated and show a negative transient signal in Fig. 2, or from a water molecule in the site, cannot be decided at present. Proton transfer to Glu L212 could represent a stabilizing effect for Q_B^- by lowering the electrostatic energy of interaction.

References

[1] Okamura, M.Y. and Feher, G. (1992) Ann. Rev. Biochemistry **61**, 861-896

[2] Michel, H., Epp, O., and Deisenhofer, J. (1986) EMBO J. **5**, 2445-2451

[3] Allen, J.P., Feher, G., Yeates, T.O., Komiya, H., and Rees, D.C (1988) Proc. Natl. Acad. Sci. USA **85**, 8487-8491

[4] El-Kabbani, O., Chang, C.-H., Tiede, D., Norris, J., and Schiffer, M. (1991) Biochemistry **30**, 5361-5369

[5] Mäntele, W. (1993) Infrared spectroscopy pf the photosynthetic reaction center in: "The Photosynthetic Bacterial Reaction Center" (Deisenhofer, J. and Norris, J., eds.) Volume II, Chapter A10, pp. 239-283, Academic Press, N.Y.

[6] Mäntele, W., Leonhard, M., Bauscher, M., Nabedryk, E., Breton, J., and Moss, D.A. (1990) in: Springer Series in Biophysics **6**, pp. 31-44

[7] Breton, J., Thibodeau, D.L., Berthomieu, C., Mäntele, W., Verméglio, A., and Nabedryk, E. (1991), FEBS Letters **278**, 257-260

[8] Breton, J., Berthomieu, C., Thibodeau, D.L., and Nabedryk, E. (1991) FEBS Lett. **288**, 109-113

[9] Bauscher, M., Leonhard, M., Moss, D.A., and Mäntele, W. (1993) Biochim. Biophys. Acta, in press

[10] Bauscher, M. and Mäntele, W. (1992) J. Phys. Chem. **96**, 11101-11108

[11] Hienerwadel, R., Thibodeau, D.L., Lenz, F., Nabedryk, E., Breton, J., Kreutz, W. and Mäntele, W. (1992) Biochemistry **31**, 5799-5808

[12] Mäntele, W., Hienerwadel, R., Lenz, F., Riedel, W.J., Grisar, R., and Tacke, M. (1990) Spectroscopy International **2**, 29-35

[13] Müller, K.H., and Plesser, Th. (1991) Biophysical Journal **19**, 231-240

[14] Hienerwadel, R., Grzybek, S., Fogel, C., Okamura, M.Y., Breton, J., Nabedryk, E. Breton, J., and Mäntele, W. (1993) manuscript in preparation

Laser-Induced Anisotropic Effects on Raman Intensities and Depolarization Ratios

J.A. Koningstein and D.F. Dawson

Ottawa-Carleton Chemistry Institute, Department of Chemistry, Carleton University, 1125 Colonel By Drive, Ottawa, Ontario, Canada K1S 5B6

1. Introduction

A major impetus for the extensive research on the water-soluble derivatives of porphyrins over the past decade, has been not only that they can be used as model chromophores for porphyrins in vivo that participate in many of the essential function of life in procaryotes and eucaryotes, but as well their clinically useful photosensitizing properties which produce singlet oxygen causing tumour cell wall damage in photodynamic therapy of cancer cells and their binding (intercalation, groove-binding) mechanisms to DNA used in chemotherapy [1]. Several techniques were used [2-4], such as circular dichroism (CD), picosecond laser induced fluorescence polarization, laser induced dichroism effect (LIDE) spectroscopy, and resonance Raman (RR) spectroscopy to probe the nature of the meso-tetrakis(N-methyl-4-pyridyl) porphine, H_2TMpyP(4), complex formed with DNA, H_2TMpyP(4)-DNA. A thorough study of the RR spectra and depolarization ratios, ρ_n, of the free H_2TMpyP(4) is required before interpreting the information obtained from ρ_n values of the H_2TMpyP(4)-DNA. In this study we present an analysis of the behaviour of ρ_n versus laser power I_l(mW) of the $\nu_2(C_\beta$-$C_\beta)$, A_{1g}, $\Delta\bar{\nu}=1558$ cm^{-1} and δ(pyr), A_{1g}, $\Delta\bar{\nu}=1643$ cm^{-1} RR peaks of the H_2TMpyP(4) in a phosphate buffer solution [3].

2. Experimental

The monobeam transmission [5] experiments were carried out (H_2TMpyP(4), 1.47×10^{-5} M, 10 mm quartz cell) using an excimer laser (Lumonics Model TE 320-2, $\approx$5 ns pulse, 25 Hz repetition rate) to pump a tunable pulsed dye laser (Lumonics Model hyperDYE-300) in the Soret B band region. The resonance Raman (RR) experiments were carried out (H_2TMpyP(4), 1.47×10^{-4} M, with a scrambler mounted in contact with the 10 mm quartz cell) using the tunable dye laser source, excited at $\lambda_l=422$ nm ($\bar{\nu}_l=23697$ cm^{-1}); and at $\lambda_l=476.5$ nm ($\bar{\nu}_l=20986$ cm^{-1}) and $\lambda_l=514.5$ nm ($\bar{\nu}_l=19436$ cm^{-1}) using a continuous wave (CW) Ar^+ laser (CCR-4 SUPERGRAPHITE, COHERENT RADIATION). All Raman intensities were normalized to the Raman band of water, $\bar{\nu}(H_2O)=3450$ cm^{-1}, internal standard. The scattered light was detected on a cooled RCA 31034 photomultiplier attached to a Jobin-Yvon double monochromator. All experiments were performed at room temperature (293 K).

3. Results and Discussion

Fig. 1 illustrates the ground state, $S_3(B)\leftarrow S_0$ (1), and excited state, $S_m\leftarrow S_3(B)$ (2), absorption spectra in the Soret B band region, respectively. On inspection of Fig. 1. one can see that at an excitation of $\lambda_l=422$ nm ($\bar{\nu}_l=23967$ cm^{-1}), the RR emissions at $\nu_2(C_\beta$-$C_\beta)$, A_{1g}, ($\Delta\bar{\nu}=1558$ cm^{-1} (451.7 nm) and δ(pyr), A_{1g}, $\Delta\bar{\nu}=1643$ cm^{-1} (453.4 nm) would not be very attenuated by the ground state (1) spectrum, but would probably be very attenuated by the excited state (2) spectrum, due to their relatively high and low transmissions in this ($\lambda\approx450$ nm) region,

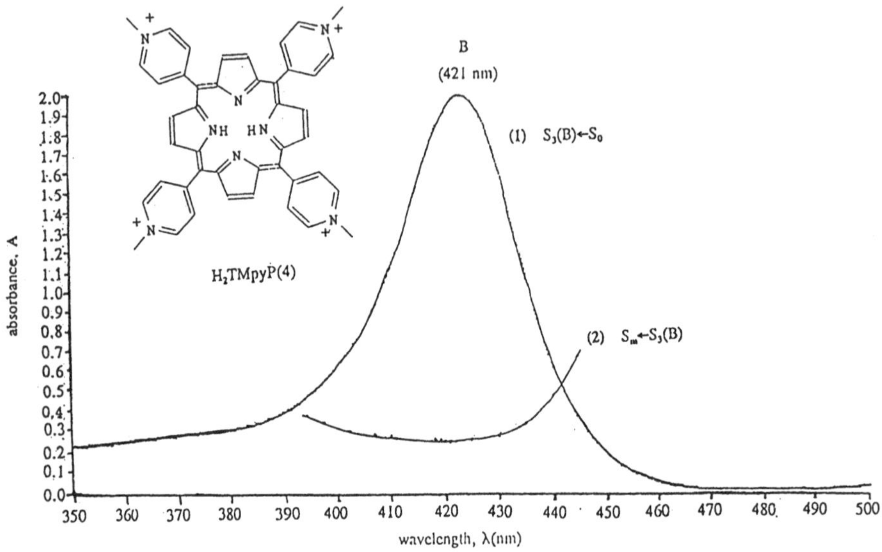

Fig. 1. Ground state (1) and excited state (2) absorption spectra of meso-tetrakis(N-methyl-4-pyridyl)porphine, $H_2TMpyP(4)$, 1.47×10^{-5} M in a 10 mm cell, in a pH=7, phosphate buffer (see text) in the Soret B band region.

respectively. Fig. 2. shows the relative Raman cross-sections, perpendicular $\sigma_R{}^{\perp}$, parallel $\sigma_R{}^{\parallel}$, and the depolarization ratio of naturally (circularly) polarized light, ρ_n, as a function of laser power I_l(mW) over a range of 0-5 mW for an excitation of $\lambda_l=422$ nm ($\bar{\nu}_l=23697$ cm^{-1}) with a pulsed dye laser. The Raman shifts of $\Delta\bar{\nu}=1558$ cm^{-1} and $\Delta\bar{\nu}=1643$ cm^{-1} correspond to $\nu_2(C_\beta\text{-}C_\beta)$,$A_{1g}$, and δ(pyr), A_{1g}, respectively.

Consider the following scattering tensor, α, with principle axes of polarizability, α_1, α_2, and α_3 expressed as,

$$\boldsymbol{\alpha} = \begin{bmatrix} \alpha_1 & 0 & 0 \\ 0 & \alpha_2 & 0 \\ 0 & 0 & \alpha_3 \end{bmatrix} \tag{1}$$

and the depolarization ratio of naturally (circularly) polarized light, ρ_n, is expresses as,

$$\rho_n = \frac{I_\perp}{I_\parallel} = \frac{6\beta^2}{45\alpha^2 + 7\beta^2} \tag{2}$$

where $I_\perp$,$I_\parallel$,α, and β denote the observed Raman intensity in the perpendicular direction, the observed Raman intensity in the parallel direction, the spherical (isotropic) component of the polarization ellipsoid, and the non-spherical (anisotropic) component of the polarization ellipsoid, respectively. On inspection of Fig. 2. one can see that the behaviour of ρ_n for $\Delta\bar{\nu}=1558$ cm^{-1} (increasing linearly) is very different from that of $\Delta\bar{\nu}=1643$ cm^{-1} (almost horizontal) with increasing I_l.

The first factor one must consider is ground state depletion. There is enough power in the pulse to saturate the $S_3(B)$ state (by consecutive stepwise absorption of two photons, $S_3(B) \leftarrow S_0 + h\nu_l$, and then $S_m \leftarrow S_3(B) + h\nu_l$ [1]) and thus cause ground state depletion. As well, if

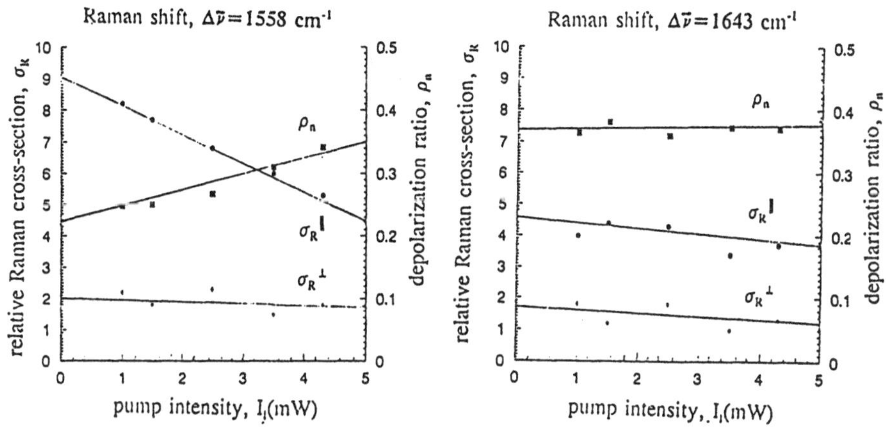

Fig. 2. Plot of relative Raman cross-section, perpendicular, $\sigma_R^{\perp}$, parallel, $\sigma_R^{\parallel}$, and depolarization ratio of naturally (circularly) polarized light, ρ_n, versus pump intensity, I_l(mW), for meso-tetrakis(N-methyl-4-pyridyl)porphine, H_2TMpyP(4), 1.47×10^{-4} M, in a 10 mm cell, in a pH=7, phosphate buffer (see text), excited at λ_l=422 nm ($\bar{\nu}_l$=23697 cm^{-1}).

some intersystem crossing $S_1 \rightsquigarrow T_1$ occurs additional ground state depletion could occur. Therefore as less molecules are residing in the ground state more laser photons, $h\nu_l$, are available for Raman scattering, both in the $I_\perp$ and $I_\parallel$ cases, and ρ_n would be unaffected.

The second factor one must consider is the induced anisotropy in the angular density distribution function of ground state molecules, n_1^G, in the scattering volume. Consider a molecule sitting at the origin of a right-handed XYZ-cartesian coordinate system. The Raman pump and probe are the same and they both come in along the X-axis, are scattered and detected along the Y-axis. Molecules whose transition moments are parallel to the exciting beam are more strongly absorbed than those at varying angles and if the rotational relaxation time, τ_{rot}(H_2TMpyP(4)), is of comparable magnitude to the duration of the pump/probe pulse then anisotropy is introduced into the previously isotropic angular density distribution of the concentration of the N ground state molecules in the scattering volume. Calculation of the isotropic, I^0, and anisotropic, I^1, contributions to the observed Raman intensities, I, under parallel (ZX-plane) [2] and perpendicular (YX-plane) [6] pumping conditions show the ρ_n curve for $\Delta\bar{\nu}$=1558 cm^{-1} is almost independent of laser power, whereas ρ_n for $\Delta\bar{\nu}$=1643 cm^{-1} is relatively much more dependant on laser power, I_l (driving ρ_n down $\approx$3x more). Thus ρ_n is either unaffected or reduced, not increased and one must consider a third factor effecting ρ_n.

Consider if the polarization of the excited state spectrum, $S_m \leftarrow S_3$(B), where to be, e.g., in the ZY-plane. When the pump comes in perpendicular (YX-plane), is scattered then scrambled, $I_\perp=6\beta^2$ is split into $I_{ZY}=3\beta^2$ and $I_{XY}=3\beta^2$. Similarly when the pump comes in parallel (ZX-plane), is scattered and scrambled, $I_\parallel=45\alpha^2+7\beta^2$ is split into $I_{ZY}=45\alpha^2+4\beta^2$ and $I_{XY}=3\beta^2$. Now if the transmission of the excited state spectrum is low in the region of the Raman emissions then the I_{ZY} components will be heavily absorbed, whereas the I_{XY} components will be unaffected. Therefore $I_\perp$ would be reduced by a relatively small amount, whereas $I_\parallel$ would be reduced by a large amount, thus driving up the value of ρ_n. This effect may be termed as an anisotropic polarized excited state absorption spectrum effect and can be seen to dominate the behaviour of ρ_n for $\Delta\bar{\nu}$=1558 cm^{-1} driving it up (whereas for $\Delta\bar{\nu}$=1643 cm^{-1}, the induced anisotropy in the density distribution of the ground state , n_1^G, almost balances out the anisotropic polarized excited state absorption effect).

Depolarization ratios, ρ_n, were calculated for $\Delta\bar{\nu}$=1558 cm^{-1} to be 0.22, 0.36 and 0.50 excited at λ_l=422, 476.5 and 514.5 nm, respectively. As well, for $\Delta\bar{\nu}$=1643 cm^{-1}

ρ_n=0.37, 0.38, and unmeasurable (lost in the baseline noise) excited at λ_l=422, 476.5, and 514.5 nm, respectively. For $\Delta\bar{\nu}$=1558 cm^{-1}, the transition in ρ_n values of 0.22 to 0.36 and then to 0.50 could be interpreted as the equal resonance enhancement of α_1 and α_2, the unequal resonance enhancement of α_1 and α_2, and the resonance enhancement of α_2 only, respectively. Thus as the value of ρ_n changes from 0.22 to 0.36, it could be interpreted as a sign of x-, and y- splitting of the Soret B band.

In conclusion, it is interesting to speculate that under certain experiment conditions where the excited state has significant, even comparable population to ground state (e.g. the $(Cu(II)(TMpyP(4))^{\bullet+}(dA\text{-}dT)^{\bullet-})$ exciplex of reference [7]) one is most likely to see the anisotropic polarized excited state absorption spectrum effect manifest itself on the calculation of ρ_n, if the excited state (exciplex) absorption spectrum has significantly low transmission values in the region of the Raman emissions.

References

[1] K. Kalyanadundaram, Photochemistry of Polypyridine and Porphyrin Complexes, Academic Press, New York, pp.. 419-426, 598-603 (**1992**).

[2] P. Myslinski. J.A. Koningstein, and Y. Shen, "Theory for laser induced dichroism and its effect on absorption, fluorescence, and Raman Spectra", J. Chem. Phys., 96(12), 8091-8098, (**1992**).

[3] Y. Shen, P. Myslinski, T. Treszcanowicz, Y. Lu, and J.A. Koningstein, "Picosecond Laser-Induced Fluorescence Polarization Studies of Mitoxantrone and Tetrakisporphine/DNA complexes, J. Phys. Chem., 96, 7782-7787, (**1992**).

[4] Y. Lu, and J.A. Koningstein, "Structure Studies of H_2TMpyP4 and ZnTMpyP4 Bound to DNA Using Laser-Induced Dichroism in Solution", J. Phys. Chem., 97, 6155-6160, (**1993**).

[5] B. Halperin, and J.A. Koningstein, "Conditions for excited state Raman and absorption processes during optical pumping", Can. J. Chem., 57, 2692-2702, (**1981**).

[6] D.F. Dawson, Y. Shen, Y. Lu, and J.A. Koningstein, to be published.

[7] G.P. Strahan, D. Lu, M. Tsubui, and K. Nakamoto, "Resonance Raman Spectra of Electronically Excited, Water-Soluble Copper (II) Porphyrins Bound to Oligonucleotides: Possibility of Translocation from a GC to an AT Site", J. Phys. Chem., 96, 6450-6457, (**1992**).

Excitation Transport in Helical Proteins

R. Lohikoski[1], *J. Timonen*[1], *J. Nieminen*[3], *V. Helenius*[2], *and J. Korppi-Tommola*[2]

[1]Department of Physics, University of Jyväskylä, P.O. Box 35, SF-40351 Jyväskylä, Finland
[2]Department of Chemistry, University of Jyväskylä, P.O. Box 35, SF-40351 Jyväskylä, Finland
[3]Department of Physical Chemistry, University of Helsinki, P.O. Box 13, SF-00014 Helsinki, Finland

Abstract. Recent results for excitation dynamics in and IR-absorption spectrum of helical polypeptides are briefly reviewed.

1. Davydov's Soliton in Helical Polypeptide Chains

In he early 1970's already A. S. Davydov and his co-workers [1] formulated a nonlinear theory which was able to give a reasonable explanation for energy transfer in helical proteins. The essence of this theory is that energy is absorbed into the α-helical protein structure in the form of intramolecular vibrations, which are localized through interaction with sound waves propagating in the protein. The transport of this self-localized energy packet ("soliton") along the protein via the three hydrogen bonded channels happens almost without dispersion, provided the coupling between the modes is suitably strong. During the past ten years this theory and its modifications have been of great intrest in biophysics [2]. Quite little attention has been paid however to the dynamics of the model. In Section 2 below we shall report briefly our recent results for the dynamics of the localized excitations.

Thus far very little experimental work has been done in search for solitons in α-helical polypeptides or proteins. Difficulties encountered in studying real proteins in their realistic environment have been circumvented by doing experiments on crystalline acetanilide (ACN). The hydrogen bonded peptide units in the chain-like structures of ACN have bond lengths and bond angles which are almost identical to those found in α-helical proteins. Careri's first experiments on ACN revealed a strongly temperature dependent side-band on the red-shifted side of the amide-I band at 1665 cm^{-1} [3]. The only consistent explanation which could be given to this band was a localized vibrational mode [3]. Recent measurements show however that the anomalous band is side-group dependent [4] and therefore its existence in true polypeptide chains is not obvious. Our IR-experiments on short helical polypeptides confirm this suspection. The current state of these studies is given in Section 3.

2. Single excitation dynamics

The equations of motion derived from the Scott's version of the Davydov Hamiltonian describe the time-evolution of one excitation interacting with sound waves on one of the three hydrogen-bonded peptide channels of the α-helical structures. We assume further that the kinetic energy of "trapped" sound waves is much less than their potential energy. In this adiabatic approximation the time-evolution of the probability amplitude a_m (m is the site index) of the excitation is given by

$$i\dot{a}_m = V(a_{m+1} + a_{m-1}) - \chi |a_m|^2 a_m \ , \qquad (1)$$

which is a discrete form of the Schrödinger equation with a nonlinear self-interaction term caused by coupling to sound waves. Here the constants have been defined such that $V = -J/\hbar$ and $\chi = \chi_a^2/(w\hbar)$, where J is the dipole-dipole interaction energy between adjacent peptide groups, χ_a is the coupling constant between intra- and intermolecular excitations and w is the spring constant between neighbouring peptide units.

We have solved eqn. (1) numerically varying the chain length (from 3 to 999), the value of the coupling constant, the initial distribution of probability and the boundary conditions.

These numerical calculations show existence of dynamically regular and chaotic regions, and self-trapping, pre-trapping and self-focusing transitions. All these features are very sensitive to the parameters and the initial conditions of the model.

Here we shall concentrate on the formation and dynamics of stable wave packets (solitons). By considering the site occupation probabilities $|a_m|^2$ as a function of time and χ, we find either dispersive, slowly dispersive, stable or trapped wave packet(s). The self-trapping value of χ for constant initial conditions and varying N remains roughly constant. However, the soliton window in χ is wider for longer chains. This is at least partly due to perturbations caused by reflections from the boundaries.

Common to all stable solitary waves eqn. (1) generates is irregular, diffusion-like dynamics. This is demonstrated in fig. 1. We have analysed the soliton trajectories in order to decide whether it describes a random process or a chaotic process. One class of random processes is Brownian motion. A particle executing Brownian motion has the Maxwell distribution of velocities. The dynamics of such a particle can be described by the Fokker-Planck equation, and in this case it can be shown [5] that the velocity distribution approaches the Maxwell distribution in a definite manner. By calculating the velocity distribution of the soliton as a function of time we in fact find that it does not approach the Maxwell distribution in the way it does in the Fokker-Planck case. This excludes the possibility of Brownian motion. We have also computed the box counting dimension of the trajectories of solitons. Typically this dimension is high for a random process. Our results show low dimension (between 1 and 2).

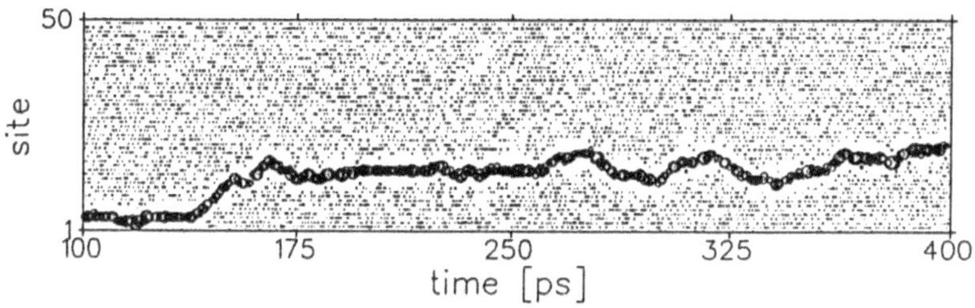

Fig. 1. Local maxima of the site occupation probability of eqn. (1) for $J = 1.55 * 10^{-22}$ J [6], $\chi \approx 5.51$ and free boundary conditions.

3. IR-spectroscopy of polypeptides

We have made standard IR-absorbtion measurements on tryptophan-(alanine)$_{15}$ and tyrosine-(alanine)$_{15}$. For the sake of comparison we have also repeated the previous IR-absortion measuments on ACN. Measurements on longer polypeptide chains are in progress.

In order to avoid the ambiguities inherent in fitting the data with Lorentzian components, we have used the Principal Component Analysis (PCA). PCA-analysis of the ACN-KBr data e.g. shows that 96.1 % of the total temperature dependence can be explained by the first principal component consisting of bands at 1650 cm^{-1} (unsymmetric), 1660 cm^{-1} and 1664 cm^{-1}.

In the amide-I region the temperature behaviour of the polypeptides differs completely from that of ACN. Both samples show a broad absortion band having a maximum at 1641 cm^{-1} (1630 cm^{-1}) for Trp-(Ala)$_{15}$ (Tyr-(Ala)$_{15}$) and a shoulder at 1662 cm^{-1} (1658 cm^{-1}). Practically no temperature dependence can be found between 15 K and 300 K. In the NH-stretch region situation is also quite different. Here the two polypeptides have two broad absortion bands at 3277 cm^{-1} and 3409 cm^{-1} for Trp-(Ala)$_{15}$, and at 3272 cm^{-1} and 3412 cm^{-1} for Tyr-(Ala)$_{15}$. Both bands have weaker, red-shifted and strongly temperature dependent sidebands at 3238 cm^{-1} and 3360 cm^{-1} for Trp-(Ala)$_{15}$, and at 3238 cm^{-1} and 3358 cm^{-1} for Tyr-(Ala)$_{15}$. Analysing these by PCA and comparing their behaviour with that of the 1650 cm^{-1} band in ACN, we find nearly identical temperature dependence for the integrated intensities. It is evident that the mechanism producing all these bands is similar. Assuming that the observed temperature dependence arises from coupling to a number of acustic or optical lattice modes, the temperature dependence of the integrated intensities of the bands should be proportional to exp(-aT^2) (a=constant) [7]. Figure 2 shows that this is indeed the case.

The appearance of two red-shifted side-bands in the two polypeptides can be explained by Fermi resonance. It is known that the two basic bands observed in the 3200-3400 cm^{-1} region result from mixing of the basic NH-stretch mode with the first overtone of amide-II or a combination of amide modes [8]. At low temperatures the coupling of the NH-stretch mode and lattice modes generates a red-shifted mode which is mixed with the same amide overtone as the basic

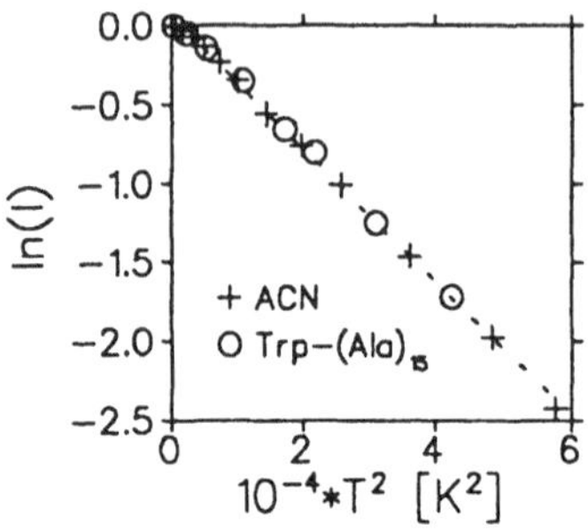

Fig. 2. Logarithm of the integrated intensity as a function of temperature squared of the 1650 cm^{-1} band of ACN and of the two red-shifted side-bands in the NH-stretch region of trp-$(ala)_{15}$. Dashed line is the least square fit to the data points.

NH-stretch mode. This explains the observed band structure in the NH-stretch region.

It seems quite possible that a localized excitation formed by the NH-stretch mode through coupling with lattice modes may be relatively common in helical polypeptides and may also play a role in energy transport.

References

[1.] A. S. Davydov, *Biology and Quantum Mechanics* (Pergamon Press), 1982.

[2.] A. C. Scott, Phys. Rep. 217, 1 (1992), and references therein.

[3.] G. Careri, U. Buontempo, F. Carta, E. Gratton and A. C. Scott, Phys. Rev. Lett. 51, 304 (1983); G. Careri, U. Buontempo, F. Galluzzi, A. C. Scott, E. Gratton and E. Shyamsunder, Phys. Rev. B 30, 4689 (1984).

[4.] M. Barthes, R. Almairac, J. L. Sauvajol, J. Moret, R. Currat and J. Dianoux, Phys. Rev. B 43, 5223 (1991).

[5.] F. Reif, *Fundamentals of Statistical and Thermal Physics* (McGraw-Hill Book Company), 1965.

[6.] N. A. Nevskaya and Yu. N. Chirgadze, Biopolymers 15, 637 (1976).

[7.] D. M. Alexander, Phys. Rev. Lett. 54, 138 (1985); D. M. Alexander and J. A. Krumhansl, Phys. Rev. B33, 7172 (1986); S. Takeno, Prog. Theor. Phys. 71, 395 (1984); 73, 853 (1985); 75, 1 (1986).

[8.] T. Miyazawa, J. Mol. Spectrosc. 4, 168 (1960); J. F. Rabolt, W. H. Moore and S. Krimm, Macromolecules 10, 1065 (1977).

Rapid Mixing Resonance-Raman Spectroscopy of Unstable π-Cation Radical Intermediates in Heme Containing Peroxidases

C.M. Hosten, A.M. Sullivan, V. Palaniappan, M.M. Fitzgerald, and J. Terner

Department of Chemistry, Virginia Commonwealth University, Richmond, VA 23283-2006, USA

The catalytic sequences of the heme containing peroxidases involve two high valent intermediates known as compounds I and II [1]. Compound I is two oxidation equivalents above the resting ferric enzyme, and contains an Fe(IV) porphyrin π-radical cation, though cytochrome c peroxidase is a notable exception [2]. Compound II is the one electron reduction product of compound I, with the electron restored to the porphyrin ring, and contains an Fe(IV) heme. Our laboratory has been involved in resonance Raman studies of these intermediates formed by rapid mixing and using laser excitation in the violet and near ultraviolet [3]. Although intermediates similar to compounds I and II are considered to occur in other physiologically important heme enzymes such as cytochrome oxidase and cytochrome P-450, their existence in the catalytic cycles of these enzymes has not been as well documented as for the peroxidases and catalases. The catalases are heme enzymes similar to the peroxidases, which disproportionate hydrogen peroxide. In this account we describe some of our work in progress on the intermediates of horseradish peroxidase and chloroperoxidase. Of the peroxidases, horseradish peroxidase is among the best known because of its commercial availability as a common biochemical reagent. Chloroperoxidase possesses important spectroscopic and structural similarities to cytochrome P-450 [4].

There has been considerable recent interest and discussion over the symmetry state assignments of the porphyrin π-radical cations in various peroxidases [5]. The radical intermediate of horseradish peroxidase has historically been considered to be A_{2u}, while those of chloroperoxidase and catalase have been considered to be A_{1u} based on electronic absorption data [6,7]. Reconstitution of horseradish peroxidase with iron mesoporphyrin or iron deuteroporphyrin has been reported to result in porphyrin π-radical cations with A_{1u} symmetry [8,9].

The symmetry state assignments of model metalloporphyrin π-radical cations have lately been under revision and reevaluation. Octaethylporphyrin has been reported to form metalloporphyrin π-radical cations of the A_{1u} symmetry type, while meso-substituted tetraphenylporphyrins form metalloporphyrin π-radical cations of predominantly A_{2u} symmetry [10-13]. The β-pyrrole substitution pattern of protoporphyrin-IX is the same as for octaethylporphyrin, as opposed to the methine substitution of the meso-tetraphenylporphyrin compounds. On this basis the protoporphyrin based peroxidases such as horseradish peroxidase and chloroperoxidase would be expected to form A_{1u} porphyrin π-cation radicals in accord with the recent symmetry state assignments of the model octaethylporphyrin π-radical cations. The long standing A_{2u} symmetry assignment for horseradish peroxidase can be rationalized to be the result of the effect of electron donation by the proximally coordinated imidazole [14]. It is therefore of interest to report that resonance Raman spectra obtained in our laboratory for horseradish peroxidase compound I, have the resonance Raman spectral pattern relative to the ferryl heme of compound II that is indicative of an A_{1u} radical, rather than A_{2u} [15]. In addition, for chloroperoxidase and horseradish peroxidase reconstituted with iron mesoporphyrin, we observe similar resonance Raman spectral patterns that are more typical of A_{1u} than A_{2u} radicals of the model metalloporphyrin π-cation radicals. It should be mentioned that there still remain considerable differences between the enzyme intermediates and the model compounds. An example is the exchange interaction between the paramagnetic centers. For the available iron based model compounds, the ferryl S=1 spin and the porphyrin π-radical cation S'=1/2 spin are ferromagnetically coupled [16]. In chloroperoxidase compound I, the interaction is strongly antiferromagnetic [17], while for horseradish peroxidase compound I the exchange interaction is very small [18]. The effects of axial ligation and protein environment on the enzymatic porphyrin π-cation radicals remain to be elucidated.

The distinctive resonance Raman spectral patterns of the A_{1u} versus A_{2u} type porphyrin π-radical cation have been attributed to the bonding and antibonding interactions of the A_{1u} and A_{2u} molecular orbitals [11]. The Cβ-Cβ interaction is bonding in the A_{2u} orbital but antibonding in the A_{1u} orbital [19]. Removal of an electron from the A_{1u} orbital is thus expected and observed [11,13] to result in an upshift of modes that involve the Cβ-Cβ stretching motion such as modes ν_2 and ν_{11} [20], while removal of an electron from the A_{2u} orbital results in substantial downshifts of these modes. Additional correlations of band shifts with symmetry state have been made via semiemipirical calculations which confirm that it is reasonable to expect modes ν_2 and ν_{11} to be distinctive indicators of radical type [21]. These bands were predicted and experimentally confirmed to undergo some of the largest shifts upon formation of the porphyrin π-radical cation. Of the two mentioned frequencies, ν_2 is the easier to identify since it is the stronger of the two and readily identifiable by its polarization. However an E_u mode, ν_{37}, and polarized vinyl modes appear in the same region. The E_u mode gains Raman activity due to symmetry lowering by the vinyl groups of the protoporphyrin, causing E_u to correlate to A and B components. Evidently the polarized A component is enhanced under the near-ultraviolet excitation used for our experiments and a second polarized mode appears in the vicinity of ν_2 [15]. In order to suppress the E_u mode activity and to eliminate the vinyl modes, we substituted mesoheme into horseradish peroxidase simplifying the ν_2 frequency region [14]. Mesoporphyrin has ethyl groups in place of the protoporphyrin vinyls, resulting in higher symmetry that is expected to suppress the E_u mode activity. As listed in Table I, we have found that mesoheme substituted horseradish peroxidase still displays an upshifted ν_2 frequency for the oxoferryl porphyrin π-radical cation of compound I relative to the ferryl heme of compound II.

In Table I are listed some of our recent data on ν_2 and ν_{11} frequencies which have been identified through polarization measurements for horseradish peroxidase (native protoheme and mesoheme substituted enzyme) and chloroperoxidase. The common property that we observe experimentally thus far and tabulated in Table I is that ν_2 and ν_{11} undergo upshifts rather than downshifts. Downshifts in ν_2 and ν_{11} are considered to be diagnostic for A_{2u} metalloporphyrin π-cation radicals.

Table I: Comparison of some Resonance Raman frequencies (in cm^{-1}) for peroxidases.

Enzyme	Compound I		Compound II		Differences	
	ν_{11}	ν_2	ν_{11}	ν_2	$\delta\nu_{11}$	$\delta\nu_2$
Horseradish Peroxidase						
Protoheme	1570	1606	1560	1587	+10	+19
Mesoheme		1616		1603		+13
Chloroperoxidase						
Protoheme	1572	1589	1551	1582	+21	+7

Abbreviations: $\delta\nu_{11}$ and $\delta\nu_2$ refer to differences in resonance Raman frequencies (in cm^{-1}) between compounds I and II.

This work acknowledges National Institutes of Health supplemental grant GM34443-06S1A1 for financial support, and Gary Pennington, Cathy Leyco and Sahana Vijay for technical assistance.

References:

1. Ortiz de Montellano, P.R. (1992) Ann. Rev. Pharmacol. Toxicol. 32, 89-107.
2. Bosshard, H.R., Anni, H. and Yonetani, T. (1991) in Peroxidases in Chemistry and Biology (Everse, J., Everse, K.E. and Grisham, M.B., eds.) CRC Press, Boca Raton, FL, pp. 51-86.
3. Palaniappan, V., Sullivan, A.M., Fitgerald, M.M., Shifflett, J.R., and Terner, J. (1992) Time-Resolved Vibrational Spectroscopy V, (Takahashi, H., ed.) Springer Proceedings in Physics, Vol. 68, 32-35.
4. Dawson, J.H. and Sono, M. (1987) Chem. Rev. 87, 1255-1276.
5. Fujii, H. (1993) J. Amer. Chem. Soc. 115, 4641-4648
6. Dolphin, D. and Felton, R.H. (1974) Acc. Chem. Res. 7, 26-32.
7. Hewson, W.D. and Hager, L.P. (1979) in The Porphyrins D. Dolphin ed, Vol. VII, Academic Press, NY., pp. 295-332.
8. DiNello, R.K. and Dolphin, D.H. (1979) Biochem. Biophys. Res. Commun. 86, 190-8.
9. DiNello, R.K. and Dolphin, D.H. (1981) J. Biol. Chem. 256, 6903-6912.
10. Spellane, P.J., Gouterman, M., Antepas, A., Kim, S., Liu, Y.C. (1980) Inorg. Chem. 19, 386-391.
11. Czernuscewicz, R.S., Macor, K.A., Li, X.-Y., Kincaid, J.R., and Spiro, T.G. (1989) J. Amer. Chem. Soc. 111, 3860-3869.
12. Skillman, A.G., Collins, J.R. and Loew, G.H. (1992) J. Amer. Chem. Soc. 114, 9538-9543.
13. Oertling, W.A., Salehi, A., Chang, C.K. and Babcock, G.T. (1989) J. Phys. Chem. 93, 1311-1319.
14. Macor, K.A., Czernuscewicz, R.S., and Spiro, T.G. (1990) Inorg. Chem. 29, 1996-2000.
15. Palaniappan, V. and Terner, J. (1989) J. Biol. Chem. 265, 16046-16053.

16. Mandon, D., Weiss, R., Jayaraj, K., Gold, A., Terner, J. Bill, E., and Trautwein, A.X. (1992) Inorg. Chem. 31, 4404-4409.
17. Rutter, R., Hager, L.P., Dhonau, H., Hendrich, M., Valentine, M., and Debrunner, P. (1984) Biochemistry 23, 6809-6816.
18. Schultz, C.E., Devaney, P.W., Winkler, H. Debrunner, P.G., Doan, N., Chiang, R., Rutter, R. and Hager, L.P. (1979) FEBS Lett. 103, 102-5.
19. Maggiora, G.M. (1973) J. Amer. Chem. Soc. 95, 6555-6559.
20. Li, X.-Y., Czernuszewicz, R.S., Kincaid, J.R., and Spiro, T.G. (1989), J. Amer. Chem. Soc. 111, 7012-7023.
21. Prendergast, K. and Spiro, T.G. (1991) J. Phys. Chem. 95, 9728-9736.

Structural and Environmental Studies of the Vitamin E (α-Tocopheroxyl) Radical Using Time-Resolved Resonance-Raman Spectroscopy

A.W. Parker[1] *and R.H. Bisby*[2]

[1]SERC, Rutherford Appleton Laboratory, Chilton, Didcot, Oxon OX11 OQX, UK

[2]Department of Biological Sciences, University of Salford, Salford M5 4WT, UK

The time-resolved resonance Raman spectrum of the vitamin E free radical, α-tocopheroxyl, has been measured. The spectrum is similar to that obtained for 4-methoxyphenoxyl, i.e. consisting of two intense bands at *ca.* 1500 and 1595 cm^{-1}, assigned to the C=C (ν_{8a}) and C-O (ν_{7a}) stretches respectively. This indicates that the vitamin E free radical is stabilised by electron delocalisation involving the chromanol ring oxygen. The ratio of these two bands increases with increase in solvent polarity. This has been used to determine the position of the α-tocopherol radical within lipid bilayer membranes.

1. Introduction

Vitamin E (α-tocopherol) has long been established as an essential nutrient in the human diet and there is growing interest in the protection it may offer against some diseases [1]. Oxidative cellular damage is associated with a number of pathological processes including carcinogenesis and aging. α-Tocopherol functions as a chain-breaking antioxidant by scavenging harmful free radicals which damage lipoproteins and cell membranes. In this process the α-tocopheroxyl radical is formed by reaction with a lipid peroxyl radical as in Scheme 1 shown below. α-Tocopherol is a minor membrane component (about 1 molecule per 3000 lipid molecules). In order to prevent depletion of α-tocopherol and to provide continued antioxidant protection, α-tocopherol must be regenerated from the free radical. One possibility, illustrated in Scheme 1, is that this reduction process is carried out

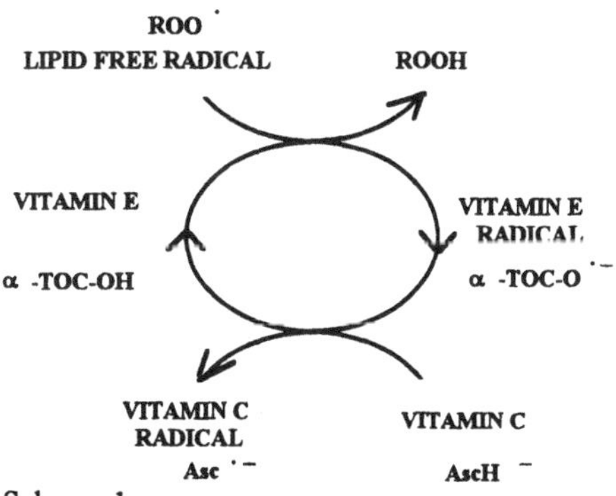

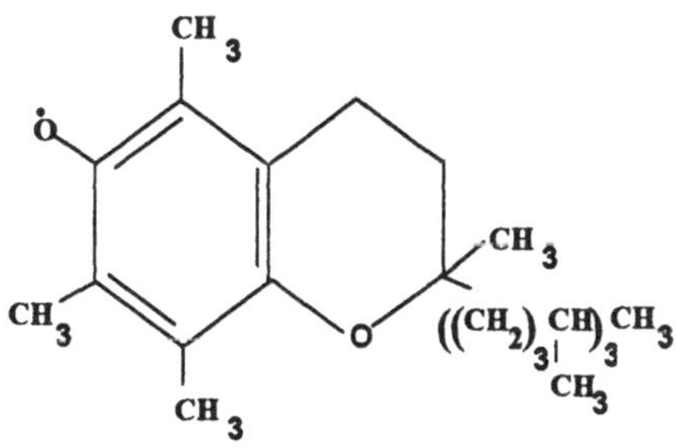

α-tocopheroxyl radical

Scheme 1

Springer Proceedings in Physics, Vol. 74 Time-Resolved Vibrational Spectroscopy VI
Editors: A. Lau, F. Siebert, W. Werncke

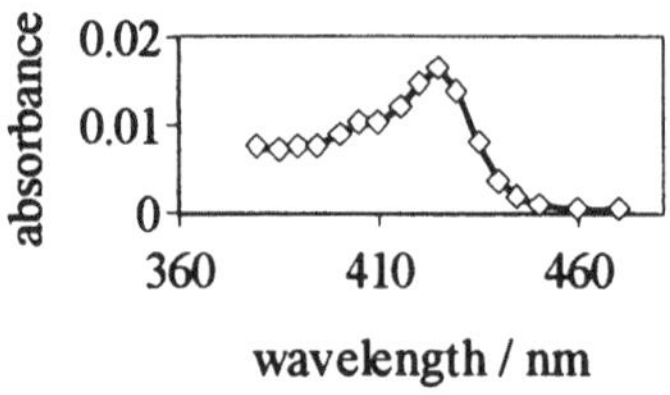

Fig. 1. Transient absorption spectrum from laser flash photolysis at 308 nm of α–tocopherol (1 x 10^{-4} mol dm^{-3}) in DPPC bilayers. Absorbances are measured 1 μs after the laser pulse.

by ascorbate (vitamin C). However, as vitamin E is fat soluble and vitamin C is water soluble this regeneration process would require that the ascorbate ($AscH^-$) in the aqueous phase has access to the α–tocopheroxyl radical in the membrane or lipoprotein particle.

We have recently shown [2] that using a 308 nm laser pulse it is possible to selectively generate, by monophotonic photoionisation, the α-tocopheroxyl radical in solution as well as in micelles and lipid bilayer membranes of dimyristoylphosphatidylcholine (DMPC) and dipalmitoylphosphatidic acid (DPPC). The uv/visible spectrum of the α-tocopheroxyl radical obtained by laser flash photolysis is shown in Fig. 1. In the presence of ascorbate the α-tocopheroxyl radical lifetime is considerably reduced both in solution phase and in the microheterogenous micellar and membrane phases, the reaction being the repair process given in Scheme 1. For the α-tocopheroxyl radical within a membrane bilayer and with ascorbate in the aqueous phase, the reaction is fast, $k \approx 10^5$ dm^3 mol^{-1} s^{-1}. The activation energy for this reaction is between 18 and 26 kJ mol^{-1}, consistent with a fast one electron transfer reaction with concerted proton exchange [2,3]. These results indicate that the reaction is not constrained by the compartmentalisation of the reactants and the location of the α-tocopheroxyl radical in such structures is therefore of interest. We report here the use of time resolved resonance Raman spectroscopy (TR^3) to investigate the stereo electronic factors helping to stabilise the α-tocopheroxyl radical and to probe the microenvironment of this radical within micellar membrane systems [4].

2. TR^3 Spectra of 4-methoxyphenoxyl and 4-methoxy-2,3,5,6-tetramethyl-phenoxyl Radicals.

Photoionisation of 4-methoxyphenol in ethanol, using a 308 nm laser pulse, is rapidly followed by loss of the phenolic proton to generate the 4-methoxyphenoxyl radical. The resonance Raman spectrum of this species is given in Fig. 2a and the assignments of these bands, based on those of Tripathi and Schuler [5] for a series of substituted phenoxyl radicals, are given in Table 1. The TR^3 spectrum of this radical is indicative of the delocalisation of the unpaired electron. The high frequency of the C-O stretching vibration indicates that this bond has nearly double bond character and the high intensity of the C=C stretch shows a strong interaction with the *para* oxygen atom with the phenoxyl π-system. For comparison, the resonance Raman spectrum for 4-methoxy-2,3,5,6-tetramethylphenoxyl radical is

Table 1-Resonance Raman bands (cm^{-1}) of tocopheroxyl radicals in solvents, micelles and membranes

Radical	Solvent	Dielectric Constant (20 °C)	C-O Stretch (ν_{7a})	C=C Stretch (ν_{8a})	C-H Bend (ν_{9a})	Intensity Ratio[a] (ν_{8a}/ν_{7a})
4-methoxyphenoxyl	ethanol	24.5	1514 (1518)[b]	1598 (1607)[b]	1156	0.92
4-methoxy-2,3,5,6-tetramethylphenoxyl	ethanol	24.5	1501	*ca.* 1593[c]		<0.15
α-tocopheroxyl	ethanol	24.5	1504	1586		0.62
	tetrahydrofuran	7.6	1501	1581		0.22
	SDS micelles		1500	1593		0.81
	Triton micelles		1498	1586		0.36
	DMPC bilayers		1500	1587		0.45

[a]Ratios are ± 0.03, [b] values in parenthesis for 4-methoxyphenoxyl are taken from reference 5 where the solvent was water, [c]very weak.

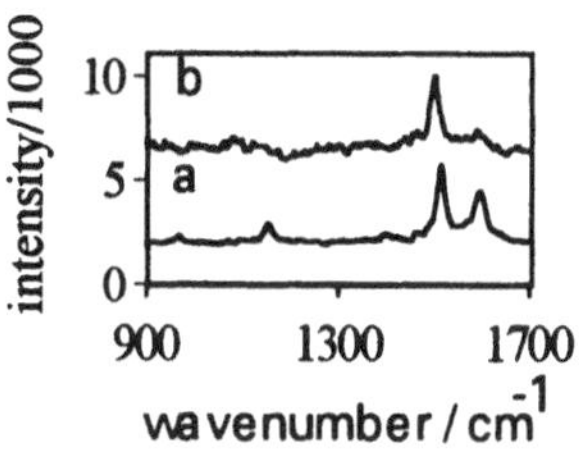

Fig. 2. TR^3 spectra produced by photoionisation of 4-methoxyphenol (5×10^{-3} mol dm^{-3}) in ethanol, time delay between pump laser (308 nm) and probe laser (425 nm) 500 ns and (b) 4-methoxy-2,3,5,6-tetramethyl-phenol (4×10^{-3} mol dm^{-3}) in ethanol, time delay between pump laser (270 nm) and probe laser (425 nm) 500 ns. Both spectra are pump/probe - probe only.

illustrated in Fig. 2b and shows only a single strong band at 1501 cm^{-1}, assigned to the C-O (ν_{7a}) stretch. The loss in intensity of the C=C stretch occurs because the 2p orbital of the 4-methoxy-oxygen atom in 4-methoxy-2,3,5,6-tetramethylphenoxyl radical liesalmost parallel to the plane of the phenyl ring due to steric interactions from the 3 and 5-methyl substituents. This causes a reduction in the electronic delocalisation leading to a lessening in radical stability and is the reason why 4-methoxy-2,3,5,6- tetramethylphenol is a poorer antioxidant than 4-methoxyphenol. This effect also explains why the absorption spectrum of 4-methoxy-2,3,4,5-tetramethylphenol is 40 nm to the blue of 4-methoxyphenol.

3. TR^3 Spectra of the α-tocopheroxyl Radical

Fig. 3 shows the TR^3 spectra of the α-tocopheroxyl radical in a variety of environments. The spectra are similar to the spectra of the 4-methoxyphenoxyl

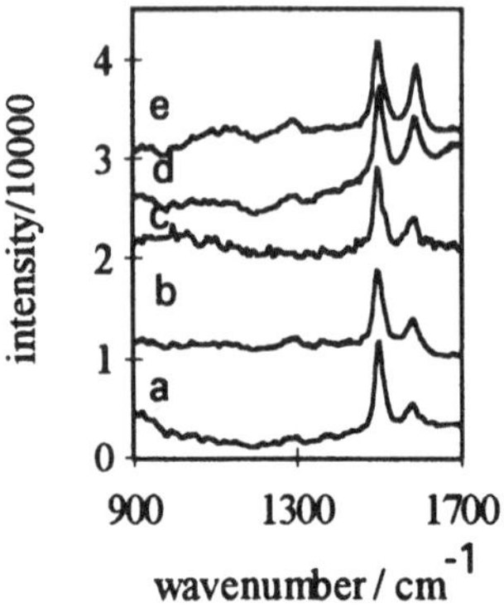

Fig. 3. TR^3 spectra of α-tocopheroxyl radical produced by photolysis, 308 nm, of α-tocopherol (5×10^{-3} to 50×10^{-3} mol dm^{-3}) and using a probe laser operating at 425 nm. All spectra are pump/probe - probe only, with a time delay of 2 μs except for the SDS micelle system where a time delay of 200 ns was used The solvent/systems are; (a) tetrahydrofuran, (b) Triton X-100R with 50×10^{-3} mol dm^{-3} phosphate, pH 7, (c) DMPC lipid bilayer, (d) ethanol and (e) SDS micelles with 0.1 mol dm^{-3} phosphate, pH 7.

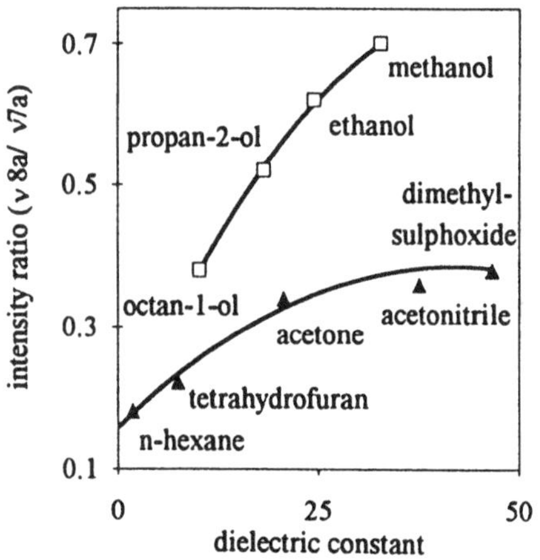

Fig. 4. Plot of intensity ratios, ν_{8a}/ν_{7a}, from the TR^3 spectra of the α-tocopheroxyl radical in solvents as a function of their dielectric constant.

radical having a high frequency C-O stretch as well as intense C=C stretching bands indicative of the delocalisation of the unpaired electron giving rise to stabilisation of the radical as discussed by Ingold and workers [6-8]. Further, the intensity ratio of the stretches reflects the polarity and ability of the solvent to act as a hydrogen bond donor as demonstrated in Fig. 4. This property can be used as a probe for the microenvironment experienced by the of α-tocopheroxyl radical. The intensity ratio of the spectrum in Fig. 3e indicates that in SDS micelles the radical is in a relatively polar environment. This reflects both the orientation of the radical, with the chromanoxyl group towards the micelle/aqueous interface, and a significant access of solvent (water) to this site. In contrast, the TR^3 intensity ratio of the radical in Triton X-100R micelles indicates that the α-tocopheroxyl radical is in a less polar environment and hence shielded from the solvent in this type of micelle. In DMPC lipid bilayers the in ν_{8a}/ν_{7a} intensity ratio is 0.45 indicating that the vitamin E radical experiences a polarity inbetween that of Triton X-100R and SDS micelles. The polar character of this environment is consistent with the vitamin E radical being close to the membrane surface. This permits the rapid regeneration of vitamin E in the membranes by ascorbate in the aqueous phase.

Acknowledgements

The authors are grateful to Dr E. W. Thomas (University of Salford) for synthesis of 4-methoxy-2,3,5,6-tetramethylphenol and to Miss S. M. Tavender for her help with the experimental setup. We also thank the SERC for Facility time.

1. See for example C. A. Rice-Evans and A. T. Diplock: Free Rad. Biol. Med. **15**, 77 (1993)
2. R. H. Bisby and A. W. Parker: FEBS Lett. **290**, 205 (1991)
3. R. H. Bisby and A. W. Parker: In *From Basic Science to Medicine,* ed. by G. Poli, E. Albano and M. U. Dianzani (Birkhauser Verlag, Basel, Switzerland 1993) p.31
4. A. W. Parker and R. H. Bisby: J. Chem. Soc. Faraday Trans. (In press 1993)
5. G. N. R. Tripathi and R. H. Schuler: J. Phys. Chem. **92**, 5129 (1988)
6. G. W. Burton and K. U. Ingold: J. Am. Chem. Soc. **103**, 6472 (1981)
7. G. W. Burton, T. Doba, E. J. Gabe, L. Hughes, F. L. Lee, L.Prasad and G. W. Ingold: J. Am. Chem. Soc. **107**, 7053 (1985)
8. G. W. Burton and K. U. Ingold: Acc. Chem. Res. **19**, 194 (1986)

Picosecond Resonance-Raman Spectroscopy of the Reaction Center of *Rhodobacter sphaeroides* R26

E. Nishizawa[1], Y. Misono[2], K. Itoh[2], and Y. Koyama[1]

[1]Faculty of Science, Kwansei Gakuin University, Uegahara, Nishinomiya 662, Japan

[2]Department of Chemistry, School of Science and Engineering, Waseda University, Shinjuku-ku, Tokyo 169, Japan

Abstract. We attempted to record, by using picosecond pulses (351 nm, 50 ps, 1 kHz), the Raman spectra of a bacterial photo-reaction center in transient redox states. Based on the results of transient absorption spectroscopy, these redox states detected were assigned to $P^+ \cdot B \cdot H^- \cdot Q_a$, $P^+ \cdot B \cdot H^- \cdot Q_a^-$ and $P^+ \cdot B \cdot H \cdot Q_a$ states. The Raman line due to the methine stretching appeared at 1601 cm^{-1} upon the formation of P^+. The Raman line around 1675 cm^{-1} to be ascribed to the hydrogen-bonded keto-carbonyl stretching of neutral H disappeared upon the formation of H^-.

In the electron-transfer reaction in bacterial photo-reaction centers (RC), an electron is transferred from the special pair BChls (P) to the accessory $BChl_L$ (B), the $BPhe_L$ (H) and then the primary-acceptor Quinone (Q_a). Resonance Raman spectroscopy using picosecond pulses is a useful tool to obtain information about the structural changes of these chromophore molecules in the transient redox states. Their Raman spectra have been recorded by using picosecond pulses (351 nm, 50 ps, 1 kHz) in this study. The assignments of Raman lines of both the methine and the carbonyl stretchings, which could be clearly observed in resonance with the Soret absorption bands of BChls and BPhes, are also discussed.

Each Raman spectrum of RC of *R. sphaeroides* R26 was recorded by placing the suspension in 0.1 M Tris (pH 8.0), 0.1 % Triton X-100 (4 x 10^{-4} M), in an ampul, which was sealed under nitrogen atmosphere and rotated during Raman measurements.

We examined the effects of the laser power and of the addition of a reductant (100 mM $Na_2S_2O_4$) or an oxidant (100 mM $K_3Fe(CN)_6$) on the Raman spectra (Fig. 1).

The methine stretching. The Raman line of the methine stretching should appear around 1600 cm^{-1}. The Raman line was observed at 1608 cm^{-1} in Spectrum (a) and at 1601 cm^{-1} in Spectra (b), (c) and (d).

The carbonyl stretchings. Two Raman lines were observed at 1655 and 1676 cm^{-1} in Spectrum (a). Raman lines corresponding to the former line (~ 1655 cm^{-1}) were observed in Spectra (b), (c) and (d) at 1659, 1649 and 1653 cm^{-1}, respectively. The latter line (~ 1675 cm^{-1}) was not observed in Spectra (b) and (c) (Group I), but in Spectra (a) and (d) (Group II).

Assignments of redox states. By using the 50 ps, 1 kHz pulses, both (1) short-lived redox states having a rise time shorter than 10^1 ps and a decay time longer than 10^2 ps, and (2) long-lived redox states having decay time longer time 10^0 ms should be detected. According to the results of transient absorption spectroscopy [1, 2], the $P^+ \cdot B \cdot H^- \cdot Q_a$ state is generated upon irradiation of a native RC, while the $P^+ \cdot B \cdot H^- \cdot Q_a^-$ state is generated when a reductant exists. On the other hand, the $P^+ \cdot B \cdot H \cdot Q_a$ state was found when the above oxidant was added. These results predict that the $P \cdot B \cdot H \cdot Q_a$, $P^+ \cdot B \cdot H^- \cdot Q_a$, $P^+ \cdot B \cdot H^- \cdot Q_a^-$ and $P^+ \cdot B \cdot H \cdot Q_a$ states should be detected in Spectra (a), (b), (c) and (d), respectively.

The Raman line of the methine stretching of P^+ appears at 1600 cm^{-1} (1590 cm^{-1}) in resonance with the Q_y (Q_x) absorption band [3, 4]. Since these frequencies seem to agree with that of Raman lines of the methine stretching detected in spectra (b), (c) and (d), the formation of P^+ in

Springer Proceedings in Physics, Vol. 74 Time-Resolved Vibrational Spectroscopy VI
Editors: A. Lau, F. Siebert, W. Werncke

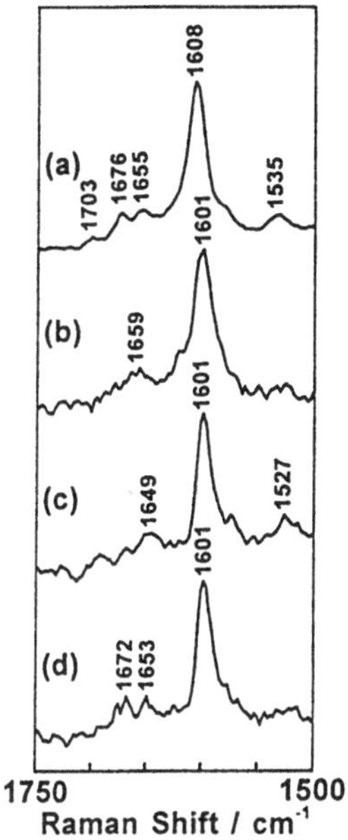

Figure 1. The picosecond Raman spectra of RC of *R. sphaeroides* R26.
Spectrum (a): the native RC recorded by using a low power to avoid photo-excitation.
Spectra (b), (c) and (d): the difference spectra obtained by subtracting Spectrum (a) from the each of spectra, (i) the native RC recorded with a high power, (ii) the RC with the reductant by using a low power and (iii) the RC with the oxidant by using a high power, respectively.

these spectra is suggested. This result agrees with the above prediction from the transient absorption spectroscopy.

The 1675 cm^{-1} Raman line of the carbonyl stretching was observed not for Group I but for Group II. Based on the predictions from the transient absorption spectroscopy, the difference between Group I and Group II originates from the redox states of $BPhe_L$, i.e. H^- or H. This difference means that the 1675 cm^{-1} Raman line disappears when the H^- is formed. (The Raman lines due to Q_a and Q_a^- should not be detected under the present resonance conditions.) The Raman line of the hydrogen-bonded keto-carbonyl stretching of H appears at 1678 cm^{-1} in resonance with the Q_x absorption band of H [5]. Therefore, the particular Raman line detected (~ 1675 cm^{-1}) is ascribed to the hydrogen-bonded keto-carbonyl stretching of neutral H.

References

[1] A.J. Hoff, Phys. Rep. 54, 75 (1979).
[2] V.A. Shuvalov and W.W. Parson, Proc. Nati. Acad. Sci. U.S.A. 78, 957 (1981).
[3] T.A. Mattioli, A. Hoffmann, B. Robert B. Schrader and M. Lutz, Biochemistry 30, 4648 (1991).
[4] M. Lutz and J. Kleo, Biochim. Biophys. Acta 546, 365 (1979).
[5] M. Lutz, in *Proceedings of the 7th International Conference on Raman Spectroscopy*, edited by W.F. Murphy (North-Holland, Amsterdam, 1980), p. 520.

Theoretical Model for the Study of CO Binding to Hemoproteins

M.G. Zakaraya[1], *J. Ulstrup*[2], *and J.B. Legarth*[2]

[1]Institute of Inorganic Chemistry and Electrochemistry of the Georgian Academy of Sciences, Jikiya 7, 380086 Tbilisi-86, Georgia
[2]Chemistry Department A, The Technical University of Denmark, Lyngby, Denmark

Abstract. We have analyzed the non-exponential kinetics, the temperature variation, and the CO isotope effects of the CO recombination reactions with myoglobin and single-chain hemoglobin.

1. Introduction

Non-exponential kinetics, nuclear tunneling and CO isotope effects are the most crucial kinetic features of CO recombination with hemoproteins. We have made an effort to account for these properties by quantum-mechanical model, which incorporates the protein/solvent environment, the iron/heme motion and the CO motion as the essential features. We notice briefly the following about the dynamics of these three system parts.

The protein/solvent environment. This part covers a continuous manifold of protein and external solvent motion of widely different coupling strengths and conformational relaxation times. the slower of these have dissipative or relaxational character, while the faster ones can be regarded as ballistic or friction free. The formalism of handling continuous manifolds of friction-free nuclear coordinates at all temperatures is available [1,2]. Relative to the dissipative coordinates we have adopted the simplified view that the slow environmental protein motion is slow enough to be represented by static inhomogeneous conformational freezing.

The iron/heme-group motion. In the initial state, immediately subsequent to flash photolysis, the iron atom is in a high-spinstate (S=2) outside the domed heme configuration. The nuclear configuration is metastable and likely to be intermediate between the equilibrated deoxy andcarbon-monoxy state of the heme group. In the final state iron is in a low-spin state (S=0) closer to or in the heme plane. We have represented this configurational change by a single nuclear coordinate, allocated a low vibra-tional wave number and a substantial nuclear coordinate displacement.

CO dynamics. The third system part is constituted by the CO motion. Structural and spectroscopic data indicate that this motion is three-dimensional, involving coordinate displacements and vibrational frequency changes in both CO stretching, bending and translation. Prompted by the kinetic isotope effectdata for the recombination, we focused, however on CO bending and stretching.

Springer Proceedings in Physics, Vol. 74 Time-Resolved Vibrational Spectroscopy VI
Editors: A. Lau, F. Siebert, W. Werncke

2. Results and Discussion

The simplest view which accomodates the observed effects [3,4] includes the following elements.

A static environmental broadening distribution function in either the structural reorganization or the reaction Gibbs free energy. The minimum width is 0.2-0.3 eV and best fits are obtained by a pronounced asymmetry with a slower fall off at higher energies. This feature reflects discrete CO configurations, each broadened by Gaussian environmental and iron/heme-groupconfigurations.

Three local nuclear coordinates: two represent iron-heme/local-proteinmotion and CO bending, both with a low vibrational frequency in the range 60-150 cm^{-1} . The third coordinate is CO stretching with a frequency of about 2000 cm^{-1} and quantum-mechanically frozen in the whole temperature range.

This model (details are given elsewhere [1,5]) is more involved than previous models [6-8], which aimed primarily towards a disentanglement of thegross temperature dependence of the rate constant and the electronic coupling.This is understandable, as neither inhomogeneous broadening nor CO motion and kinetic isotope effects were explicitly addressed in these reports. The present analysis extends previous analysis to these other features.

References

[1] M.G.Zakaraya, T.A.Marsagishvili and D.E.Khoshtariya, Elementary processes in condenced media, (Metsniereba, Tbilisi, 1989).

[2] R.R.Dogonadze, A.M.Kuznetsov, M.A.Vorotyntsev and M.G.Zakaraya, J. Electroanal. Chem. **75**, 315 (1977).

[3] R.H.Austin, K.W.Beeson, L.Eisenstein, H.Frauenfelder and I.Gunsalus, Biochemistry **14**, 5355 (1975).

[4] N.Alberding, R.H.Austin, K.W.Beeson, S.S.Chan, L.Eisenstein, H.Frauenfelder and T.M.Nordlund, Science **192**, 1002 (1976).

[5] J.B.Legarth, J.Ulstrup & M.G.Zakaraya, Eur.J.Biochem. **205**,621(1992).

[6] J.Jortner and J.Ulstrup, J. Am. Chem. Soc. **101**, 3744 (1979).

[7] M.Redi, B.S.Gerstman and J.J.Hopfield, Biophys. J. **35**, 471 (1981).

[8] E.Bukhs and J.Jortner, J. Chem. Phys. **83,** 4456 (1985).

How Catalizes Azide Proton-Pumping in D96N Mutated Bacteriorhodopsin?

J. le Coutre and K. Gerwert

Max-Planck-Institut für molekulare Physiologie, Rheinlanddamm 201, D-44139 Dortmund, Germany

Bacteriorhodopsin (bR) from *halobacterium salinarium* converts light energy into chemical energy by pumping protons across the cytoplasmic membrane and thereby creating a proton gradient. The chromophor, all-trans-retinal, is covalently linked to one of the seven transmembrane helices via a protonated Schiff-base. After lightexcitation the chromophore isomerizes and the protein undergoes a photocycle during which one proton is transferred to the extracellular side. Using timeresolved FTIR-techniques two aspartic residues have been identified to be crucial for proton-pumping [1]. D85 works as internal proton-acceptor receiving a proton from the Schiffbase and D96 works as internal proton-donor to the Schiffbase.

In D96N and D96G mutated Bacteriorhodopsins azide is known to accelerate the M-decay at which the Schiffbase is reprotonated from the cytoplasmic side [2]. Also other small anions like formiate and cyanate are able to accelerate this reaction.

Furthermore azide accelerates the L to M transition at which the Schiff-base is deprotonated. Azide and hydrogen-azide show distinct bands at 2050cm-1 and 2150cm-1 respectively. Depending on the solvent these bands shift towards lower frequencies in an increasing hydrophobic environment [3].

Samples for timeresolved experiments were prepared at pH 5 since IR bands due to either azide or hydrogen azide can best be detected at a pH close to the pK of azide (pK=4,7). In order to detect the M-rise, temperature during the measurements was cooled down to -10°C.

Timeresolved measurements of D96N with azide exhibit an transient protonation of the azide-anion after M-decay. Fig 1 shows the timecourse of this protonation with a maximum after the M-intermediate. We can exclude that the proton which is taken up from azide comes from an internal aminoacid-residue (data not shown).

From these observations the following conclusions can be made: 1. Azide does not imitate the internal proton-donor D96 as originally assumed. 2. Azide enhances proton-conductivity by taking up a proton which is not offered by an internal amino-acid residue.

In fact the proton might be offered from an internal H-bonded-network. This can lead to a rearrangement of the network, which then causes acceleration of proton-transport.

Springer Proceedings in Physics, Vol. 74 Time-Resolved Vibrational Spectroscopy VI
Editors: A. Lau, F. Siebert, W. Werncke

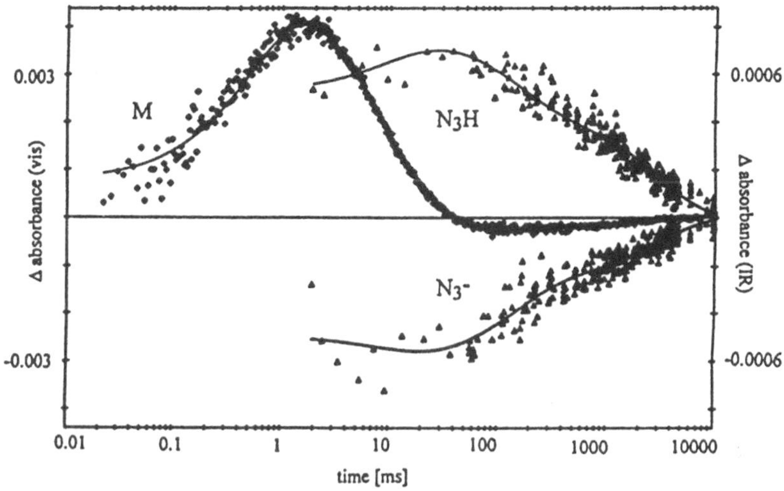

Fig. 1: Transient protonation of azide ($2040 cm^{-1}$) to hydrogen-azide ($2132 cm^{-1}$) occuring after reprotonation of the Schiff-base (M-decay measured at 410 nm).

References:

[1] Gerwert K. (1992), *Biochimica et Biophysica Acta* 1101: 147-153

[2] Tittor J., Soell C., Oesterhelt D. Butt H.-J. and Bamberg E. (1989), *The EMBO Journal* 8: 3477-3482

[3] le Coutre J. and Gerwert K. (1992), *Proceedings of the Vth International Conference on Retinal Proteins* (J.L. Rigaud ed.): 127-130

The Photocycle of the Bacteriorhodopsin Mutant ASP96→ASN Studied by Time-Resolved Resonance-Raman and Optical Transient Spectroscopy

W. Eisfeld and M. Stockburger

Max-Planck-Institut für biologische Chemie, D-37018 Göttingen, Germany

The retinal-binding protein bacteriorhodopsin acts as a light-driven proton pump in the cell membrane of *Halobacteria*. Under illumination, the retinylidene Schiff base chromophore BR_{570} goes through a series of intermediate states (K, L, M, N, O), before it is reconstituted within ~8 ms (pH 7, 20°C). During this cyclic reaction ("photocycle") a proton is released to the extracellular side, and a proton is taken up by the protein from the cytoplasmic side. The transfer of the Schiff base proton to the counterion (carboxylate side chain of residue Asp85), i.e. the formation of M, and the reprotonation of the Schiff base (M→N transition) are thought to play a key role in proton pumping.

When the aspartic acid residue at position 96 is replaced by asparagine (mutant Asp96→Asn, or D96N), the decay of M is dramatically prolonged (~ 100 times at pH 6). From this behavior it was concluded that Asp96 is the proton donor in the reprotonation pathway of the Schiff base group (for a review, see Lanyi (1992) [3]). It was argued by Váró and Lanyi (1991) [6] that in D96N the intermediate N is not formed until the decay of M and even then is not accumulated to a detectable level. Our time-resolved resonance Raman experiments with aqueous suspensions of D96N purple membrane patches at pH 6 and 20°C lead to a completely different view.

RR spectra were recorded using the spinning cell technique and laser lines at 647 nm (Kr^+) and 514 nm (Ar^+) for pump and probe radiation. Overview spectra of the mutant chromophores BR, L (single beam, residence time of a sample element in the beam $\Delta t = 120\,\mu s$) and N (pump-probe, delay time $\delta = 1$ ms) are shown in Figure 1. They are almost identical with those in wild type (WT), with slight differences in the C=C stretching frequencies (shifts of the band positions in the range $-2\,cm^{-1}$ to $+3\,cm^{-1}$). The N intermediate can be clearly distinguished from L by a different intensity distribution of the two characteristic C=C bands (Figure 1). On this basis a rise time of a few hundred microseconds is estimated for N from a series of pump-probe experiments in which δ is varied from $50\,\mu s$ to 1.2 ms (Figure 2; the pure L spectrum on top is from a single beam experiment with $\Delta t = 30\,\mu s$). The rise of N thus is not correlated with the extremely slow decay of M (950 ms). This is in close analogy to the "fast-rising" N component in WT bR which was attributed to a direct L→N transition (without a detectable de- and reprotonation of the Schiff base) (Diller and Stockburger, 1988 [1]; Eisfeld et al., 1993 [2]). This transition is caused by a conformational change of the protein backbone (Ormos, 1991 [4]; Ormos et al., 1992 [5]).

The relative amount of N and M in the photocycle of D96N was obtained from a pump-probe experiment with $\lambda_{Probe} = 476$ nm. For $\delta = 1$ ms a value of [N]/[M] = 0.7 was found which is similar to WT bR. A special capillary flow RR experiment was designed in order to determine the lifetimes of N and M. First results indicate similar values in the range 500 ms – 1 s for both species (25°C, pH 6), with N decaying slightly slower than M.

Optical transient studies with a conventional laser flash photolysis apparatus were performed at pH 6 and 20°C. The K→L transition was found to take place in $0.7\,\mu s$ ($1.2\,\mu s$ in WT). The L→M transition is biexponential ($55\,\mu s$: 44%, $350\,\mu s$: 56%) and thus on the

Springer Proceedings in Physics, Vol. 74 Time-Resolved Vibrational Spectroscopy VI
Editors: A. Lau, F. Siebert, W. Werncke

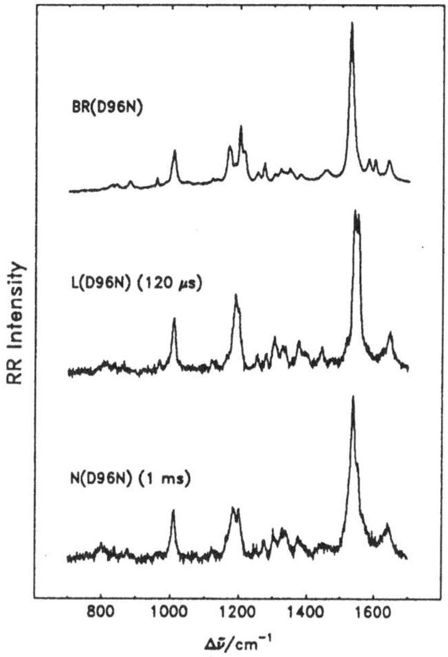

Figure 1: Overview RR spectra of the green-absorbing chromophores BR, L and N in the mutant D96N ($\lambda_{Probe} = 514\,nm$).

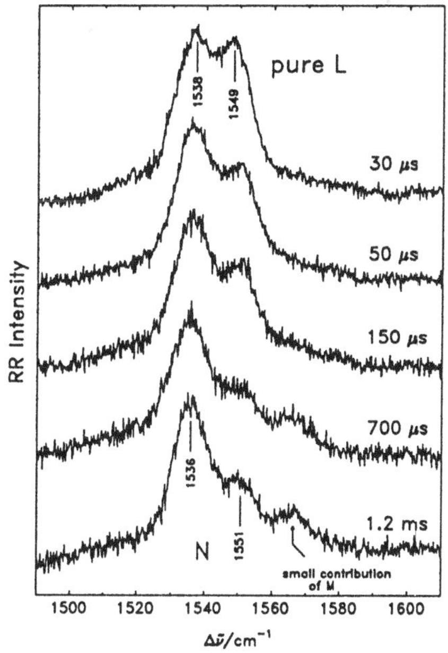

Figure 2: L→N transition, monitored in the C=C stretching region, takes place in several hundred microseconds ($\lambda_{Probe} = 514\,nm$).

average is slower by a factor of ~3 than the monoexponential function in WT (80 μs). The decay of M is monoexponential and extremely long (950 ms) compared to the bi-exponential decay in WT (3 ms, 7 ms, nearly equal amplitudes). The reconstitution of the parent chromophore (BR) in the mutant exactly matches the decay of M, i. e. no intermediate N or O can be identified in between.

In summary, for the first time an N-type intermediate in the photocycle of D96N could be characterized by RR spectroscopic data. Compared to WT bR, its lifetime under similar conditions is longer by a factor of ~100. This behavior strongly suggests that the presence of Asp96 is crucial for the reisomerization of retinal in the N→BR reconstitution step. Since this step is triggered by the uptake of a proton from the cytoplasmic side (Eisfeld et al., 1993 [2]), it can be concluded that Asp96 controls the conduction of a proton from the surface to a reactive site in the interior of the protein which in its protonated state catalyzes the reisomerization process. This view supports our earlier proposal for the proton pump mechanism (Eisfeld et al., 1993 [2]).

REFERENCES

[1] Diller, R., and Stockburger, M. (1988) *Biochemistry 27*, 7641–7651.
[2] Eisfeld, W., Pusch, C., Diller, R., Lohrmann, R., and Stockburger, M. (1993), accepted for publication in *Biochemistry*.
[3] Lanyi, J. K. (1992) *J. Bioenerg. Biomembr. 24(2)*, 169–179.
[4] Ormos, P. (1991) *Proc. Natl. Acad. Sci. USA 88*, 473–477.
[5] Ormos, P., Chu, K., and Mourant, J. (1992) *Biochemistry 31*, 6933–6937.
[6] Váró, G., and Lanyi, J. K. (1991) *Biochemistry 30*, 5008–5015.

TRS-FTIR Study of N Phosphoryl Histidine and Isopropanol Reaction System

Xinyao Hu, Yingqun Mao, and Xiaolin Tao

Department of Chemistry, Tsinghua University,
Beijing 100084, People's Rep. of China

Phosphoryl proteins & phosphoamino acids play a great part in biochemical process and play a critical role in adjusting activity of enzyme. Furthermore the role of phosphoamino acids in center of enzyme must not be underestimated. Therefore studying the characterics of phosphoamino acids is meaningful to reveal the secret of living phenomena.

In this work the FTIRs of standard samples such as phosphoryl histidine, L–histidine, isopropanol, triisopropyl phohate are measured. The temperature of phosphoryl histridine from 25℃ to rises 100℃, and the spectra are measureed in every interval of 5℃. As a result, the characteristics of these FTIR change little. This indicates that phosphoryl histidine is almost thermolly stable. In the sealed sample champer, the temperature of phosphoryl histidine and isopropanol system is kept constant and the TRS–FTIR are measured every 20 minutes. Under other temperatures, different sets of TRS–FTIR are gained.

The FTIRs from the reaction processes have peaks at 1733, 1630, 1465, 1378, 1224, 1063, 1022, 995, 896, 778cm^{-1} etc, which are found in the FTIR of phosphoryl histidine and decrease gradually. In the meantime, the FTIR of L–histidine has peaks at 1639, 1582, 1500, 1412, 1452, 1492, 1268, 1189cm^{-1} etc. which appeare and increase gradually in the FTIR from the reaction processes. Besides, some other peaks are found in the FTIR from the reaction processes. From above, some conclusions following can be seen. During the reaction phosphoryl histidine takes part in reaction and decreases gradually. L–histidine is produced and increases gradually. In the meantime some other compounds are produced. Triisopropyl phosphate is not produced in the reaction.

Treating the Absorbance as a function of reaction time and wavenumber of FTIR, plot the 2DIR. As shown in figure 1. Namely, when the temperature rises from room temperature to predetermined temperature, the absorbance at 1733cm^{-1}, which can represent the quantity of phosphoryl histidine, increases greatly; from then on, decreases rather fast and finally changes little.

The transmitivity of the 1733cm^{-1} peak from the reaction process is measured, and A, −lnA are caculated. As shown in figure 2. Their shapes

Springer Proceedings in Physics, Vol. 74 Time-Resolved Vibrational Spectroscopy VI
Editors: A. Lau, F. Siebert, W. Werncke

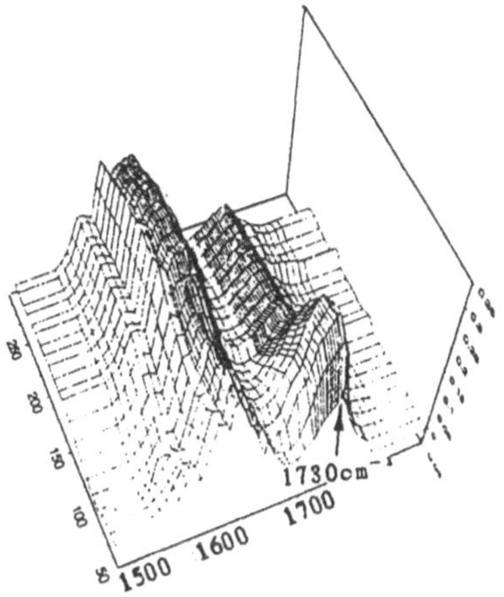

Figure 1 2DIR (45℃) of phosphoryl histidine & isopripanol system

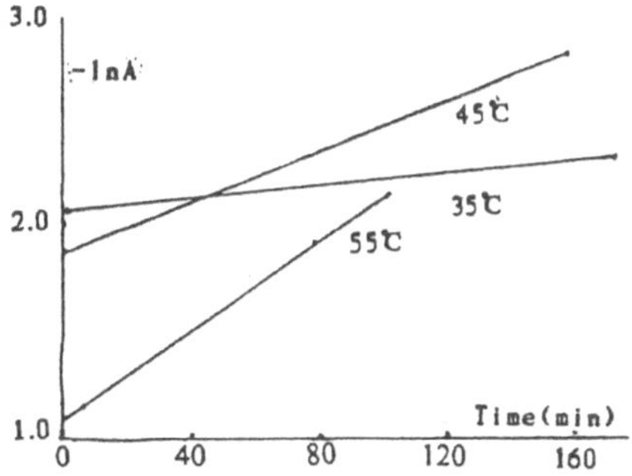

Figure 2 Kinetic curve of phosphoryl histidine & isopropanol system (35℃, 45℃, 55℃)

are the first order reaction rate constants of the reaction. Those follow: 0.0155 at 55℃, 0.0056 at 45℃, 0.0013 at 35℃.

The calculated activity energy is 103.8KJ / mol^{-1}, relate error is ± 15%.

All Spectra were obtained on a Perkin Elmer Co. 2000 FTIRs.

REFERENCE

〔1〕 G. Thomas et al, ″Protein Phosphorylation and Bio−Regulation″, S. karger, Basel, 1980.

〔2〕 G. Taborsky, Adv. protein, Chem., 28, 1 (1974).

〔3〕 S. Sakakibara et al, european Patent, EP 0085488, A2.

〔4〕 W.H. Glspen et al, ″Brain Phosphorproteins Characterization and Funtion″ Progress in Brain Research, 56, 1982.

Protein Response to Ligation Reactions in Myoglobin

T.P. Causgrove and R.B. Dyer

Photochemistry and Photophysics Group (CLS-4, Mail Stop J567), University of California, Los Alamos National Laboratory, Los Alamos, NM 87545, USA

1. Introduction

A central question in Mb ligand binding is the dynamics and energetics of motion of the protein in response to ligation. Time-resolved infrared (TRIR) spectroscopy is uniquely suited as a probe for these processes, particularly the behavior of the protein, which generally is not observable by other spectroscopies [1].

It has long been known that infrared spectra in the amide region are sensitive to protein secondary conformation [2]. Recent advances in equipment and techniques have permitted researchers to quantitatively predict secondary structures from infrared spectra [3-7], particularly in the amide I region [4,5]. Therefore, it is now possible to study protein motion in time-resolved experiments on dynamics and function.

The ligation reactions of small molecules such as CO with the heme site of Mb exemplify the mechanisms available to O_2. CO is an ideal candidate for initial time-resolved IR experiments in the amide I region because it is easily photolyzed, exhibits little geminate recombination [8,9], and the structure of both MbCO and unligated Mb have been studied by crystallographic methods [10,11]. TRIR has already been applied to the stretching vibrations of the bound and free CO ligand [12-14]; dynamics of the protein, however, have yet to be probed by TRIR spectroscopy of the protein vibrations. Here we report results on the motions of the protein in response to ligation reactions, probed in the amide I region centered about 1650 cm^{-1}.

2. Experimental

Real-time TRIR measurements were obtained using an apparatus that has been described previously [1], except that the InSb detector was replaced with a HgCdTe detector and biased amplifier (300 ns risetime). The laser diode used as a probe source was temperature tunable between 1610 and 1725 cm^{-1} with a mode spacing of about 4.5 cm^{-1}. Each mode was tunable by about 1 cm^{-1}.

Ultrafast TRIR measurements were obtained by means of a pump-probe experiment as previously described [15]. The IR probe pulse was generated by using a two step nonlinear mixing process. In the first step, the difference frequency between the 1064 nm pulse from the regenerative amplifier and the amplified dye pulse was generated in a KTP crystal, producing a pulse at 1.3 μm; this pulse was then differenced again with the 1064 nm pulse in $AgGaS_2$. Tuning the dye laser yields IR pulses (~4 ps, 100 nJ) tunable from 1200 to 2100 cm^{-1}.

Carbonmonoxy myoglobin (MbCO) was prepared by dissolving lyophilized Mb in 50 mM phosphate buffer, pD 7.4 (in D_2O). The heme was reduced with 5 mM $Na_2S_2O_4$ and liganded by several cycles of degassing and back-filling with CO.

Springer Proceedings in Physics, Vol. 74 Time-Resolved Vibrational Spectroscopy VI
Editors: A. Lau, F. Siebert, W. Werncke

Final samples were ~1.5 mM MbCO in 50 μm sealed cells with CaF_2 windows for real-time experiments or ~4 mM in 100 μm cells for picosecond experiments.

3. Results

The decay of the infrared absorption changes at 1943 and 1662 cm^{-1} following laser photodissociation of CO show identical recovery times if taken under identical conditions. Analysis of the recovery was done using the so-called A+B second order kinetics. The resulting fits gave second-order rate constants of 3.8 (1943 cm^{-1}) and 4.1 (1662 cm^{-1}) x 10^5 $M^{-1}s^{-1}$, within experimental error of each other, but slightly lower than the value of 5.0 x 10^5 $M^{-1}s^{-1}$ determined by UV-Vis spectroscopy [9].

Transients were also taken using faster amplifiers in an attempt to resolve kinetics due to the motion of the protein. Transients taken in both the amide I region and at 1943 cm^{-1} showed instrument-limited risetimes of 300 ns, indicating that protein response is complete on this time scale.

The initial amplitude of IR transients across the amide I band from 1608 cm^{-1} to 1723 cm^{-1} was recorded and plotted against frequency to create a Mb-MbCO difference spectrum (Figure 1). All of the major features in the TRIR spectrum are reproduced in the static difference spectrum (Mb-MbCO) obtained by subtracting FTIR spectra of carbonmonoxy and deoxy Mb. This is convincing evidence that the TRIR transients are a measure of conformational change of the protein backbone.

Picosecond TRIR measurements of photodissociation of CO indicate that the time-resolved bleaching of the Fe-CO absorption at 1943 cm^{-1} is indistinguishable from the instrument response, determined using a Si wafer. This is consistent with UV-Vis [16] and TRIR [13] measurements which indicate that CO photodissociation occurs in less than 150 fs. The protein response immediately following the photolysis of CO was probed at 1666 cm^{-1} (Figure 2). The convolution of the 6 ps $sech^2$ instrument response function with an 8 ps exponential function is also plotted in Figure 2. Positive transients (absorptions) were observed at 1680 cm^{-1}, but the signal-to-noise was insufficient to permit fitting the data to extract dynamics.

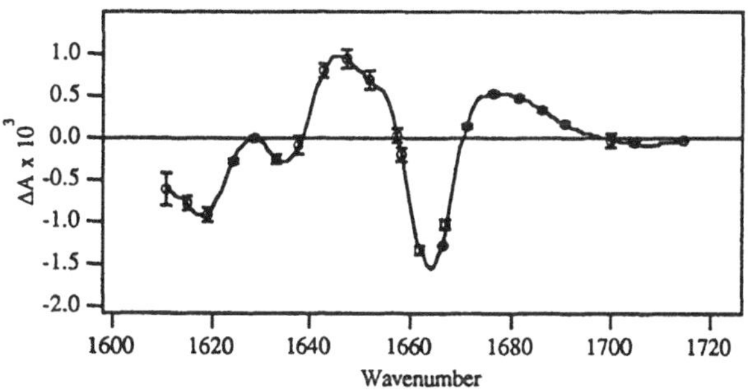

Figure 1. Difference spectrum generated from initial amplitudes of IR transients of 1.6 mM sperm whale Mb in the amide I region. The symbols are the actual data; the line is a cubic spline interpolation.

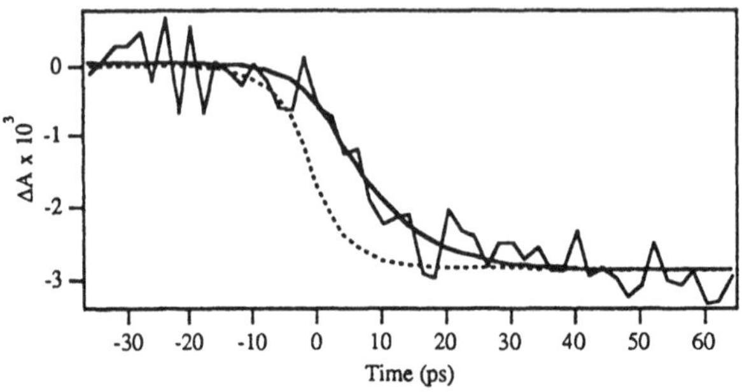

Figure 2. Picosecond TRIR absorption at 1666 cm^{-1}, together with the measured instrument response function (dashed line) and the best fit to the data of a convolution of this function with an exponential function (τ = 8.1 ps).

4. Discussion

Recently, development of Fourier self-deconvolution [7,17] and second derivative spectra [3,5] and criteria for the subtraction of water and water vapor bands have allowed tables of amide I band assignments to be made for proteins in H_2O [3] and D_2O [7] solution. The most straightforward assignment is that for α-helix, which shows only a single narrow band at 1653-1656 cm^{-1}. The Mb difference spectrum (Fig. 1) has an isosbestic point near this frequency, consistent with the finding that the helix structures of Mb are quite similar between the CO and deoxy forms, but only shift relative to each other [10]. The bleach at 1664 cm^{-1} in the difference spectrum is in the region assigned to turn structures; bands at 1676 cm^{-1} (a transient absorption in Fig. 1) have been assigned to either turn or extended chain conformations. The other main absorption at 1645 cm^{-1} is in the area assigned to unordered structure. Bands at lower frequency have been assigned to general β-type structures [7]. Specific structural assignments to the spectral features in Fig. 1 will require pH-dependent studies, selective isotopic labeling or site-directed mutagenesis. It is clear, however, that the dynamics measured are due to changes in the protein backbone conformation and possibly the motion of specific side chains which have absorptions in this spectral region.

Picosecond TRIR experiments give information on the initial stages of protein motion, and provide a connection between molecular dynamics simulations and actual protein motion. These results for Mb at 1666 cm^{-1} suggest that at least part of the conformational change occurs in 8 ps. These dynamics can be compared to resonance Raman results [18], which showed that the Fe-N(his) stretching frequency reaches its equilibrium deoxy value within the 30 ps pulsewidth of the resonance Raman probe pulse. Transient grating spectroscopy [19] suggests that a global change in the protein structure is occurring in less than 30 ps. In contrast, relaxation of the circular dichroism spectrum probed at 355 nm of photodissociated MbCO to the equilibrium deoxy spectrum requires 300 ps [20]. The differences among these measurements may be due to the differences in what is being probed. In our TRIR measurements at 1666 cm^{-1} we are probing a specific part of the protein backbone structure. Measurements at other frequencies may reveal different dynamics.

In conclusion, we have demonstrated that TRIR in the amide I region gives structural information regarding protein conformational changes in real time.

Assignment of many of the amide I peaks to specific amide or side chain structures will require much additional effort. It is clear, however, that enormous potential exists for elucidating structural relaxation dynamics and energetics with a high degree of structural specificity using this approach.

5. References

1. Dyer, R. B.; Einarsdóttir, Ó.; Killough, P. M.; López-Garriga, J. J.; Woodruff, W. H. *J. Am. Chem. Soc.* **1989**, *111*, 7657-7659.
2. Elliott, A.; Ambrose, E. J. *Nature* **1950**, *165*, 921-922.
3. Dong, A.; Huang, P.; Caughey, W. S. *Biochemistry* **1990**, *29*, 3303-3308.
4. Dousseau, F.; Pézolet, M. *Biochemistry* **1990**, *29*, 8771-8779.
5. Dong, A.; Huang, P.; Caughey, W. S. *Biochemistry* **1992**, *31*, 182-189.
6. Kaiden, K.; Matsui, T.; Tanaka, S. *Appl. Spectros.* **1987**, *41*, 180-184.
7. Susi, H.; Byler, D. M. *Meth. Enzymol.* **1986**, *130*, 291-311.
8. Henry, E. R.; Sommer, J. H.; Hofrichter, J.; Eaton, W. A. *J. Molec. Biol.* **1983**, *166*, 443-451.
9. Gibson, Q. H.; Olson, J. S.; McKinnie, R. E.; Rohlfs, R. J. *J. Biol. Chem.* **1986**, *261*, 10228-10239.
10. Kuriyan, J.; Wilz, S.; Karplus, M.; Petsko, G. A. *J. Mol. Biol.* **1986**, *192*, 133-154.
11. Takano, T. *J. Mol. Biol.* **1977**, *110*, 569-584.
12. Dixon, A. J.; Glyn, P.; Healy, M. A.; Hodges, P. M.; Jenkins, T.; Poliakoff, M.; Turner, J. J. *Spectrochim. Acta* **1988**, *44A*, 1309-1314.
13. Jedju, T. M.; Rothberg, L.; Labrie, A. *Opt. Lett.* **1988**, *13*, 961-963.
14. Anfinrud, P. A.; Han, C.; Hochstrasser, R. M. *Proc. Natl. Acad. Sci. USA* **1989**, *86*, 8387-8391.
15. Stoutland, P. O.; Dyer, R. B.; Woodruff, W. H. *Science* **1992**, *257*, 1913-1917.
16. Petrich, J. W.; Poyart, C.; Martin, J. L. *Biochemistry* **1988**, *27*, 4049-4060; Petrich, J. W.; Martin, J. L. *Chem. Phys.* **1989**, *131*, 31-47.
17. Surewicz, W. K.; Mantsch, H. H. *Biochim. Biophys. Acta* **1988**, *952*, 115-130.
18. Findsen, E. W.; Friedman, J. M.; Ondrias, M. R.; Simon, S. R. *Science* **1985**, *229*, 661-665.
19. Richard, L.; Genberg, L.; Deak, J.; Chiu, H.-L.; Miller, R. J. D. *Biochemistry* **1992**, *31*, 10703-10715.
20. Xie, X.; Simon, J. D. *Biochemistry* **1991**, *30*, 3682-3692.

Index of Contributors

MIX
Papier | Fördert gute Waldnutzung
FSC® C083411

Kolibri 360 GmbH
Ferdinand-Jühlke-Straße 7
99095 Erfurt, Deutschland
produktsicherheit@kolibri360.de